T0074315

Studies in Computational Intelligence

Volume 536

Series editor

Janusz Kacprzyk, Polish Academy of Sciences, Warsaw, Poland
e-mail: kacprzyk@ibspan.waw.pl

For further volumes:
http://www.springer.com/series/7092

About this Series

The series "Studies in Computational Intelligence" (SCI) publishes new developments and advances in the various areas of computational intelligence—quickly and with a high quality. The intent is to cover the theory, applications, and design methods of computational intelligence, as embedded in the fields of engineering, computer science, physics and life sciences, as well as the methodologies behind them. The series contains monographs, lecture notes and edited volumes in computational intelligence spanning the areas of neural networks, connectionist systems, genetic algorithms, evolutionary computation, artificial intelligence, cellular automata, self-organizing systems, soft computing, fuzzy systems, and hybrid intelligent systems. Of particular value to both the contributors and the readership are the short publication timeframe and the world-wide distribution, which enable both wide and rapid dissemination of research output.

Anthony Lewis Brooks · Sheryl Brahnam
Lakhmi C. Jain
Editors

Technologies of Inclusive Well-Being

Serious Games, Alternative Realities, and Play Therapy

 Springer

Editors
Anthony Lewis Brooks
Department of Architecture, Design and
    Media Technology
Aalborg University
Esbjerg
Denmark

Lakhmi C. Jain
Faculty of Education, Science, Technology
    and Mathematics
University of Canberra
Canberra
Australia

Sheryl Brahnam
Computer Information Systems
Missouri State University
Springfield
USA

ISSN 1860-949X   ISSN 1860-9503   (electronic)
ISBN 978-3-642-45431-8   ISBN 978-3-642-45432-5   (eBook)
DOI 10.1007/978-3-642-45432-5
Springer Heidelberg New York Dordrecht London

Library of Congress Control Number: 2013958234

Printed on acid-free paper

Springer is part of Springer Science+Business Media (www.springer.com)

Foreword

Over the last decade we have witnessed a dramatic increase in the global adoption of innovative digital technologies. This can be seen in the rapid acceptance and growing demand for high speed network access, mobile devices/wearable displays, smart televisions, social media, hyperrealistic computer games, and novel interaction and behavioral sensing devices (e.g., MS Kinect, Fitbit). Consumer-driven interactive technologies that were considered to be visionary just 10 years ago have now become common to the current digital landscape. At the same time, the power of these technologies to both automate processes and create engaging user experiences has not gone unnoticed by healthcare researchers and providers. This has led to the emergence and growing adoption of clinical applications that both leverage off-the-shelf technology and push the boundaries of new technologic development. As electric typewriters gave way to word processors and handwritten letters to email, we are now witnesses to a wave of technological innovation that is driving how healthcare is accessed and delivered, and how clinical users engage with it.

It was during the "computer revolution" in the 1990s that promising technologically driven innovations in interactive behavioral healthcare had begun to be considered and prototyped. Primordial efforts from this period can be seen in R&D that aimed to use computers to enhance productivity in patient documentation and record-keeping, to deliver "drill and practice" cognitive rehabilitation, to improve access to care via internet-based teletherapy, and in the use of virtual reality simulations to deliver exposure therapy for specific phobias. Since that time, continuing advances in computation speed and power, 3D graphics and image rendering, display systems, body tracking, interface technology, haptic devices, authoring software, and artificial intelligence have supported the creation of low-cost and usable interactive clinical technology systems now capable of running on commodity level personal computational devices.

Other factors beyond the rapidly accelerating advances in enabling technologies and concomitant cost reductions have driven interest in healthcare technology applications. In part due to the ubiquitous presence of technology in everyday society, there is a growing attitude of acceptance by clinical researchers and practitioners (i.e., reduced suspicion and "technophobia"), particularly with the growing numbers of digital generation providers and patients. Moreover, there is now an emerging scientific literature reporting positive outcomes across a range of clinical applications. Such scientific support for the clinical efficacy and safe delivery of care has also fostered the view that technologic innovation may both improve care while reducing the escalating healthcare costs that have become one of the hallmarks of post-industrial western society. Thus, the convergence of the exponential advances in underlying enabling technologies with a growing body of research and experience has fueled the evolution of healthcare technology applications to near mainstream status. And this state of affairs now stands to transform the vision of future clinical practice and research for addressing the needs of those with clinical health conditions.

It is in this context, that the present volume of chapters has so much to offer. While 20 years ago the title, *"Technologies of Well-Being: Serious Games, Alternative Realities, and Play Therapy"*, would raise the specter of Star Trek, Lawnmower Man, and Super Mario Brothers, in the current context it instead evokes a sense that new possibilities are within our reach as we harness technology to create user experiences that promote human well-being. The use of games and simulation technology and play for engaging users with health care has passed through its initial phase of scientific doubt and quizzical skepticism. These concepts are now seen as vibrant options that bring together both art and science for a useful human purpose. No longer seen as harebrained schemes, we see respected scientific journals like *Nature, American Psychologist,* and *JAMA* publishing research that probes these concepts. Papers in this area are routinely presented at long-established scientific venues in addition to the more specialized homegrown conferences that our community has now evolved. Major funding agencies are now earnestly supporting R&D in these areas. And, when you describe what you do for a living to your neighbor, they get it right away and seem genuinely impressed! In essence, the science and technology has caught up with the vision in clear and demonstrative ways.

So, what might a reader consider as they study this book?

We sometimes observe that great insights into the present turn up in the words of those who lived and died in a not-too-distant past. Such insights, derived from a relatively recent yet sufficiently disjointed past, can deliver a vision of the future that illuminates our present in surprising ways. This can be nicely illustrated in the words of the French author, poet, and aviator, Antoine de Saint-Exupery (1900–1944) with his comment, *"The machine does not isolate man from the great problems of nature but plunges him more deeply into them."* While interpretations may vary, in one sentence from a writer who lived exclusively in the first half of the twentieth century, I see the exquisite juxtaposition of apprehension and engagement that always looms in our pursuit of technology solutions that address the problems of nature. This is not a bad thing. With whatever technology tools we have available, we plunge deeper into the nature of problems, and hopefully come close to where the solutions lie. I see this book in much the same fashion; a fascinating collection of visionary works by a diverse collection of scientists and practitioners who implicitly acknowledge the same struggle. The many ideas presented in these pages for using digital technology to help change the course of challenged lives in ways unthinkable in just the last century is bold and provocative. And to do this requires a team of scientists, artists, programmers, clinicians, users, among others, who are willing to plunge deeply into the struggle, rather than to use technology to become isolated from the reality of the challenges that we aim to address. The authors in this book have successfully done this and these writings will play a significant part in further illuminating the bright future in this area.

USA, December 2013   Skip Rizzo

Preface

This book is the first single volume that brings together the topics of serious games, alternative realities, and play therapy. The focus is on the use of digital media for the therapeutic benefit and well-being of a wide range of people—spanning those with special needs to the elderly to entire urban neighborhoods. The editors of this book believe it timely to bring together these topics to demonstrate the increasing trans/inter/multidisciplinary initiatives apparent today in science, medicine, and academic research—interdisciplinary initiative that are already profoundly impacting society.

Esbjerg, Denmark
Springfield, USA
Canberra, Australia

Anthony Lewis Brooks
Sheryl Brahnam
Lakhmi C. Jain

Contents

1 Technologies of Inclusive Well-Being at the Intersection
   of Serious Games, Alternative Realities, and Play Therapy   1
   Anthony Lewis Brooks, Sheryl Brahnam and Lakhmi C. Jain
   1.1 Introduction   1
   1.2 Contributions in This Book   3
   1.3 Conclusion.   10

Part I Technologies for Rehabilitation

2 Design Issues for Vision-Based Motor-Rehabilitation
   Serious Games   13
   Antoni Jaume-i-Capó, Biel Moyà-Alcover and Javier Varona
   2.1 Introduction   14
   2.2 Related Work.   15
   2.3 Design Issues   15
        2.3.1 Development Paradigm   16
        2.3.2 Interaction Mechanism   16
        2.3.3 Interaction Elements   16
        2.3.4 Feedback.   16
        2.3.5 Adaptability.   17
        2.3.6 Monitoring.   17
        2.3.7 Clinical Evaluation   18
   2.4 Case Study: Vision-Based Rehabilitation Serious Game   18
        2.4.1 Development Paradigm   19
        2.4.2 Interaction Mechanism   19
        2.4.3 Feedback.   20
        2.4.4 Interaction Elements   21
        2.4.5 Adaptability.   21
        2.4.6 Monitoring.   22
        2.4.7 Clinical Evaluation   22
   2.5 Conclusions   23
   References   23

3   Development of a Memory Training Game   25
    Kristoffer Jensen and Andrea Valente
    3.1   Introduction   25
    3.2   Memory...   26
            3.2.1   History of Memory Ideas......................   27
            3.2.2   Memory Overview   27
            3.2.3   Memory Assessments   29
            3.2.4   Memory and Games   30
    3.3   Game Design and Development   32
            3.3.1   Levels.......................................   33
            3.3.2   Development   33
    3.4   Scientific Uses......................................   35
            3.4.1   Self-Assessment   36
            3.4.2   Other Scientific Uses   37
    3.5   Conclusions..   37
    References   38

**4   Assessing Virtual Reality Environments as Cognitive Stimulation
    Method for Patients with MCI**   39
    Ioannis Tarnanas, Apostolos Tsolakis and Magda Tsolaki
    4.1   Introduction   40
            4.1.1   Individuals with Mild Cognitive Impairment   40
            4.1.2   MCI Subtypes   41
            4.1.3   Amnestic-Type Mild Cognitive Impairment.......   43
    4.2   Virtual Environments and aMCI........................   44
            4.2.1   Spatial and Visual Memory   45
            4.2.2   Virtual Reality, Spatial Memory and aMCI   46
    4.3   The Virtual Reality Museum   47
            4.3.1   The Virtual Reality Museum
                     Technical Components   48
            4.3.2   The Virtual Reality Museum Cognitive Theory   49
            4.3.3   The Virtual Reality Museum Cognitive Exercises...   53
    4.4   Research Methodology   54
            4.4.1   Design.......................................   54
            4.4.2   Participants   54
            4.4.3   Procedures...................................   55
            4.4.4   Data Recordings and Analysis   58
    4.5   Results   60
            4.5.1   Neuropsychological Variables Outcome..........   60
            4.5.2   Electrophysiological Measures Outcome   60
    4.6   Discussion   65
    4.7   Conclusion...   66
    References   69

5   Adaptive Cognitive Rehabilitation   75
Inge Linda Wilms
5.1   Introduction   75
5.2   The Challenges   76
       5.2.1   Fundamental Learning and Adaptation   76
       5.2.2   Brain Injury: The Damage Done   78
       5.2.3   Recovering from Brain Injury Through Training . . .   79
5.3   Advanced Technology and Cognitive Rehabilitation   81
       5.3.1   A Brief History   81
       5.3.2   Assessment   82
       5.3.3   Treatment   82
       5.3.4   Computer-Based Rehabilitation Training Today   83
       5.3.5   The Challenge of Individuality in Injury
               and Treatment   83
       5.3.6   The Same is not the Same   84
       5.3.7   The Prism Example   85
5.4   Artificial Intelligence and Rehabilitation   87
       5.4.1   Adjusting Level of Difficulty
               in Cognitive Rehabilitation   87
5.5   Concluding Comments   89
References   89

**6   A Body of Evidence: Avatars and the Generative
    Nature of Bodily Perception**   95
Mark Palmer, Ailie Turton, Sharon Grieve,
Tim Moss and Jenny Lewis
6.1   Introduction   96
6.2   Complex Regional Pain Syndrome   96
6.3   The Rubber-Hand Illusion   98
6.4   The Somatosensory System   103
6.5   The Body and Space   105
6.6   Changes in Body Perception   107
6.7   A Need to Communicate Painful Contradictions   109
6.8   An Erie Realisation and a Form of Acceptance   112
6.9   The Nature of Pain   113
6.10   Discussion   117
References   119

**7   Virtual Teacher and Classroom for Assessment
    of Neurodevelopmental Disorders**   121
Thomas D. Parsons
7.1   Introduction   121
7.2   Neurodevelopmental Disorders: Differentiation
       of Their Cognitive Sequelae   123

7.3   Virtual Environments for Neurocognitive Assessment   126
7.4   Assessment of Neurodevelopmental Disorders
        Using Virtual Environments   127
7.5   Virtual Reality for Assessment and Treatment
        of Social Skills   128
7.6   Virtual Teacher/Classroom Environment
        for Assessment/Treatment of Attention   129
7.7   Conclusions   131
References   132

8   Engaging Children in Play Therapy: The Coupling
    of Virtual Reality Games with Social Robotics   139
    Sergio García-Vergara, LaVonda Brown, Hae Won Park
    and Ayanna M. Howard
    8.1   Introduction   140
    8.2   Related Work   141
    8.3   A Virtual Reality Game for Upper-Arm Rehabilitation   142
            8.3.1   Introduction   142
            8.3.2   Objective   143
            8.3.3   Description of Overall System   143
            8.3.4   Description of Real-Time Kinematic Assessment . . .   147
            8.3.5   Pilot Study with Children   148
    8.4   Integration of Social Robotics in Gaming Scenarios   150
            8.4.1   Engagement Through Behavioral Interaction   151
            8.4.2   Pilot Study with Children   152
            8.4.3   Learning from Gaming Demonstration   155
            8.4.4   Pilot Study with Children   157
    8.5   Discussion and Future Work   160
    References   161

Part II   Technologies for Music Therapy and Expression

9   Instruments for Everyone: Designing New Means
    of Musical Expression for Disabled Creators   167
    Rolf Gehlhaar, Paulo Maria Rodrigues, Luis Miguel Girão
    and Rui Penha
    9.1   Introduction   168
    9.2   The Evolution of Robotic and Technology
            Assisted Musical Instruments   169
    9.3   A Description of the Process   171
            9.3.1   Context   171
            9.3.2   Designing the Instruments   173

9.4 Developing Musical Activities with the Newly
Created Instruments 183
9.4.1 Composing for and Performing with I4E 184
9.4.2 The Final Composition: Viagem 185
9.5 Feedback from Those Involved 187
9.6 Advice to Practitioners 192
9.6.1 Problems Encountered and Methods
for Arriving at Solutions 193
9.7 Concluding Remarks 194
References 195

10 Designing for Musical Play 197
Ben Challis
10.1 Introduction 197
10.2 Snoezelen and the Evolution of the Sensory Space 199
10.3 Music and Sound in Therapeutic Contexts 201
10.4 Sensory Spaces in Practice 203
10.4.1 Perceptions and Attitudes 204
10.4.2 The Spaces 205
10.4.3 Specialist Technologies 206
10.4.4 Repurposed Technologies 208
10.4.5 Activities (Music and Sound) 209
10.5 Technologies for Musical Play 212
10.5.1 Pressure-Sensitive 212
10.5.2 Touch-Sensitive 213
10.5.3 Movement-Sensitive 214
10.5.4 Switch-Access 215
10.6 Reconsidering the Design and Use of Sensory Spaces 215
References 217

Part III Technologies for Well-Being

11 Serious Games as Positive Technologies for Individual
and Group Flourishing 221
Luca Argenton, Stefano Triberti, Silvia Serino,
Marisa Muzio and Giuseppe Riva
11.1 Introduction 222
11.2 Positive Psychology 222
11.3 The Hedonic Perspective: Fostering Positive
Emotional States 223
11.3.1 Using Technology to Foster Positive
Emotional States 224
11.3.2 Can Serious Games Foster Positive
Emotional States? 225

11.4   The Eudaimonic Perspective: Promoting Individual
        Growth and Fulfillment   226
        11.4.1   Using Technologies to Promote Individual
                 Growth and Fulfillment   227
        11.4.2   Can Serious Games Promote Individual
                 Growth and Fulfillment?   228
11.5   The Social Perspective: Enhancing Integration
        and Connectedness   229
        11.5.1   Using Technologies to Enhance Integration
                 and Connectedness   230
        11.5.2   Can Serious Games Enhance Integration
                 and Connectedness?   231
11.6   Mind the Game™: A Serous Game to Promote
        Networked Flow   232
        11.6.1   Technology   232
        11.6.2   Sharing Goals and Emotional Experiences:
                 Sport as a Narrative Tool   232
        11.6.3   Creating a Space of Liminality   235
        11.6.4   Identifying a Common Activity to Overcome
                 the Space of Liminality   235
        11.6.5   A Pilot Study   236
11.7   Conclusion   240
References   240

12   Spontaneous Interventions for Health: How Digital Games
     May Supplement Urban Design Projects   245
     Martin Knöll, Magnus Moar, Stephen Boyd Davis
     and Mike Saunders
     12.1   Introduction: Morning Stroller Clubs and an iPhone
             Stair Climbing Game   246
     12.2   Active Design Guidelines and Community Games   247
     12.3   Spontaneous Interventions for Health   248
             12.3.1   Lack of Expertise to Explore ICT
                      in Urban Design   249
             12.3.2   Lack of Data Evaluating Medical
                      and Social Impact   249
             12.3.3   Not Stimulating User Co-design   250
             12.3.4   Lack of Cooperation and Business Models
                      for Further Development   251
     12.4   Context-Sensitive Games as a Spring Board
             for Communication   251

12.5 Roadmap Towards ICT Supported Spontaneous
Interventions 254
12.5.1 Encouraging Interdisciplinary Educational
and Research Projects 254
12.5.2 Developing New Strategies to Evaluate the Effects
of Mobile Health Games in the Wild 255
12.5.3 Improve Access to Co-design Projects
with Playful Approaches 255
12.5.4 Working on New Business Models that also Target
Local Neighborhoods 256
12.6 Conclusion and Outlook 256
References 257

13 Using Virtual Environments to Test the Effects of Lifelike
Architecture on People............................... 261
Mohamad Nadim Adi and David J. Roberts
13.1 Introduction................................... 262
13.1.1 Using Virtual Reality for Testing and Evaluating . . . 265
13.1.2 Crossover and Relations 267
13.2 Research Direction 269
13.3 Experiments................................... 270
13.3.1 Experiment 1 272
13.3.2 Experiment 2 273
13.3.3 Experiment 3 274
13.3.4 Experiment 4 276
13.4 Discussion.................................... 278
13.5 Conclusion.................................... 280
References 281

Part IV Technologies for Education and Education
for Rehabilitative Technologies

14 An Overview of Virtual Simulation and Serious Gaming
for Surgical Education and Training...................... 289
Bill Kapralos, Fuad Moussa and Adam Dubrowski
14.1 Introduction................................... 290
14.1.1 Alternative Educational Models 290
14.1.2 Open Problems with Virtual Simulation
and Serious Games........................ 292
14.1.3 Paper Overview 294
14.2 Virtual Simulation and Gaming in Surgical Education...... 294
14.2.1 Transfer of Skills to the Operating Room 295
14.3 Serious Games: Fidelity and Multi-Modal Interactions 297

14.4   The Serious Games Surgical Cognitive Education
        and Training Framework   298
        14.4.1   Graphical Rendering   300
        14.4.2   Sound Rendering   300
        14.4.3   Multi-cue Interaction and Cue Fidelity   301
14.5   Conclusions   302
References   303

**15   The Ongoing Development of a Multimedia Educational
       Gaming Module**   307
       Elizabeth Stokes
       15.1   Introduction   307
       15.2   The Coursework   308
       15.3   The Lifecycle of the Gaming Coursework   309
               15.3.1   Case Studies (Profiles)   309
       15.4   Pupils' Profile   309
               15.4.1   Gender   309
               15.4.2   Age   310
               15.4.3   Medical Conditions   310
               15.4.4   Communication   310
               15.4.5   Computer Ability   311
               15.4.6   Reading Level   312
               15.4.7   Comprehension   312
               15.4.8   Educational Area to Consider   312
               15.4.9   Likes   312
       15.5   The Development and Production   312
       15.6   Feedback and the Next Cohort of Students   313
       15.7   The Responsibility of the Academic   314
       15.8   Students' Useful Contribution and the Assets They Bring . . .   314
       15.9   Students' Gain from the Coursework   315
       15.10   Partnership and Collaboration   315
       15.11   Theoretical Knowledge, Practical Skills
               and Design Techniques   316
       15.12   Relating Academia with Society and Humanity   316
       15.13   Gains from the Experience Itself   316
       15.14   Students' Gains   316
       15.15   Schools, Practitioners and Pupils Gains   317
       15.16   Conclusion   318
       References   319

Part V   Disruptive Innovation

16   Disruptive Innovation in Healthcare and Rehabilitation........   323
       A. L. Brooks
       16.1   Introduction.....................................   323
               16.1.1   Disruptive Innovation   323
       16.2   Cloud-Based Disruptive Innovation..................   325
       16.3   SoundScapes: Serious Games, Alternative
               Realities and Play..............................   327
               16.3.1   Serious Games   328
       16.4   Alternative Reality   329
               16.4.1   Analogizing from a Games Perspective..........   330
               16.4.2   Training Trainers Retreat....................   333
               16.4.3   A Cloud-Based Archive Architecture   335
               16.4.4   Justifying the Transcending of Performance
                        Art-Related Quotes.........................   336
               16.4.5   Aesthetic Resonance   338
               16.4.6   Neuroaesthetics/Neuroesthetics................   342
               16.4.7   Games, Interactive Technology, and 'Alternative'
                        Environments in Healthcare..................   345
               16.4.8   Presence   346
       16.5   Conclusions....................................   348
       References   349

Glossary...   353

About the Editors   359

Chapter 1
Technologies of Inclusive Well-Being at the Intersection of Serious Games, Alternative Realities, and Play Therapy

Anthony Lewis Brooks, Sheryl Brahnam and Lakhmi C. Jain

Abstract This chapter introduces the intersection between serious games, alternative realities, and play therapy as it promotes well-being. A summary of the chapters included in this book is also presented.

Keywords Alternative realities · Serious games · Play · Therapy · ICT · Healthcare · Inclusive well-being · Quality of life (QOL)

1.1 Introduction

As future demographic trends point to increased pressures and demands on service industry providers addressing growing needy communities, such as children, the aged, and those challenged through impairment, many predict that digital technologies will play an increasing role in supplementing intervention practices and methods. Substantiating this claim is the rising awareness illustrated by major research funding activities directed at these groups throughout developed countries. These activities have the mission of contributing to knowledge and of realizing emerging enterprise and industrial developments in the area, as well as of

A. L. Brooks (✉)
School of ICT, AD:MT, Medialogy, Aalborg University, Niels Bohrs,
vej 8 6700, Esbjerg, Denmark
e-mail: tb@create.aau.dk

S. Brahnam
Missouri State University, 901 S. National, Springfield, MO 65804, USA
e-mail: sbrahnam@missouristate.edu

L. C. Jain
Faculty of Education, Science, Technology & Mathematics, University of Canberra,
Canberra, ACT 2601, Australia
e-mail: Lakhmi.jain@unisa.edu.au

A. L. Brooks et al. (eds.), *Technologies of Inclusive Well-Being,*
Studies in Computational Intelligence 536, DOI: 10.1007/978-3-642-45432-5_1,
© Springer-Verlag Berlin Heidelberg 2014

encouraging and of informing new educational programs involving technology that proactively look to contribute to a societal "wealth through health" regime.

The goal of this book is to introduce and to describe some of the latest technologies offering therapy, rehabilitation, and more general well-being care. Included along with the work of researchers from the serious games, alternative realities (incorporating artificial reality, virtual reality (VR), augmented reality, mixed reality, etc.), and play therapy disciplines are the writings of digital artists who are increasing working alongside researchers and therapists to create playful and creative environments that are safe and adaptive and that offer tailored interventions. The chapters in this book illustrate how complementary overlapping between topics has become the accepted norm. Such acceptance contributes to a readdressing and a questioning of associated values resulting in new themes and topics. Accordingly, the topics of this volume are selected to be wide in scope and to offer academics an opportunity to reflect on intersections in their work. These can be specific to the concepts of serious games, alternative reality, and play therapy, or to any number of related topics. Such intersections have embedded weightings from the distinct and individual toward the extreme, as seen, for example, in the common area in Fig. 1.1, an example where serious games are conducted in alternative realities and where the goal is play therapy utilizing the plasticity of digital media.

Unlike entertainment systems, the goals of alternative realities therapy and serious play demand the addition of sophisticated feedback systems that monitor user progress. These systems must encourage progress and intelligently and progressively adapt to users' individual needs within an environment that is challenging, engaging, and user friendly for patients and health care professionals. Such systems require the evolution of new paradigms in test battery creation that take advantage of the controllable digital framework, embodied data feedback, and other opportunities uniquely offered by virtual interactive spaces.

Much literature on play therapy (and therapeutic play) focuses on interactions between a professional therapist and children where the use of toys and other objects, i.e., physical artifacts, are expressive channels for communicating and interpreting a person's condition. In this book the additional opportunities to supplement such traditional practices via the use of digital media are posited. Serious games link to games (and gameplay) but are being used toward a serious outcome to solve a defined problem. The games are used 'seriously' in alternative realities, i.e., in computer generated environments that are interactive through embedded virtual artifacts.

These computer-generated alternative realities are commonly referred to as mixed, augmented, or artificial reality. Virtual reality in therapy and rehabilitation is not a new subject: a large number of papers reporting research advancing the field with transfer to activities of daily living have appeared over the last decade. This book contributes to the field by acknowledging the impact of digital media and by questioning potentials offered in traditionally *non-digitized* practices. Through the use of digital media in play therapy with children, for instance, the destruction, breakages, and damages to physical artifacts witnessed in traditional play therapy are eradicated. Instead, computer graphic environments are safe,

Fig. 1.1 Venn diagram illustrating the three fields of serious games, alternative realities, and play therapy. At the extreme is the triadic intersection where gameplay is conducted in computer generated realities and where therapeutic outcomes from play therapy are targeted

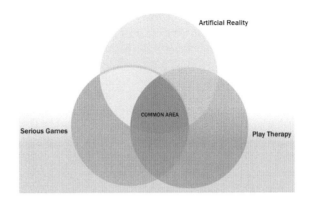

adaptive, and interactive, providing a world where things can be "virtually broken" any number of times and inexhaustively repaired, offering to clinicians both qualitative and quantitative aspects of evaluation alongside a flexibility for creating new tools for developing the clinical outcomes required by play therapy and other medical and educational interventions.

The aim of this book is to offer a platform for researchers in and across these disciplines to share with peers and to critique and to reflect on each other's studies. The authors, as well as the editors, come from various corners of the globe. This diversity is seen from a positive perspective, as it highlights how interpretations of subject matter and the use of culturally sensitive terminologies differ in and across cultures as well as within these fields.

In closing, it is important to state that this book does not attempt to cover all areas (serious games, alternative realities, and play therapy) equally, as the title might suggest, nor does it attempt to define these areas. Rather this book was conceived to be a catalyst for debate on interactive computer technology used in a manner whereby, for instance, creative industries, health care, human computer interaction, and technology sectors are encouraged to communicate with each other and to stimulate thinking about application design and intervention practices that are needed to supplement and to satisfy the societal demographic service needs of the future, needs that are predicted to grow as the population ages, needs that many predict only technology can provide the scaffolding to service. From this position we anticipate your reading of and reactions to these chapters as a step toward such thinking.

1.2 Contributions in This Book

The chapters in this book are divided into five parts that reflect major themes currently at the intersection of serious games, alternative realities, and play therapy for rehabilitation, intervention, assessment, education, and the promotion of well-being:

Part 1: Technologies for Rehabilitation
Part 2: Technologies for Music Therapy and Expression
Part 3: Technologies for Well-Being
Part 4: Technologies for Education and Education for Rehabilitative Technologies
Part 5: Disruptive Innovations

The first two chapters in Part 1 describe serious games for rehabilitation. These chapters are followed by one chapter that discusses problems adapting traditional therapies to games, one chapter that describes using a virtual environment for cognitive rehabilitation, and three chapters that employ social agents (virtual and robotic) either in therapy games or in virtual environments. The two chapters in Part 2 focus on projects centered on the design of novel musical instruments for people with special needs. The three chapters in Part 3 describe using virtual reality, mobile technology, and interactive architecture for personal and community well-being. The first chapter in Part 4 addresses virtual reality simulations for educating surgeons. This is followed by a chapter describing coursework developed to train designers to develop serious educational games for students with special games. Finally, the chapter in Part 5 addresses a theme echoed in several chapters: the disruptive potential of innovative digital games developed for people with special needs as these technologies expand to wider populations.

Below we provide a synopsis of each of the remaining chapters in this book.

Part 1: Technologies for Rehabilitation

In Chap. 2, "Design Issues for Vision-Based Motor-Rehabilitation Serious Games" Antoni Jaume-i-Capó, Biel Moyà-Alcover, and Javier Varona contend that a problem in game-based rehabilitation systems is that the focus is too much on the serious components of these games (task customization and measurements, for example) at the expense of user interactions, perceptions of ability, and levels of engagement. In this chapter the authors outline a game development paradigm composed of seven design elements that incorporate user interactions that motivate users to continue their rehabilitation program. These elements are described in some detail and include a development paradigm, an appropriate interaction mechanism (they advocate vision-based systems), interactive elements, feedback, adaptability, monitoring, and clinical evaluations. The authors then describe their vision-based rehabilitation game for patients with Cerebral Palsy (CP) and their implementation of these seven design elements. In a clinical study of patients with CP who had all formerly failed to adhere to conventional physiotherapy treatment programs, it was found that these same patients were motivated to complete their programs when using the rehabilitation game. Moreover, patients showed an interest in continuing the rehabilitation process even after the study was completed. This chapter provides a good overview of the developmental process of rehabilitation games (see also Chaps. 9 and 15).

In Chap. 3, "Development of a Memory Training Game," Kristoffer Jensen and Andrea Valente present a multiplatform game called Megame for memory training and assessment. The authors provide detailed descriptions of two versions of Megame, concentrating on the game parameters that can be changed to provide levels of feedback on memory capacity and the software platform used to implement the game. Also included in this chapter is a short history of some of the main theories of working memory, its importance in learning and attention, and a review of methods for memory assessment, especially focused on contemporary assessment methods using games.

In Chap. 4, "Assessing Virtual Reality Environments as Cognitive Stimulation Method for Patients with MCI," Ioannis Tarnanas, Apostolos Tsolakis, and Magda Tsolaki describe a virtual reality museum used as an intervention tool for improving navigation, spatial orientation, and spatial memory in patients with Amnestic-type Mild Cognitive Impairment (aMCI). The authors begin the chapter with an overview of aMCI, defining typical cognitive profiles of patients with subtypes of this disorder, outlining assessment methods, and discussing problems with current interventions. Also included in this chapter is a section that justifies from a neurological perspective the use of virtual environments in the treatment of aMCI. This discussion is followed by a detailed description of their virtual reality museum, including the technologies used in its development, cognitive exercises, clinical protocol, research methodology, and results using their virtual reality museum as an aMCI intervention. Older patients are shown to improve in a number of cognitive functions (for instance, working memory) after cognitive training, lending support to the idea that playing virtual reality training games improves untrained cognitive functions in aMCI.

In Chap. 5, "Adaptive Cognitive Rehabilitation," Inge Linda Wilms addresses some of the advantages and challenges inherent in translating into software the therapeutic activities necessary for successful rehabilitation of persons suffering from brain injury. The advantages and challenges primarily revolve around the following: assessment, generating adaptive levels of difficulty, individualizing rehabilitation, and providing feedback that is missing due to brain injury. It is Wilms's contention that the potentials offered by new technologies in rehabilitation have yet to be actualized. After explaining brain injury and discussing theories of rehabilitation, the author provides an historical overview of computer technology applied to brain injury rehabilitation, including games technology, focusing sections on assessment, individualization, and the use of artificial intelligence as a means of automatically adjusting levels of difficulty in cognitive rehabilitation, each section describing the specific merits of example systems and providing many suggestions for improving and expanding their effectiveness.

In Chap. 6, "A Body of Evidence: Avatars and the Generative Nature of Bodily Perception," Mark Palmer, Ailie Turton, Sharon Grieve, Tim Moss, and Jenny Lewis describe an Avatar that functions as a tool enabling patients suffering from Complex Regional Pain Syndrome (CRPS) to describe their bodily perceptions. The authors devote one section to discussing CRPS, a chronic pain condition in the extremities that alters patient perceptions of the affected limbs (perceived by some as entirely missing and by others as dramatically enlarged), perceptions so perplexing and disconcerting that many patients find it very difficult to talk about their experiences. Palmer et al. relate CRPS to the rubber hand illusion (RHI), the experience of perceiving a rubber hand as one's own, and extensions of the RHI to experiences in immersive virtual reality (IVR) with virtual limbs and bodies. Included in this chapter is a detailed discussion (both scientific and phenomenological) of the somatosensory system, i.e., the receptors and processing involved in the sense of touch, temperature, proprioception, and pain, and the central nervous system, as disruptions within these systems may account for the symptoms experienced by patients with CRPS. The remainder of the chapter describes their proof of concept study with CRPS patients using an avatar to enable them to realize and to communicate their abnormal bodily perceptions.

In Chap. 7, "Virtual Teacher and Classroom for Assessment of Neurodevelopmental Disorders," Thomas D. Parsons proposes evaluating children with High Functioning Autism (HFA) and differentiating this population from children with Attention Deficit Disorder (ADHD) by immersing them in virtual classroom environments with a virtual docent that acts as a social orienting system. Most research in the area of HFA assessment relies on paper and pencil psychometric testing of executive functions, an assessment method that has several drawbacks, including limited ecological validity, defined as the degree of similarity that a test or training method has to real world situations. Computer-based neuropsychological assessments, in contrast, offer increased standardization of administration and accuracy of timing presentation and response latencies as well as ease of administration and data collection. Included in this chapter is a detailed account of the neurodevelopmental disorders of HFA and ADHD, a brief overview of virtual reality systems related to assessment of children in virtual classrooms, including the effect these virtual reality environments have on individuals with Autism Spectrum Disorder (ASD), and descriptions of systems that use embodied conversational agents to improve social skills. Parson presents convincing arguments for building assessment systems that unite these technologies.

In Chap. 8, "Engaging Children in Play Therapy: The Coupling of Virtual Reality Games with Social Robotics" Sergio García-Vergara, LaVonda Brown, Hae Won Park, and Ayanna M. Howard focus on virtual reality play therapy with robotic systems that increases motivation and engagement for children with impaired motor skills. Systems that are motivational for adults are not necessary motivational for children. In this chapter, the authors provide an overview of the state of the art in gaming and robotics in rehabilitation, an overview that highlights important problems in existing virtual reality systems and interactive robotic devices. This is followed by a detailed description of two serious games for children that were developed by the authors: a virtual reality system that provides feedback and tracking of user performance during game play and a pilot study that integrates robotics into the virtual reality gaming scenario, with the robot serving to increase the quality of a child's interaction in a virtual environment by adaptively engaging with the child through the use of verbal and nonverbal (gesturing) cues. Studies are presented that validate the effectiveness of achieving the stated goals of both systems.

Part 2: Technologies for Music Therapy and Expression

In Chap. 9, "Instruments for Everyone: Designing New Means of Musical Expression for Disabled Creators" Rolf Gehlhaar, Paulo Maria Rodrigues, Luis Miguel Girão, and Rui Penha describe their project to enable musical expression for a group of physically and mentally challenged people by designing a special set of tools intended to match their predilections, capacities, ambitions, and tastes—a set of musical tools that were eventually used in a public performance of a musical work composed by one of the authors and that continues to be used as part of institutional occupational activities. The goals in developing the tools were not primarily therapeutic but rather geared towards enabling the new musicians to produce high quality sound given a reasonable learning curve and to promote communal musical interaction while acknowledging the special circumstances of each performer's capacities. Included in this chapter is a short history of the evolution of robotic and technology assisted musical instruments, spanning from the 3rd century BCE to the present time. The authors also provide a detailed description of their developmental process, which included administering a questionnaire to ascertain the musical interests, specific skills, and disabilities of each musician. These questionnaires were used to design the instruments, which are described in detail along with diagrams and many photographs. Also provided are complete descriptions of the musical activities that enabled the new musicians to master the instruments, of the composition process using the new instruments, and of the arrangements for the final performance. Feedback from all who participated was solicited, and two sections offer advice and a discussion of problems encountered for those interested in developing their own instruments and performances for people with disabilities. The authors feel that the constraints in the design of the novel instruments for this project became for them a catalyst for the emergence of new ideas that are relevant for artists and musicians generally.

In Chap. 10, "Designing for Musical Play," Ben Challis explores the use of music and sound in technologically assistive improvised play in sensory spaces. The author reports his visits to nine Special Needs Education (SNE) schools with the goal of evaluating their sensory spaces and determining what combinations of technology, layouts, and activities provide good sensory spaces for special needs students. This chapter includes a short history of sensory spaces, with special attention paid to the advantages and short comings of the Snoezelen model, which involves placing a person with autism or special needs in a soothing yet stimulating environment, along with some background information on music and sound therapy. The remainder of the chapter focuses on the nine SNE schools and their sensory spaces, which the author evaluates in terms of the general characteristics of the spaces themselves, the activities (music and sound) available within the space, and the technologies that utilize touch, movement, pressure, and other actions triggering midi events to produce sound. The chapter ends with a critique of current sensory spaces and activities and with suggestions for improving them by making these spaces more interactive and less passive.

Part 3: Technologies for Well-Being

In Chap. 11, "Serious Games as Positive Technologies for Individual and Group Flourishing," Luca Argenton, Stefando Triberti, Silvia Serino, Marisa Muzio, and Giuseppe Riva explore how serious games can empower personal experience by nurturing positive emotions and by promoting engagement and social connections. The authors argue that serious games can be considered "positive technologies," a field of study that assumes technology has the capacity of increasing emotional, psychological, and social well-being and that investigates how Information and Communication Technologies (ICT) empower and enhance the quality of personal experiences in these areas. The chapter begins with an overview of the concept of well-being, not so much from a philosophical perspective, but rather as this concept has been understood and investigated in positive psychology, especially by Csikszentmihalyi. The chapter then moves on to discuss serious games, conceptualized as flow activities that naturally evoke positive emotional states: the sensorial pleasure in the game aesthetics, the epistemophilic pleasure in the way games stimulate curiosity and satisfy desires for novelty in environments that offer occasions for developing mastery and control, and the pleasure of victory experienced in interacting and collaborating with others in multiuser games. Specific games are discussed in terms of promoting well-being and a pilot study is described that explored the impact a serious game had on the optimal functioning of different groups.

In Chap. 12, "Spontaneous Interventions for Health: How Digital Games May Supplement Urban Design Projects" Martin Knöll, Magnus Moar, Stephen Boyd Davis, and Mike Saunders explore the possibilities of using mobile games to inform the relationship between health and the external environment. The authors describe how doctors, sanitarians, and town planners have historically worked together, starting as early as the late 19th century, to build housing, streets, and parks and how the citizenry in the form of stroller clubs, for example, worked to increase physical activity in an era when more people were becoming sedentary. These clubs, however, did not stimulate members to explore, question, and reshape their environment. Contemporary analogs of stroller clubs take the form of iPhone apps that invite players to climb monuments and to take the stairs, but they too have the potential to do more. To illustrate possibilities, the authors review some urban design projects where citizens work to reshape their local environments to stimulate physical activity, a phenomenon referred to as "spontaneous interventions." The authors also review persuasive games that encourage healthy behaviors throughout the environment. The authors then persuasively argue that health games could be used in the future not only to increase healthy behaviors but also to supplement urban research and design by raising attention to new complexes, by stimulating participation, by identifying locales for potential improvement, and by evaluating impact. A road map is provided for developing interactive communication technologies that support spontaneous interventions. Another important contribution of this chapter is to show how the disparate disciplines of urban design, preventive healthcare, and serious gaming could be merged into a new field of research that benefits the health of citizens.

In Chap. 13, "Using Virtual Environments to Test the Effects of Lifelike Architecture on People," Mohamad Nadim Adi and David J. Roberts describe their work using virtual reality to investigate how interactive elements in architectural designs (animated, reactive, organic, and intelligent architecture) influence people's emotional states (that can range from feelings of well-being to feelings of uncanniness) and productivity (the experience of flow) using interactive scenarios. In this chapter the authors provide a brief history of interactive, or lifelike, architecture along with a detailed discussion of some of the benefits and problems using virtual reality for testing and evaluating interactive buildings. The authors then report the results of several experiments using an immersive large screen projection system they call OCTAVE for evaluating interactive building designs. Their findings suggest that immersive systems are of value in evaluating the impact that lifelike architectural designs are likely to have on occupants.

Part 4: Technologies for Education and Education for Rehabilitative Technologies

In Chap. 14, "An Overview of Virtual Simulation and Serious Gaming for Surgical Education and Training," Bill Kapralos, Fuad Moussa, and Adam Dubrowski address a number of issues in the application of serious games for surgical education and training, specifically focusing on issues of fidelity, multimodal cue interaction, immersion, and knowledge transfer and retention. The chapter begins with an overview of the literature on medical and surgical educational methods, first by exploring problems and ethical considerations using cadavers and animals and then by moving on to discuss issues involved with simulations, such as the lack of fidelity and limited sensory modalities. The authors admit that both traditional and simulated methods of training offer residents the opportunity of developing the specific levels of expertise needed to perform surgeries on human beings, but simulations have the marked advantages of being more cost effective and of encouraging more experimentation in purposively making and then learning to handle mistakes. This overview of training methods is followed by a discussion of fidelity and multimodal cue interaction, including a discussion of visual, auditory, olfactory, and haptic modalities, as these elements function more generally in virtual simulations and in serious games, but all with an eye towards considering how best to incorporate these important modalities in surgical simulations. The chapter ends with a presentation of the authors' Surgical Cognitive Education and Training Framework (SCETF), a framework that explicitly addresses fidelity and multimodal cue interaction, the authors providing a detailed description of the graphical and sound rendering capabilities of SCETF as well as their methods for handling multi-cue interaction.

In Chap. 15, "The Ongoing Development of a Multimedia Educational Gaming Module," Elizabeth Stokes provides a detailed description of an ongoing multimedia educational module that facilitates the design of games for students with special needs, especially for those students on the autistic spectrum who vary widely in their abilities, medical conditions, computer skills, and interests. The author describes coursework developed for undergraduate students who were assigned the task of developing personalized educational games that were based on actual individualized profiles of real learners with disabilities (not hypothetical subjects), games that had the potential of actually benefiting these special needs students.

Part 5: Disruptive Innovation

In Chap. 16, "Disruptive Innovation in Healthcare and Rehabilitation," Anthony Brooks discusses disruption and innovation in terms of his work. Disruption is a powerful body of theory that describes how people interact and react, how behavior is shaped, how organizational cultures form and influence decisions. Innovation is the process of translating an idea or invention into a product or service that creates value or for which customers will pay. Disruptive Innovation in context of Brooks' work in healthcare and rehabilitation relates to how development of a cloud-based converged infrastructure resource, similar to that conceived in a national (Danish) study entitled Humanics, can act as an accessible data and knowledge repository, virtual consultancy, networking, and training resource to inform and support fields of researchers, practitioners and professionals. High-speed fiber networking, smart phone/tablet apps, and system presets can be shared while AI and recommendation engines support directing global networks of subscribers to relevant information including methods and products. Challenges and problems to fully realize potentials are speculated. A review of his research acts as the vehicle to illustrate such a concept.

1.3 Conclusion

This book is a continuation of the previous volumes in our series on *Advanced Computational Intelligence Paradigms in Healthcare*. The fact that this book on serious games, alternative realities, and play therapy is published under the series of *Studies in Computational Intelligence* (Springer SCI) reflects the widening interdisciplinary scope of applied computation. From this perspective it is anticipated that this first volume on serious games, alternative realities, and play therapy will promote and provoke peer debate across various cultures and fields.

Part I
Technologies for Rehabilitation

Chapter 2
Design Issues for Vision-Based Motor-Rehabilitation Serious Games

Antoni Jaume-i-Capó, Biel Moyà-Alcover and Javier Varona

Abstract When rehabilitation sessions are for maintaining capacities, the demotivation of patients is common due to the difficulty in improving their situation. Recent experiments show that rehabilitation results are better when users are motivated and serious games can help to motivate patients in rehabilitation processes. We developed a rehabilitation serious-game for a set of patients who had abandoned therapy due to demotivation in the previous years. The serious-game was developed following desirable features for rehabilitation serious-games presented in the related research works. From this development, we present implementations guidelines for developing serious-games as motivational tool for rehabilitation therapies. The experiments performed in previous works validate that the interacting design issues defined help motivation in therapy and that they are adequate in rehabilitation therapy.

Keywords Serious games · Rehabilitation · Vision-based interfaces

A. Jaume-i-Capó (✉) · B. Moyà-Alcover · J. Varona
Departament de Ciències Matematiques i Informaticà, Universitat de les Illes Balears, Cra. de Valldemossa km 7.5, 07122 Palma, Spain
e-mail: antoni.jaume@uib.es
URL: http://www.uib.es

B. Moyà-Alcover
e-mail: gabriel.moya@uib.es

J. Varona
e-mail: xavi.varona@uib.es

A. L. Brooks et al. (eds.), *Technologies of Inclusive Well-Being*, Studies in Computational Intelligence 536, DOI: 10.1007/978-3-642-45432-5_2, © Springer-Verlag Berlin Heidelberg 2014

2.1 Introduction

A serious game [15] is defined as a game that allows the player-user to achieve a specific purpose through the entertainment and engagement component provided by the experience of the game. Recent research [4, 16, 20] shows that the cognitive and motor activity required by video games engage the user's attention. In addition, users focus their attention on the game and this helps them in forgetting that they are in therapy [20, 23]. This motivation is particularly important in long-term rehabilitation for maintaining motor capacities. Demotivation is frequent in chronic patients, because therapy usually consists of repetitive and intensive activities that become boring. As a result, the user may not concentrate on the therapy program which thereby loses its effectiveness.

Therefore, in order to build appropriate serious games for rehabilitation purposes, therapy studies have included the motion-based input devices for serious games, such as $EyeToy^{TM}$ [8, 19] or $Wiimote^{TM}$ [7]. A common conclusion in previous works is that existing commercial video games for these devices are difficult to use in rehabilitation therapy, because they were designed for patients with full capabilities. Specifically, [2] enumerates the problems associated with commercial video games when used in rehabilitation: targeting mainly upper body gross motor function, lack of support for task customizations, grading, and quantitative measurements. In addition, Sandlund et al. [22] state that patient's interest in gaming faded somewhat over time indicating that there is a need for flexible games that adapt to the changing ability of the user and offer a continuous challenge to maintain the interest. In order to solve these problems, researchers have been developing their own games.

Unfortunately, game-based rehabilitation system designers frequently over-emphasize the serious game rather than the user interaction. When these games are designed for disabled people, the interaction design issues are fundamental to achieve a high patient motivation. In addition, game interaction design is usually defined without taking into account user's perceptions with regard to their actions in order to achieve the rehabilitation goals.

Motor rehabilitation usually consists of body movements and patients with motor disabilities can have difficulties holding physical devices. Computer vision technology allows capture body movements and it is non-invasive. For this reason, we built interactive games using computer vision techniques, which can be summarized by capturing the visual information of the performance of user actions.

The experiments performed in previous works [10, 16, 21] validate that the interacting design issues defined help motivation in therapy and that they are adequate in rehabilitation therapy. In this work we present a set of design issues, from our previous work implementing vision-based serious-games for motor rehabilitation.

This work is organized as follows. In Sect. 2.2, we present the previous work. Next, in Sect. 2.3 are presented the design issues. Section 2.4 is devoted to presenting a case study focusing on the design issues. Section 2.5 is dedicated to conclusions.

2.2 Related Work

In previous work, we can find different serious games for different types of rehabilitation. Video games for balance rehabilitation have been presented in [3] where therapy is performed by a center-of-pressure, and the game is designed to older adults in order to incorporate appropriate balance exercises. Regarding upper limb rehabilitation, [14] presented a serious-game based movement therapy which aims to encourage stroke patients with upper limb motor disorders to practice physical exercises, [5] showed Virtual Reality (VR) system for stroke patients, [4] designed several serious games which use low-cost webcams as input technology to capture data of users movements, [6] created a simple game in which the patient tried to move a coloured circle from an initial position to a goal position using a robotic device designed for arm rehabilitation, in [9] implemented a haptic glove serious game for finger flexion and extension therapy, [1] presented several home-based serious games which use a webcam and a *Wiimote*^{TM}, and [2] designed a low-cost VR-based system using *Wiimote*^{TM}.

Recent research studies have proposed what features are desirable for rehabilitation serious games. [11] proposed target audience, visibility and feedback as important human factors, [4] identified two principles of game design theory which have particular relevance to rehabilitation: meaningful play, the relationship between player's interactions and system reaction, and challenge, maintaining an optimum difficulty is important in order to engage the player. [20] identified as important main criteria for the classification of serious games in the rehabilitation area: application area, interaction technology, game interface, number of players, game genre, adaptability, performance feedback, progress monitoring and game portability. Alankus et al. [1] concluded that serious games must ensure that patients are correctly performing and must provide a motivating context for therapy, in order to have maximum impact on the rehabilitation process.

2.3 Design Issues

From our experience of implementing vision-based motor-rehabilitation serious games [10, 16, 21], defining an interaction model adapted to the user's capabilities and following the desirables features for rehabilitation serious games presented, we can state that success of using this type of serious game depends on 7 design issues: the development paradigm, the interaction mechanism, the interactive elements, the feedback, the adaptability, the monitoring, and the clinical evaluation. These design issues are presented in more detail in the next section.

2.3.1 Development Paradigm

After first meetings with physiotherapists, we discovered that engineering technical language is totally different to physiotherapy. To ensure objectives, we decided to develop the game using the prototype development paradigm [18], which facilitated communication between engineers and physiotherapists. This paradigm ensures all the necessary information to perform the different tasks is provided in a clear and understandable way.

2.3.2 Interaction Mechanism

A serious game should not develop a new rehabilitation therapy. It is more suitable to transfer existing therapy to a serious game, where the selected rehabilitation therapy is the means of interaction with the serious game. As a major number of patients with motor disease cannot hold a device, a camera can be used as an input device in order to define the interaction model adapted to the user's capabilities [17]. Recent technological advances have created the possibility to enhance naturally and significantly the interface perception by means of visual inputs [24], the so-called Vision-Based Interfaces (VBI).

In general, vision-based interaction systems aim to provide reliable computer methods to detect and analyse human movements. The process is repeated over time, allowing for monitoring of the users interacting actions. According to the computer vision technique used it is possible to achieve different levels of detail. In addition, due to the dependence on real conditions (lighting, distances, clothes...) the interaction environment limits the techniques that can be used. Figure 2.1 depicts one example of vision-based interaction which can be implemented by detecting the users silhouette, the skin colour or the hand motion.

2.3.3 Interaction Elements

Interaction objects must be selected in order to show the users images that achieve an optimal level of motivation, by choosing themes of particular interest to each user. When the interaction objects were related to some of these interests [10], the patients performed the rehabilitation activity faster.

2.3.4 Feedback

The game must respond to the actions of the user through different types of feedback, in order to user be aware of their current state. In general, when using VBI with a system [4, 11], providing feedback is critical for users to feel in control

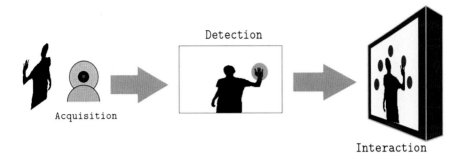

Fig. 2.1 Vision-based interaction which can be implemented by detecting the users silhouette, the skin colour or the hand motion

and helping them to understand what is happening. Especially if there is not contact with the interface by means of an interaction device. A significant problem of vision-based interaction is that users have no interaction device of reference. The user, therefore, always should know when interaction is taking place using visual and audible feedback.

2.3.5 Adaptability

Rehabilitation sessions must adapt to the characteristics of the different users [4, 11, 20]. As the difficult level of therapy depends on the user interaction (motions), physiotherapists should create a set of templates which define the position where interaction elements must have, in order to define different levels in the game depending on the skills and the evolution of each user. In addition, the game should define different configurations parameters to customize games and adapt them to different users.

2.3.6 Monitoring

The system must be able to archive different information about each user, configuration parameters and a dataset for each session consisting of patients performance, in order to simplify a patient's progression as monitored by the specialists [1, 11, 20]. Therefore, the system has two class of users, the patient and the specialist. Each are in pursuit of different objectives of the interaction with the system, see Fig. 2.2.

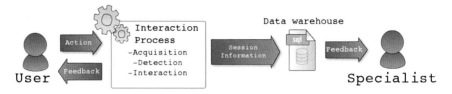

Fig. 2.2 Feedback system depending on user role

2.3.7 Clinical Evaluation

The clinical evaluation aims to quantify the improvement of the rehabilitation according to the kinds of functional exercises. In order to perform a successful clinical evaluation, it should define the experiment, the participants, and the measurements depending on the final goals and type of therapy. However, from our experience, we recommend to design the whole intervention through a pre-assessment and post-assessment of every measurement, see Fig. 2.3. Optimally, the success of clinical evaluation increase when one can include a control group and the largest number of measurements.

2.4 Case Study: Vision-Based Rehabilitation Serious Game

We present a rehabilitation serious game that we implemented for a Cerebral Palsy (CP) rehabilitation center [16]. CP is a term used to describe a group of chronic conditions affecting body movement and muscle coordination. CP is the most common physical disability in children [13]. The number of adults with cerebral palsy is increasing due to increased survival of low birth weight infants, and increased longevity of the adult population. Many children and adults with CP have poor walking abilities and manipulation skills. One of the factors that contributes to their problems with gait and reaching movements is poor balance control. The objectives of medical intervention and physical therapy are to prevent and mitigate the degradation of balance and postural control, in order to conserve autonomy of movement, and to improve muscle coordination when performing voluntary movements.

Each year, part of the center's patients ceased the therapy, after years of rehabilitation sessions to maintain the capacities. The habitual users of the rehabilitation center, are aware that rehabilitation sessions are focused on maintaining capacities, and so they experience demotivation due to the difficulty in improving their situation, and also due to the repetitive nature of the exercises performed in every session. Therefore, the rehabilitation center was interested in a balance rehabilitation serious game to motivate their users, in order to favour coordination and trunk control, to stimulate cognitive and communicative aspects, and improve their activity Activities of Daily Living (ADL).

Fig. 2.3 Visual feedback

2.4.1 Development Paradigm

The serious game was developed using the prototype development paradigm, following requirements indicated by physiotherapists, for a year. During this period, in order to improve the game, the system was tested with real users once a week.

2.4.2 Interaction Mechanism

We selected balance exercises from standard therapy in the rehabilitation room [12], to transfer to a serious game in order to improve balance, to reduce demotivation in patients, and to obtain more adherence to this long-term therapy. The selected balance rehabilitation therapy is the means of interaction with the serious game.

We designed a game which covers the objective required by physiotherapists, which consists in changing the user gravity center, where users stand in front of the standard monitor (or large projector) and, using their movements, they interact with the video game, see Fig. 2.4. In addition, since users may have difficulty in holding devices, the designed game is markerless and device free. With this configuration, users can see the serious game while they are interacting with it. The serious game environment is defined as follows:

- Users must be facing the screen to see the serious game in order to interact using the balance rehabilitation therapy
- Users must see themselves on the screen in order to orient themselves with respect to the interaction objects.

The designed video-game tries to cause a specific body movement in order to change the user gravity center. To do this, the users must interact, by means body movements, with objects they cannot reach without changing their center of mass, see Fig. 2.5. Concretely, the user must delete a set of items that appear on the monitor.

In this way, a user focuses their attention in the serious game instead of their posture, because the goal was not to improve the rehabilitation process, but rather to improve user's motivation.

Fig. 2.4 Game environment configuration

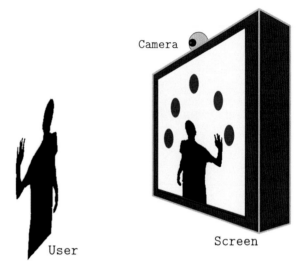

Camera

User   Screen

Fig. 2.5 User magnitude without changing center of mass, in *grey*

2.4.3 Feedback

Visual feedback consists of deleting or changing of the object of interaction. Auditory feedback is added in order to reinforce the action result. For example when the interaction object is completely deleted from the screen, an audio feedback is played. And when the game ends, the user receives different types of visual and audio feedback, depending on the end game conditions.

Moreover, we used another type of feedback in VBI, represented by the provision of the visual representation of the users body on the screen, see Fig. 2.6. Users must see themselves on the screen in order to orient themselves with respect to the interaction objects. This configuration allows the user to view the serious game and themselves while performing the interaction.

Fig. 2.6 Visual feedback

2.4.4 Interaction Elements

Interaction elements can be changed in order to show the users' images to achieve an optimal level of motivation. This is by choosing themes of particular interest to each user (see Fig. 2.7). In the study, patients were asked what hobbies they had in order for the designer to seek interaction elements related to these interests.

2.4.5 Adaptability

In order to make rehabilitation sessions adaptable to the characteristics of the different users, a set of templates were created, that define the size and the position that interaction objects must have (see Fig. 2.7), so we can define different levels in the game depending on the skills and the progression of each user. In addition to the patterns, we defined the following configuration parameters to customize games and adapt them to different users:

- *Maximum playing time*: The specialist can set a limit time for each session depending on the users characteristics
- *Mirror effect*: To increase the game difficulty, the game screen can be reversed. So when users move their right hand, they view as if they move their left hand
- *Contact time*: The specialist can customize how long a player must be in contact with an element in order to delete that object
- *User distance*: The distance between user and screen in meters. The larger the distance from the screen, the larger the center-of-mass change needed.

With these implemented adaptability we ensured the *challenge* [4].

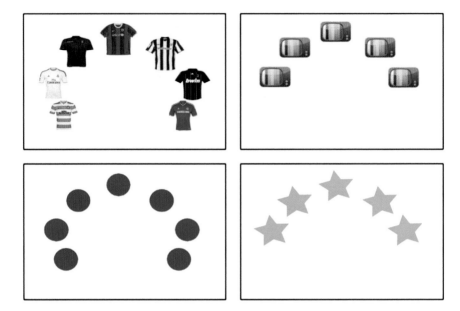

Fig. 2.7 Interaction screens with themes for motivating users and different templates

2.4.6 Monitoring

The serious-game saves and maintains an xml file for each user, which is easy to parse and analyse, where configuration parameters and a dataset for each session is stored. This consists of: date of the session, level pattern, playing time, removed percentage, user distance from monitor device and contact time. This way the monitoring of users by the therapists is simplified. This design has the potential to be used in other rehabilitation systems. It is important to remark that there are two types of users: the user and the specialist, with different interaction objectives with the system.

2.4.7 Clinical Evaluation

The research team made a request to all adults in the rehabilitation center who met the inclusion criteria and 90 % agreed to participate in the study. The study population finally included 9 adults, 2 women and 7 men. Their families signed an informed consent.

One of the inclusion criteria was having no adherence to physiotherapy treatment after attending the conventional program of the centre as a long-term therapy. Therefore, patients in this study were undergoing rehabilitation only with our experimental system. This rehabilitation program started after 24 weeks of training, and it consisted of one session per week, during 24 weeks. Before and after the 24 week therapy period, users were pre- and post-assessed using Berg Balance Scale, Functional Reach Test, and Balance Tinetti Test.

Results showed [10, 16] that patients improved their balance slowly; improvements were also detected in individual items. With regards to motivation, in previous years the set of users had abandoned their therapeutic plans. Using the presented video games, no patients abandoned and, on completion of the study, they showed interested in continuing the rehabilitation process with the video games.

2.5 Conclusions

From our previous work implementing vision-based serious-games for motor-rehabilitation, in this work we have presented implementations guidelines for developing vision-based motor-rehabilitation serious games, based on related work and our experience, in order to help others researchers in this field. We can state that the successful of this type of serious games depends on 7 design issues: the development paradigm, the interaction mechanism, the interactive elements, the feedback, the adaptability, the monitoring, and the clinical evaluation.

As a case study, we showed a serious game that we implemented for a CP rehabilitation center, in order to improve patients' balance that consists in changing the user gravity center. The technology employed were vision-based interaction, due to the difficulties of some patients have for holding physical devices. Results showed that patients improved their balance slowly. Regarding motivation, the set of users had resigned their therapeutic plan in previous years. Using the presented video games, no users abandoned and, on completion of the study, they showed interested in continuing the rehabilitation process with the video games.

Acknowledgments This work was partially supported by the projects MAEC-AECID A2/037538/11 and TIN2012-35427 of the Spanish Government, with FEDER support.

References

1. Alankus, G., Lazar, A., May, M., Kelleher, C.: Towards customizable games for stroke rehabilitation. In: Proceedings of the 28th International Conference on Human Factors in Computing Systems, pp. 2113–2122, ACM (2010)
2. Anderson, F., Annett, M., Bischof, W.F.: Lean on wii: physical rehabilitation with virtual reality and wii peripherals. Annu. Rev. Cybertherapy Telemedicine **8**, 181–184 (2010)
3. Betker, A.L., Desai, A., Nett, C., Kapadia, N., Szturm, T.: Game-based exercises for dynamic short-sitting balance rehabilitation of people with chronic spinal cord and traumatic brain injuries. Phys. Ther. **87**(10), 1389–1398 (2007)
4. Burke, J.W., McNeill, M.D.J., Charles, D.K., Morrow, P.J., Crosbie, J.H., McDonough, S.M.: Optimising engagement for stroke rehabilitation using serious games. Vis. Comput. **25**(12), 1085–1099 (2009)

5. Cameirao, M.S., Bermudez, I.B.S., Duarte Oller, E., Verschure, P.F.: The rehabilitation gaming system: a review. Stud Health Technol. Inform. **145**, 65–83 (2009)
6. Colombo, R., Pisano, F., Mazzone, A., Delconte, C., Micera, S., Carrozza, M.C., Dario, P., Minuco, G.: Design strategies to improve patient motivation during robot-aided rehabilitation. J. Neuro-engineering Rehabil. **4**, 1–12 (2007)
7. Deutsch, J.E., Borbely, M., Filler, J., Huhn, K., Guarrera-Bowlby, P.: Use of a low-cost, commercially available gaming console (wii) for rehabilitation of an adolescent with cerebral palsy. Phys. Ther. **88**(10), 196–207 (2008)
8. Flynn, S., Palma, P., Bender, A.: Feasibility of using the Sony Playstation 2 gaming platform for an individual poststroke: a case report. J. Neurol. Phys. Ther. **31**(4), 180 (2007)
9. Jack, D., Boian, R., Merians, A.S., Tremaine, M., Burdea, G.C., Adamovich, S.V., Recce, M., Poizner, H.: Virtual reality-enhanced stroke rehabilitation. IEEE Trans. Neural Syst. Rehabil. Eng. **9**(3), 308–318 (2001)
10. Jaume-i-Capo, A., Martinez-Bueso, P., Moya-Alcover, B., Varona, J.: Interactive Rehabilitation System for Improvement of Balance Therapies in People With Cerebral Palsy. IEEE Trans. Neural Syst. Rehabil. **PP**(99), (2012). doi:10.1109/TNSRE.2013.2279155
11. Jung, Y., Yeh, S.C., Stewart, J.: Tailoring virtual reality technology for stroke rehabilitation: a human factors design. In: CHI'06 Extended Abstracts on Human Factors in Computing Systems, pp. 929–934, ACM (2006)
12. King, L.A., Horak, F.B.: Delaying mobility disability in people with parkinson disease using a sensorimotor agility exercise program. Phys. Ther. **89**(4), 384–393 (2009)
13. Krigger, K.W.: Cerebral palsy: an overview. Am. Fam. Physician **73**(1), 91–100 (2006)
14. Ma, M., Bechkoum, K.: Serious games for movement therapy after stroke. In: IEEE International Conference on Systems, Man and Cybernetics, 2008. SMC 2008, pp. 1872–1877, IEEE (2008)
15. Michael, D.R., Chen, S.L.: Serious games: games that educate, train, and inform. Muska & Lipman/Premier-Trade (2005)
16. Moyaa-Alcover, B., Jaume-i-Capoo, A., Varona, J., Martinez-Bueso, P., Mesejo Chiong, A.: Use of serious games for motivational balance rehabilitation of cerebral palsy patients. In: The Proceedings of the 13th International ACM SIGAC—CESS Conference on Computers and Accessibility, pp. 297–298, ACM (2011)
17. Norman, D.A.: The invisible computer: why good products can fail, the personal computer is so complex, and information appliances are the solution. MIT press, Cambridge (1999)
18. Overmyer, S.P.: Revolutionary vs. evolutionary rapid prototyping: balancing software productivity and hci design concerns. In: Proceedings of the Fourth International Conference on Human-Computer Interaction, pp. 303–307 (2002)
19. Rand, D., Kizony, R., Weiss, P.L.: Virtual reality rehabilitation for all: Vivid GX versus Sony Playstation II EyeToy. In: 5th International Conference on Disability, Virtual Environments and Association Technologies, pp. 87–94 (2004)
20. Rego,P., Moreira, P.M., Reis, L.P.: Serious games for rehabilitation: a survey and a classification towards a taxonomy. In: 5th Iberian Conference on Information Systems and Technologies (CISTI), pp. 1–6, IEEE (2010)
21. Reyes-Amaro, A., Fadraga-Gonzalez, Y., Vera-Perez, O.L., Dommguez-Campillo, E., Nodarse-Ravelo, J., Mesejo-Chiong, A., Moyaa-Alcover, B., Jaume-i-Capó, A.: Rehabilitation of patients with motor disabilities using computer vision based techniques. JACCES: J. Accessibility Design All **2**(1), 62–70 (2012)
22. Sandlund, M., Lindh Waterworth, E., Hager, C.: Using motion interactive games to promote physical activity and enhance motor performance in children with cerebral palsy. Dev. Neurorehabilitation **14**(1), 15–21 (2011)
23. Sandlund, M., McDonough, S., Hager-Ross, C.: Interactive computer play in rehabilitation of children with sensorimotor disorders: a systematic review. Dev. Med. Child Neurol. **51**(3), 173–179 (2009)
24. Varona, J., Jaume-i Capo, A., Gonzalez, J., Perales, F.J.: Toward natural interaction through visual recognition of body gestures in real-time. Interact. Comput. **21**(1), 3–10 (2009)

Chapter 3
Development of a Memory Training Game

Kristoffer Jensen and Andrea Valente

Abstract This paper presents Megame *[me-ga-me]*, a multiplatform game for working memory training and assessment. Megame uses letters and words to amuse, train and assess memory capacity. Based on memory research, the main parameters of the working memory have been identified and some improvement possibilities are presented. While it is not clear that the working memory in itself can be improved, other cognitive functions are identified that may be improved while playing Megame. Other uses of Megame include spelling and vocabulary training and learning modality assessment. Megame is written in Python using an agile and iterative approach, taking advantage from rapid prototyping and allowing for user-driven development.

Keywords Memory · Learning · Attention · Interference · Game

3.1 Introduction

Memory is a central component of cognition and thinking in general. According to Baddeley [1] reasoning, comprehension, and learning capacity are all correlated with the capacity of the short-term memory. The concept of short-term memory is now largely superseded by that of the working memory, active in all aspects of memory, the encoding, storage and retrieval. For instance, the working memory is active in the consolidation process, at least when actively thinking about the information being consolidated. It is therefore of importance to utilize ones

K. Jensen (✉) · A. Valente
Department of Architecture, Design and Media Technology, Aalborg University Esbjerg,
Niels Bohrsvej 8, 6700 Esbjerg, Denmark
e-mail: krist@create.aau.dk

A. Valente
e-mail: av@create.aau.dk

A. L. Brooks et al. (eds.), *Technologies of Inclusive Well-Being*,   25
Studies in Computational Intelligence 536, DOI: 10.1007/978-3-642-45432-5_3,
© Springer-Verlag Berlin Heidelberg 2014

memory in the best possible way. While it is not clear that one can improve the memory using memory-training methods [2], many general brain and memory enhancement games exists. Still, some of the games may be fun to play, and at least allow improvement in tasks related to the game played. In addition specific elements related to memory, such as attention, interference, chunking etc. can be supported, stimulated or trained by playing simple games. We present here an approach to such a game, with a focus on training the working memory while directly estimating its capacity, thus motivating the player to improve the memory while playing.

The game itself is inspired by classic memory games, like *memory*,[1] and we envision it as a single-player, multi-platform game. The gameplay is simple: various symbols, images or shapes are presented to the player, then hidden; the player has to remember them later, while composing a sequence (of letters for example). Megame has many configurable parameters, for instance we design it so that we are able to increase the number of letters, and in that way assessing the limit with respect to the number of elements in working memory. Memory capacity assessment is the central goal of the Megame game but other uses are also envisioned. Finally, it is not our goal to create an addictive game, but rather a good occasional challenge that can be used for players' self-assessment and that will provide us with a tool for testing different memory models.

This chapter is organized as follow, Sect. 3.2 presents the main theories on memory, in particular the working memory, and the relationships of importance between this memory and learning, attention and memory assessment. Section 3.3 presents the design and development of Megame, with details of the two versions of the game and the parameters that can be changed when changing level in the game, and the details of the feedback on the memory capacity. Section 3.3 also covers the software development methodology. In Sect. 3.4 different uses and settings are shown to be of interest for scientific experimentation about the memory and related areas, and the chapter ends with a conclusion.

3.2 Memory

The processes involved in memory are encoding, storing and retrieving information, and this processing of information may be considered to be central to cognition [1]. Memory is considered to take place in three stages [3], the initial sensory memory of span up to 300 ms approximately, the working memory with a span of a few seconds and up to a minute using attention, and the long-term memory (LTM) with a time span of years. The sensory store and the working memory are modality-specific, which means that they treat in particular visual, auditory and touch

---

[1] http://en.wikipedia.org/wiki/Concentration_(game).

independently. This may have implications for Megame, and future work include using items in different modalities, in order to train and assess these.

3.2.1 History of Memory Ideas

Memory is today very related to information and we take for granted that machines can remember and find data for us. However, as discussed in Rose [4] remembering was a complex and central skill in pre-writing societies. Some professions developed techniques for enhancing individual's memory: for instance poets could memorize entire epic poems by re-structuring them into rimes, using a certain *metre*, and by the means of singing. Writing as a form of external, persistent memory was opposed by great figures like Socrates (for whom we know through the writings of Plato), since:

> Writing destroys memory and weakens the mind, relieving it of work that makes it strong. Writing is an inhuman thing.[2]

However, apart from poets, other ancient professions required memorization of long presentations and arguments, lawyers for instance. Therefore, various mnemotechnic methods were devised; an example is the *method of loci* described by Cicero in his *De Oratore*.[3] Its key idea is to remember unstructured information by association with specific physical locations or objects in a room. The person that wants to memorize the items can, for example, visualize a room that is familiar to her (or a fictitious room) and associate a word or an item to each object in that room. To remember, one takes a mental tour of the room and retrieve words by association with the objects that are encountered. Other memorization techniques suggest the creation of a story, where the details of the narrative help recalling items.

These and many other memorization techniques, described throughout antiquity and the middle ages,[4] show that memory was always considered something that need to be trained and that assumption is clearly that memory can be improved following the right methods.

3.2.2 Memory Overview

It is clear that any interaction, such as Megame, that does not involve all modalities in the early memory systems, will not have an influence on all parts of

---

[2] In *Phaedrus*, Plato, around 500 B.C.

[3] Literally "On the Orator", written circa 55 B.C.

[4] As in *Ars Memoriae* ("The Art of Memory") by Giordano Bruno, 1582.

the working memory. It is interesting in this context to determine the principal target group of the game. This can be done, for instance, by looking at the use for dyslexic children, or by choosing the target group using a questionnaire, such as the VARK questionnaire [5]. In the VARK model, learning is supposed to use one modality principally, Visual, Aural, Read/write or Kinesthetic. As a large part of learning in schools take place in the read/write category, children with less preference for this modality benefit from activities, such as playing Megame, that strengthens the preference for reading and writing.

While Megame mainly regards the working memory, the identification/recognition of correct word is related to the long-term memory and other cognitive skills. In general, the working memory can be said to be active in all three memory processes, as it is central in the encoding stage, where each new information element is temporarily stored, but also active in the other stages, as information stored or retrieved necessarily passes through the working memory. Thus, improvement of the working memory is bound to be beneficial to all aspects of memory, and thus of cognition in general. The proposed game is also likely to be influential in increasing the vocabulary due to the generation effect [6], that states that information that is generated, such as the words to be created from the letters in Megame, is better memorized.

While the game remains in the letter/word category, no serious issues should arise from the complicated level of representation processing [3]. This processing enables us to remember the meaning of information (that resides in higher levels of processing), while forgetting the exact wording that resides of lower levels. This occurs, for instance, when you read something and the say it in different wording. The information has passes to a higher level of processing, where the meaning, and not the wording is central, and then back again to be said out loud. Assumingly, there is no such level of representation in spelling, as this is an exact, deterministic process.

If the target word is visible, the player does not need any cues. However, if the target word is hidden, the results may be dependent on the player's ability to cue the correct word. Cueing or priming is a common process in memory retrieving, but it is not believed to be of strong relevance in these games, that is regarding letters only in this phase of development. Still priming [7] is seen as a potential means of increasing or affecting the difficulty of the game.

The two main factors that influence memory seem to be attention, which increases the time the information remains in the memory (by increasing the activation strength of the element under attention), and interference that weakens the activation strength of the information due to conflicting memory traces. Without attention, memory weakens within a minute, and it is believed that the development of attention combined with efficient coding schemes are the main possible memory enhancement. Better and more efficient encoding reduce interference, as new information is placed at proximity to relevant but not interfering information. However, Wixted [8] resumes the research on forgetting (which may take place at the time of encoding, or at the time of retrieval), and advances the role of the limited capacity of the brain, that induces retroactive interference, i.e.

that new memory encoding taking place after the first memory will interference with the consolidation process of this memory trace. This would also explain why sleep and drugs are helpful in retrieving memory traces, and it could be explaining the use of power-naps, and other structured methods of relaxation within e.g. mediation. Another different aspect of memory is chunking [9]. For instance, the capacity of reading involves chunking, as the letters that constitute a word are chunked into that word in the cognitive processes.

Memory traces are stored through consolidation [10] and also through repetitions. Consolidation consist of an initial synaptic stage that takes place within minutes and up to a few hours, and the system consolidation that takes place within weeks up to years. After the synaptic stage, the memory trace is resistant to interference. After the system consolidation, the information trace is no longer bound to the main memory cortical part, the hippocampus. Consolidation occurs when information is repeated. Repetitions can be either massed or distributed, and according to Glenberg and Lehmann [11], distributed repetitions are more efficient in memory recall. This is generally called the spacing effect.

3.2.3 Memory Assessments

Megame can assess the capacity of the working memory of the participant. In order to do so, the model presented in Jensen [12] is used. This model considers two components of the working memory: the number of elements, and the duration of each element. The activation strength of the working memory is decaying exponentially with the number of elements, $A_n = 1 - ln(N + 1)$, and with the duration of the element, $A_t = 1 - ln(t + 1)$. The total activation strength is $A = A_n + A_t$. In that way, the model can either contain many elements for a short time, or few elements a long time, as shown in Fig. 3.1. When the activation strength of an element becomes negative, $A < 0$, the element is purged from memory.

Another aspect where attention comes into play is the serial position effect, where earlier elements, which potentially have received more repetitions and thus entered the realms of the long term memory, are better remembered than the middle elements. It does not seem to be a problem in our game, as obviously the earlier letters may receive more attention and rehearsals, but it must be understood that these letters are already present in the long term memory (at least for everybody except very young children), and it does not affect the assessment of the memory capacity to any degree.

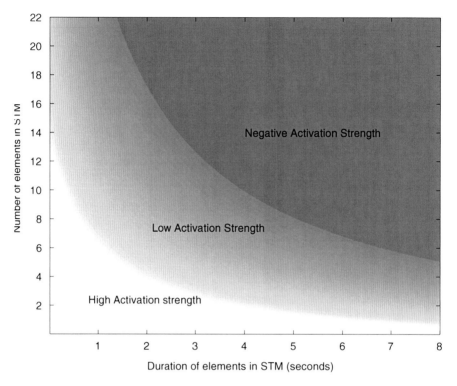

Fig. 3.1 Activation strength as a function of duration and number of elements. Elements are purged when activation strength is negative (*dark grey area*). (From [20])

3.2.4 Memory and Games

Many games have been developed with relations to improving the memory capacity. Here should be mentioned the popular Memory game,[5] and Kim's Game[6] which enables children (and others) alone or together to improve their concentration, by trying to remember where the cards are in order to form pairs and guess the content of a hidden object collection. Other popular memory games include the electronic Simon[7] in which the gameplay include listening and watching to a sequence of notes and corresponding colored lights, and then reproducing the sequence on the buttons in the same position as the light and

---

[5] Memory, also known as Concentration or Pairs can be played with a standard deck of playing cards or with special decks. Source: http://en.wikipedia.org/wiki/Concentration_(game).

[6] This game is described in *Kim* by Rudyard Kipling, 1901, in Chap. 9. The book is freely available at Project Gutenberg (http://www.gutenberg.org/files/2226/2226-h/2226-h.htm).

[7] http://www.accessmylibrary.com/coms2/summary_0286-18399548_ITM.

sound. These games were designed to depend on good memory skills in order to succeed. Many brain development and trainer computer games have also been developed, for example for the Nintendo DS console, such as the Nintendo's Brain age[8] and *The Professor's Brain Trainer: Memory*.[9] Today, companies such as CogMed,[10] Lumocity,[11] and many others, including game and platform developers, introduce programs via games and internet, and also propose programs in schools and workplaces.

It is clear that in general, people learn more when they get older, but it is not sure they learn better. This is what is termed crystalized and fluid intelligence [13]. While previous studies has shown improvement in memory when performing tasks that supposedly improve the working memory, recent studies [14] and meta-studies [2] has shown that this is mainly an effect of learning the task and that the working memory capacity does not improve, nor is it clear that any transfer of improvements to other tasks occur.

This aspect of far-transfer, here meaning to be able to use the improvement in cognitive skills in daily life is important to keep in mind, when designing the flow [15] of the game. Therefore, in the level design of Megame we take an ethical standpoint in order to ensure that the flow element is not too addictive; crippling flow would simply require tuning the rate of increase in difficulty throughout the levels.

An interesting, and perhaps counter-intuitive argument can be made about the role of repetition and acquisition of perceptual patterns. In [16] the mind in general is discussed as:

> The mind is not a machine, however, but a special environment which allows information to organize itself into patterns. This self-organizing, self-maximizing, memory system is very good at creating patterns and that is the effectiveness of mind.

Thinking in patterns can however hinder creativity (according to de Bono, who is interested in creative thinking, problem solving and lateral thinking). This extends also to memory and in particular visual memory and painting [17], where methods are devised to break habits and re-learn how to "see". All this points to memory plasticity and to the possibility that exercise, play and memory techniques can effectively alter (in a positive or negative sense) one's skills in memorization and perception of patterns. Therefore we are aware of the importance of testing our game in responsible ways.

Finally, it is not our goal to create a strongly addictive game, but rather a good occasional challenge that players can use for self-assessment and that will provide a tool for testing various models and assumptions about memory.

---

[8] http://www.nintendo.com/games/detail/Y9QLGBWxkmRRzsQEQtvqGqZ63_CjS_9F.

[9] http://www.amazon.co.uk/The-Professors-Brain-Trainer-Nintendo/dp/B000LITROQ.

[10] http://www.cogmed.com/.

[11] http://www.lumosity.com/.

Fig. 3.2 Gameplay of HTW version of game. Once selection is done, success or failure can affect the level characteristics

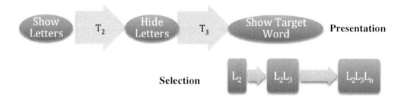

Fig. 3.3 Gameplay for the VTW version of the game

3.3 Game Design and Development

There currently exist two gameplays in Megame. One version, the Hidden Target Word (HTW) version Fig. 3.2 presents a number of letters one after one, with a brief pause between each letter. After each letter is presented, it is hidden, and the player should remember it. The goal is to form a word of a given number of letters out of a subset of the presented letters. This is done by clicking at the letters one by one, in the correct order, which can be done after any letter, assuming enough letters exist to form the word. If the word is formed correctly, the player may continue and the level difficulty may increase. The second version, the Visible Target Word (VTW) version Fig. 3.3, first presents all letters for a brief moment. When the time is out, all letters are turned upside down, hiding the letters. After another brief moment, the target word is presented, and the player should click on the letters of the target word, in the correct order, in order to win the game.

The gameplay is very simple, consisting of two stages, the presentation stage, and the selection stage. There is a pause after the letters are shown (for T_2 s), and another pause before the target word is shown (T_3 s); the time between to letters are shown is T_1 s, and all three time parameters, T_1, T_2 and T_3, can be 0.

If the target word is hidden, it is difficult to identify other modalities that could create true sequences distinguishable from other false sequences. Only words are such sequences for non-experts. On the contrary, if the target word is visible, then it is possible to use other presentation modalities, such as sequences of colours, shapes, sounds etc. It is also possible to present nonsense words.

3.3.1 Levels

Megame permits the modification of the difficulty level, through several means. Because of the delays between presentation of letters (T_1), and between letter presentation and hiding (T_2), and presentation of target word (T_3), there is a possibility to change the level difficulty while at the same time measuring the duration of the players working memory.

The changes in the letters that relates to level difficulty include the number of letters in the target word (N), the number of letters on the table (K). K must be equal or larger than N (unless letters are repeated in the target word), since the target word needs N letters. The difficulty is believed to be related to the additional unused letters (M = K–N), that interferes with the choice of the target word letters. Therefore, increasing M increases the difficulty.

If the table contains several copies of the same letter there could be two effects, either it becomes easier, because there are several possibilities to find the target letter, or it becomes confusing, because you loose the localization of the letter, since it exists in several locations. It is also possible to alter the order of the target word on the table, by simple inversions, mirroring (backwards), circular shifting, and scramble all letters except the first and the last.

3.3.2 Development

We decided to develop our game in Python, following an agile and iterative approach. Python is a very good language for rapid prototyping and coupled with the pygame library, it offers great productivity. As soon as the specification of the Megame was in place, it was possible to quickly design a mock-up of the game (visible in Fig. 3.4). We used Pygame, a widely adopted python graphic library based on SDL[12]; the adoption of Python and Pygame makes our scripts simple to write and highly portable, since standard implementations of both language and library are available for the most common operating systems (including Windows, Mac OS and Linux).

Moreover, Python runs on Android devices (see Fig. 3.5): as discussed in [18], detailing the Scripting Layer for Android (SL4A), a python applications can be developed on a laptop, then transferred to an Android device where a local Python interpreter will run it. The main problem running Python on tablet and similar devices is that all the Android-specific services are usually accessed via a Java Virtual Machine: the solution employed by SL4A is to setup remote communication between the Python runtime and the Java runtime, using remote procedure calls (RPCs) and JSON-based serialization[13] to exchange data back and forth

---

[12] Single Directmedia Layer—http://www.libsdl.org/.

[13] JavaScript Object Notation—http://www.json.org/.

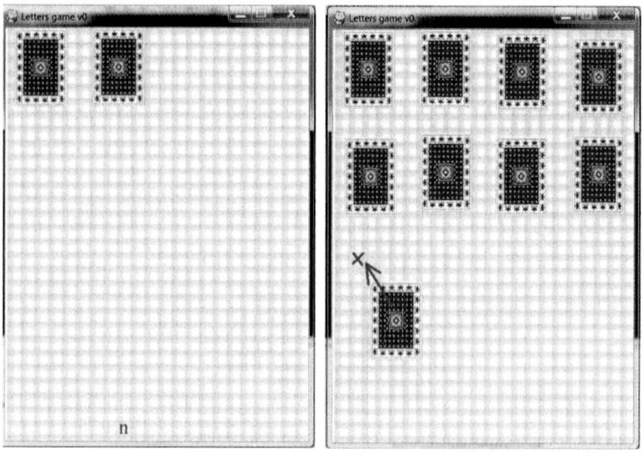

Fig. 3.4 - Version 0 of Megame. On the *left* the letters visible at the *bottom* of the application window, turn into a card one by one. The card will move (*face down*) towards its position in the main (*central*) area of the window. The player can click instead on a card to turn it. On the *right* further on in the game, a card moving towards its final position (the *blue* "x")

Fig. 3.5 Python running on Android. Python runs on *top* of the linux kernel, and it uses standard libraries. The main complication is that most of the Android-specific API are available only through Java and the Dalvik virtual machine (special Java virtual machine present on Android)

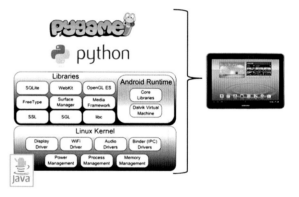

between the 2 runtimes. Thanks to this mechanism it is very simple to, for example, read the state of the accelerometers onboard an Android tablet, and use them in a Python script.

The SL4A does not contain a porting of the Pygame library, but another project is under development, that ports a significant portion of Pygame in Android: PGS4A.[14] Using this partial porting of Pygame, it is possible to develop a standard Python Pygame application, using (for instance) the mouse as main input device; the application would run (and can partially be debugged) on the laptop or stationary machine where it was developed. The application can further be packaged, signed and transferred to an Android device, where it can run and re-map touch gestures to work as mouse events.

This is perhaps the closest Python development can get today to seamless application deployment on Android. Taking advantage of SL4A and PGS4A we are able to rapidly design, run and test our game prototypes on Windows, Macintosh and Linux machines, as well as on portable devices. This way of

---

[14] Pygame Subset for Android—http://pygame.renpy.org/.

Fig. 3.6 The application is given to multiple users, each playing single-player sessions at Megame. The data can be stored locally on the Android device, then periodically the device will communicate to a central server (managed by the authors) and download player statistics and progress

working allows for the creation of multiple versions of the game within the same hour, and it provides functional prototypes that could support early user testing.

Both implementations, PC and Android device, save player statistics on the local machine (see Fig. 3.6); periodically the statistics are also transferred on a central server, administered by the authors. This makes data collection quite simple, and enables players to use the game for memory capacity assessment. In future versions the game can easily be extended with an occasional/asynchronous multiplayer mode, allowing players to engage in tournaments. Social gaming can be supported by the addition of shared high-score boards, hence providing more extrinsic motivation to players.

3.4 Scientific Uses

While the literature does not seem to favor the possibility of improvement of the working memory capacity, there are other scientific uses of Megame that seem interesting. This includes the self-assessment of memory capacity, the assessment of memory capacity related to different difficulties and different presentation modalities, and the use of this game to improve spelling and vocabulary.

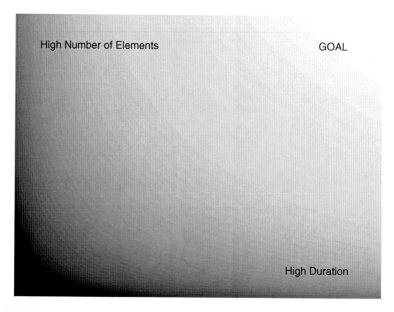

Fig. 3.7 Memory capacity feedback. Better memory capacity is to the *right* for duration and *up* for number of elements

3.4.1 Self-Assessment

The Megame permits to estimate the number of elements in the game, by increasing the number of letters that forms a word, or by randomly presenting words of different number of letters, and in that way assessing the limit of the memory capacity with respect to the number of elements. In a similar manner, the presentation rate of the letters can be varied such that the first letters may be eliminated because it extends the time limit of the working memory and the duration of the participants working memory may be assessed. The estimation of these limits in working memory capacity is presented to the participant in the Fig. 3.7, where the goal is to approach the upper right corner (the sun, denoted Goal) as much as possible.

It is interesting to observe the difference in memory capacity improvement with and without the feedback. In addition, it is also possible to assess this with regards to the different difficulty levels that are included in the game. While it is unlikely that the memory and the difficulty are directly related, it is more probably to find results linked to the notion of flow [15], in that good performance is found when the difficulty is high enough so as to avoid boredom, but not so high it creates anxiety.

3.4.2 *Other Scientific Uses*

While the memory capacity assessment is the central goal of the Megame, other uses are also envisioned. This relates to the use of this game to improve spelling and vocabulary, as it is believed that the repeated use of the game by children will expose them to the spelling and the words in a game environment that provide additional motivating. However, the assessment of spelling and vocabulary skills is not currently targeted.

The next version of Megame under development contains different modalities in addition to letters and words, for the visible target-word version. It is the plan to assess the learning modality preferences [19] by measuring the memory capacity for different modalities. Thus, the new Megame version should provide an alternative method to establish the learning modality preference.

3.5 Conclusions

This work presents the design and implementation of Megame, a multi-platform game (including PCs and Android tablets) that can be used to train and assess memory, both by instant self-assessment and by measuring and assessing long-term effects.

Megame is based on research in memory, in particular the working memory. While it is not clear that the working memory, in terms of duration or number of elements can be improved outside the task (winning at Megame), the main improvement possibilities identified here are attention and interference. If attention is improved, or interference decreased, the activation strength of the element in memory is increased, which also increases the probability of that element entering the long-term memory.

While Megame is made principally for training and assessing the capacity of working memory, other uses of the game include fundamental research in learning modality preferences, spelling and vocabulary training, and assessment of flow.

Megame has been made using an agile and iterative approach. The game is written in Python (and pygame) a very good language for rapid prototyping that offers great productivity. It was possible to go through few design-develop-test cycles in few hours, and use the prototypes to quickly explore various gameplay. Being able to deploy on major operating systems as well as on android OS, we can reach many types of users and our game can be played in different situations (e.g. while commuting), extending the base of our potential users. We are currently working at finishing a major release that encompasses both gameplay described in this work.

References

1. Baddeley, A.: Working memory. Science, New Series **255**(5044), 556–559 (1992)
2. Melby-Lervåg, M., Hulme, C.: Is Working Memory Training Effective? A Meta-Analytic Review. Developmental Psychology, Np (2012)
3. Radvansky, G.: Human Memory. Allyn and Bacon, Boston (2011)
4. Rose, S.: The Making of Memory: From Molecules to Mind. Anchor Books/Doubleday (NY) (1992)
5. Fleming, N.D., Mills, C.: Not Another Inventory, Rather a Catalyst for Reflection. To Improve the Academy, vol. 11, p. 137 (1992)
6. Bertsch, S., Pesta, B.J., Wiscott, R., Mcdaniel, M.A.: The generation effect: A meta-analytic review. Mem. Cogn. **35**(2), 201–210 (2007)
7. Meyer, D.E., Schvaneveldt, R.W.: Facilitation in recognizing pairs of words: Evidence of a dependence between retrieval operations. J. Exp. Psychol. **90**(2), 227–234 (1971)
8. Wixted, J.T.: A theory about why we forget what we once knew. Curr. Dir. Psychol. Sci. **14**(1), 6–9 (2005)
9. Miller, G.: The magical number seven plus or minus two: Some limits on our capacity for processing information. Psychol. Rev. **63**(2), 81–97 (1956)
10. Dudai, Y.: The neurobiology of consolidations, or, how stable is the engram? Annu. Rev. Psychol. **55**, 51–86 (2004)
11. Glenberg, A.M., Lehmann, T.S.: Spacing repetitions over 1 week. Mem. Cogn. **8**(6), 528–538 (1980)
12. Jensen, K.: On the use of memory models in audio features. In: Symposium of Frontiers of Research on Speech and Computer Music Modeling and Retrieval (FRSM/CMMR—2011), Bhubaneswar, India, pp. 100–107 (2011)
13. Cattell, R.B.: The measurement of adult intelligence. Psychol. Bull. **40**(3), 153–193 (1943)
14. Redick, T.S., Shipstead, Z., Harrison, T.L., Hicks, K.L., Fried, D.E., Hambrick, D.Z., Kane, D.Z., Engle, R. W.: No evidence of intelligence improvement after working memory training: A randomized, placebo-controlled study. J. Exp. Psychol. Np (2012)
15. Csíkszentmihályi, M.: Flow: The Psychology of Optimal Experience. Harper and Row, New York (1990)
16. De Bono E.: Lateral Thinking, Viking (2009)
17. Edwards, B.: Drawing on the Right Side of the Brain: The Definitive, 4th edn, Tarcher (2012)
18. Ferrill, P.: Android applications using Python and SL4A, Part 1: Set up your development environment. IBM developer Works (2011)
19. Rayner, S., Riding, R.: Towards a categorisation of cognitive styles and learning styles. Educ. Psychol: Int. J. Exp. Educ. Psychol. **17**(1–2), 5–27 (1997)
20. Jensen, K., Hjortkjær, J.: An Improved dissonance measure based on auditory memory. J. Audio Eng. Soc. **60**(5), 350–354 (2012)

Chapter 4
Assessing Virtual Reality Environments as Cognitive Stimulation Method for Patients with MCI

Ioannis Tarnanas, Apostolos Tsolakis and Magda Tsolaki

Abstract Advances in technology in the last decade have created a diverse field of applications for the care of persons with cognitive impairment. This chapter is an attempt to introduce a virtual reality computer-based intervention, which can used for cognitive stimulating and disease progression evaluation of a wide range of cognitive disorders ranging from mild cognitive impairment (MCI) to Alzheimer's disease and various dementias. Virtual reality (VR) environments have already been successfully used in cognitive rehabilitation and show increased potential for use in neuropsychological evaluation allowing for greater ecological validity while being more engaging and user friendly. Nevertheless a holistic approach has been attempted, in order to view the research themes and applications that currently exist around the "intelligent systems" healthcare given to the cognitively impaired persons, and thus looking at research directions, systems, technological frameworks and perhaps trends.

Keywords Computerized cognitive training · Computerized testing · Cognitive reserve · Dementia · Psychometrics

I. Tarnanas (✉) · M. Tsolaki
School of Medicine, Aristotle University of Thessaloniki, 3rd Neurological Clinic,
Papanikolaou Avenue 570 10 Exohi, Thessaloniki, Greece
e-mail: i.tarnanas@alzheimer-hellas.gr

M. Tsolaki
e-mail: tsolakim1@gmail.com

A. Tsolakis
School of Electrical and Computer Engineering, Aristotle University of Thessaloniki,
Egnatia Street 54124 Thessaloniki, Greece
e-mail: aptsolak@gmail.com

A. L. Brooks et al. (eds.), *Technologies of Inclusive Well-Being*,
Studies in Computational Intelligence 536, DOI: 10.1007/978-3-642-45432-5_4,
© Springer-Verlag Berlin Heidelberg 2014

4.1 Introduction

Virtual Reality (VR) and Augmented Reality (AR) are some of the most promising and at the same time challenging applications of computer graphics. Virtual Reality (VR) is stimulating the user's senses in such a way that a computer generated world is experienced as real. In order to get a true illusion of reality, it is essential for the user to have influence on this virtual environment. All that has to be done in order to raise the illusion of being in or acting upon a virtual world or virtual environment, is providing a simulation of the interaction between human being and this environment. This simulation is—at least—partly attained by means of Virtual Reality interfaces connected to a computer. When considering VEs for context-sensitive rehabilitation, it is important to first evaluate the limitations and potential of the underlying VR technology.

Over the past decades VR technology has been used in many different domains such as education [1], simulation for expert training [2] and therapy. Looking at medical uses in particular, Rizzo and Kim [3] and Rizzo et al. [4] discuss the advantages and disadvantages of VR systems in a therapeutic context. Even though both reviews have been conducted 6 and 7 years ago respectively, most of what the authors discuss still appears to be of relevance. In Rizzo and Kim's overview the following aspects were among the key characteristics for VR systems and therefore should be taken into account when developing VEs for individualized rehabilitation. In this chapter we are going to present the benefits of using a particular type of virtual reality interface, named the virtual reality museum. The VR Museum was used as an intervention tool for patients with Amnestic-type Mild Cognitive Impairment (aMCI) in order to see if it can improve the task domains of navigation, spatial orientation and spatial memory. Those tasks were chosen for their relevance for patient with aMCI for whom it is essential to be spatially oriented in order to live independently [5].

Firstly we are going to focus on MCI and the basic characteristics of aMCI patients. It is essential to define the exact cognitive profile of those patients in order to understand the difficulties that non-invasive methods of intervention may encounter. We are also going to describe related attempts to address those patients with virtual reality.

On the second part we will describe the VR Museum, the clinical protocol, the research methodology and the final results.

At the end of the chapter, we will sum up our findings with general conclusions and implications on the use of VR Museum as an intervention tool.

4.1.1 Individuals with Mild Cognitive Impairment

The concept of Mild Cognitive Impairment (MCI) was derived from milder cases of dementia and not Alzheimer's disease (AD). MCI encompasses patients with and without memory impairment. Of those with memory loss, some have memory

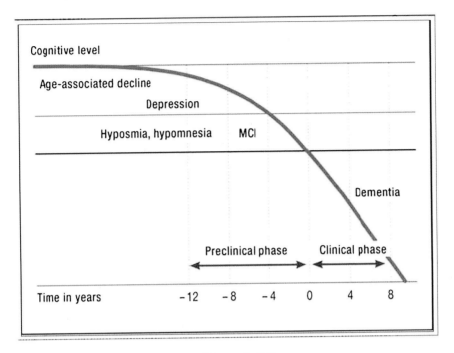

Fig. 4.1 The typical progressive course of Dementia [10]

impairment as their only deficit [amnestic MCI single domain (aMCIs)], whereas others have impairments of memory loss plus changes in other cognitive domains [amnestic MCI multiple domain (aMCImd) [6]. Multiple-domain MCI is more common than pure amnestic type MCI and is characterized by slight impairment in more than one cognitive domain but of insufficient severity to constitute dementia [7]. Of those without any memory loss, some patients have deficits in one domain only, such as executive functions, apraxia or aphasia. Or they may have deficits in several domains, excluding memory [8]. These prodromal states may progress to non-AD dementias, such as vascular dementia, frontotemporal dementia, Lewy body dementia, primary progressive aphasia, or corticobasal degeneration [9] (Fig. 4.1).

4.1.2 MCI Subtypes

Recently research by Winblad et al. [9], revealed the heterogeneity in the clinical description of MCI leading to the classification of four subtypes. Based on the number: single or multiple and type: memory, non-memory or both, of impaired cognitive domains we have:

1. Amnestic MCI—memory impairment only.
2. Multi-domain MCI-Amnestic (memory plus one or more non-memory domain).

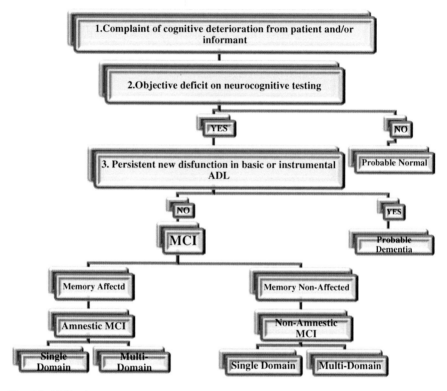

Fig. 4.2 MCI subtypes

3. Multi-domain MCI-Non-Amnestic (more than one non-memory domain).
4. Single Non-Memory MCI (one non-memory domain).

Building on this classification Hanfelt et al. [11] posited cognitive, functional and neuropsychiatric traits that can distinguish individuals with MCI. This set the basis for improved diagnosis because it provided a common "language" among research centers for future research (Fig. 4.2).

Recently, the criteria for the presence of MCI have been defined as [6, 12]:

• Subjective memory complaints, preferably validated by a third person.
• Memory impairment, non-characteristic for given age and education level.
• Preserved general cognitive function.
• Intact activities of daily living.
• Absence of dementia.

More importantly, impairment of Activities of Daily Living (ADL) has been observed in some MCI subtypes and therefore Instrumental ADL (IADL) questionnaires [13] or recent video assisted observation tools [14] have been used for their ability to act as a diagnostic marker for the MCI subtypes.

In summary, in order to have a MCI subtype diagnosis a variety of medical and neuropsychological examinations is required. A thorough physical examination, blood sample studies, imaging (MRI, RiB-PET), genetic tests (APOE, TREM2) as well as biomarkers in CSF (beta-amyloid, tau and phospho-tau protein) [15]. Occasionally, a condition, such as vitamin B12 deficiency or thyroid disease, can be identified as a cause for MCI [16]. However, one general conclusion to be drawn is that none of the above mentioned tools should be used alone, on the contrary the combination of different tools results in a more precise diagnosis. Lastly, the most recent research findings showed that the pathophysiologic findings in MCI may predict Alzheimer's Disease (and perhaps other diseases) and therefore the sooner the diagnosis the more effective the intervention [17].

4.1.3 Amnestic-Type Mild Cognitive Impairment

Decline in episodic memory is one of the hallmark features of Alzheimer's disease (AD) and is also a defining feature of amnestic Mild Cognitive Impairment (aMCI), which is posited as a potential prodrome of AD. While deficits in episodic memory are well documented in MCI, the nature of this impairment remains relatively under-researched, particularly for those domains with direct relevance and meaning for the patient's daily life. Recently in order to fully explore the impact of disruption to the episodic memory system on everyday memory in MCI, clinicians examine participants' episodic memory capacity using a battery of experimental tasks with real-world relevance [18]. They investigated episodic acquisition and delayed recall (story-memory), associative memory (face-name pairings), spatial memory (route learning and recall), and memory for everyday mundane events in 16 amnestic MCI and 18 control participants. Furthermore, they followed MCI participants longitudinally to gain preliminary evidence regarding the possible predictive efficacy of these real-world episodic memory tasks for subsequent conversion to AD.

It has been reported for patients with aMCI and more frequently for patients with AD that they have difficulties with spatial orientation in everyday activities [19]. Patients often fail to find their way in unfamiliar environments when facing entirely new spatial settings during urban transportation, traveling or shopping. In mild to more severe stages of the disease, they may be disoriented even within their familiar neighborhood or inside their own flat. The standard way to study disorientation and spatial memory is with tests consisting of navigation inside a hospital [20], sometimes as orientation in a circular arena [21] and remembering object position [22]. To this day there are only two studies, to our knowledge, which addressed spatial orientation in MCI [23, 24]. The Mapstone et al. [23] study correlated motion flow perception with results in a table-top Money Road Map (MWM) test and the Hort et al. [24] study investigated allocentric and egocentric navigation in an analogue of the MWM.

It is generally accepted that spatial navigation deficit is particularly pronounced in individuals with hippocampus-related memory impairment, such as aMCI and may signal preclinical AD [5, 24]. Laczo et al. [5] analyzed several types of errors made by the subjects' during the task to investigate which of them contributed to their impairment. They used the Hidden Goal Task, a human analogue of the Morris Water Maze, to examine spatial navigation either dependent (egocentric) or independent of individual's position (allocentric). Overall, the aMCI group performed poorer on spatial navigation than the non-aMCI group, especially in the latter trials when the aMCI group exhibited limited capacity to learn and the non-aMCI group exhibited a learning effect. Finally, the aMCI group performed almost identically as the AD group. Hort et al. [24] examined aMCI, AD and healthy controls using a four-subtests task that required them to locate an invisible goal inside a circular arena, analogues to the MWM test. Each subtest began with an overhead view of the arena showed on a computer monitor and then entered a real navigation inside of the actual space, an enclosed arena 2.9 m in diameter. They found that the AD group and amnestic MCI multiple-domain group were impaired in all subtests. The amnestic MCI single-domain group was impaired significantly in subtests focused on allocentric orientation and at the beginning of the real space egocentric subtest, suggesting impaired memory for allocentric and real space configurations. These results suggest that spatial navigation impairment occurs early in the development of AD and can be used for monitoring of the disease progression or for evaluation of presymptomiatic AD.

4.2 Virtual Environments and aMCI

Virtual Reality is a relatively new technology regarding its use for neuropsychological research. Publications to date provide evidence of some cases where a virtual environment creates the desired conditions and the necessary triggers for amnestic MCI patients to be classified and assessed. To be more precise, applications of virtual reality in neuroscience can provide experiments in a controlled environment where normal and impaired patient behavior, perception, control of movement, learning, memory and emotional aspects can be observed [25]. VR creates interactive, multimodal sensory stimuli that offer unique advantages over other approaches to neuroscientific research and applications. VR compatibility with imaging technologies such as functional MRI allows researchers to present multimodal stimuli with a high degree of ecological validity and control while recording changes in brain activity. Therapists, too, stand to gain from progress in VR technology, which provides a high degree of control over the therapeutic experience.

Normally, a real-time interaction is required in order to observe and analyze human reactions of any kind, event or task. Otherwise, computer generated experimental tasks designed for specific variables and aspects of human response are required. During the last decade and a half, research towards that direction

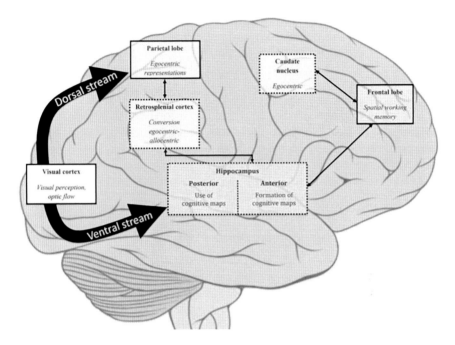

Fig. 4.3 The neural network involved in spatial navigation [32]

provided information about the use of VR in Neuroscience [26–30]. For the purposes of this chapter we are going to analyze the way VR can be used for evaluating spatial perception and memory aspects in individuals with MCI.

4.2.1 Spatial and Visual Memory

Visual Memory is responsible for retaining visual shapes and colors whereas spatial memory is responsible for information about locations and movement. It could be described as cognitive imaging and cognitive mapping. This distinction is not always clear since part of visual memory involves spatial information and vice versa [31]. When it comes to MCI, impairment to both visual and spatial memory could indicate memory deficits [23] (Fig. 4.3).

Navigation combines the two types of memory. Successful navigation requires a variety of thoughts and actions; planning, selection of an appropriate strategy and possible alterations, prospective memory and remembering previously visited locations. In particular, navigation is connected with the hippocampal function, a brain area already impaired in individuals with MCI [32]. Thus, deficits on navigational skills and spatial memory could be a solid cognitive indicator for MCI or early forms of Dementia.

The discovery of place-specific firing in the hippocampus [33] and spatial navigation impairment after hippocampal lesion in the water maze [34] gave strong support to the theory of a cognitive map. This theory dissociates hippocampal navigation, based on a configuration of distal landmarks, from navigation to and from landmarks. This concept has evolved into the dissociation between allocentric navigation, using flexible representation of an ensemble of distal landmarks and independent of actual subject positions, and egocentric navigation, using distances and angles to or from individual landmarks. In humans, the allocentric mode of navigation was shown to be connected with the hippocampal function in analogues of the Morris water maze (MWM) [35, 36], in place navigation inside a virtual town [37], and in remembering the location of objects on a table [38].

4.2.2 Virtual Reality, Spatial Memory and aMCI

Immersive virtual reality environments can provide information and some times rehabilitate spatial working memory [39]. Ensuring that the desired conditions are ecologically valid, it is possible to use VR as a tool to evaluate spatial memory in individuals with MCI by tracking their behavior inside the virtual environment in real-time [40]. As described above spatial memory can be impaired in aMCI. In one study, aMCI participants encountered two virtual environments; the first, as the driver of a virtual car (active exploration) and the second, as the passenger of that car (passive exploration). Subjects were instructed to encode all elements of the environment as well as the associated spatiotemporal contexts. Following each immersion, we assessed the patient's recall and recognition of central information (i.e., the elements of the environment), contextual information (i.e., temporal, egocentric and allocentric spatial information) and lastly, the quality of binding. The researchers found that the AD patients' performances were inferior to that of the aMCI and increasingly so when compared to that of the healthy aged groups, in line with the progression of hippocampal atrophy reported in the literature [28]. Spatial allocentric memory assessments were found to be particularly useful for distinguishing aMCI patients from healthy older adults. Active exploration yielded enhanced recall of central and allocentric spatial information, as well as binding in all groups. This led aMCI patients to achieve better performance scores on immediate temporal memory tasks. Finally, the patients' daily memory complaints were more highly correlated with the performances on the virtual test than with their performances on the classical memory test.

Taken together, these results highlight specific cognitive differences found between these three populations that may provide additional insight into the early diagnosis and rehabilitation of pathological aging. In particular, neuropsychological studies benefit when using virtual tests and a multi-component approach to assess episodic memory, and encourage active encoding of information in patients suffering from mild or severe age-related memory impairment. The beneficial effect of active encoding on episodic memory in aMCI and early to moderate AD is discussed in the context of relatively preserved frontal and motor brain functions implicated in self-referential effects and procedural abilities.

In another study, a virtual navigation based reorientation task (VReoT) was used [41] and again healthy subjects were compared with aMCI subjects regarding their performance on the reorientation test. The performance of the aMCI was significantly worse than the controls suggesting that VReoT detects spatial memory deficits. A subsequent receiver-operating characteristics analysis showed a sensitivity of 80.4 % and a specificity of 94.3 %.

4.3 The Virtual Reality Museum

Virtual Reality (VR), Augmented Reality (AR) and Web3D technologies in conjunction with database technology may facilitate the preservation, dissemination and presentation of cultural artifacts in museum's collections and also educate the public in an innovative and attractive way. *Virtual Reality* signifies a synthetic world, whereas *Augmented Reality* refers to computer generated 2D or 3D virtual worlds superimposed on the real world. *Web3D* is used to represent the application of XML (eXtended Markup Language) and VRML (Virtual Reality Markup Language) technologies to deliver interactive 3D virtual objects in 3D virtual museums. Precedents made use of 3D multimedia tools in order to record, reconstruct and visualize archaeological ruins using computer graphics and also provide interactive AR guides for the visualization of cultural heritage sites [42]. These new emerging technologies are used not only because of their popularity, but also because they provide an enhanced experience to the virtual visitors. Additionally, these technologies offer an innovative, appealing and cost effective way of presenting cultural information. Virtual museum exhibitions can present the digitized information, either in a museum environment (e.g., in interactive kiosks), or through the World Wide Web.

Our Virtual Museum system has been developed in XML and VRML and is described in detail [43, 44]. The system allows museum curators to build, manage, archive and present virtual exhibitions based on 3D models of artifacts. The innovation of our system is that it allows end-users to explore virtual exhibitions implemented using very simple everyday interfaces (e.g. joystick, mouse) (Fig. 4.4).

The cultural artifacts are digitized by means of a custom built stereo photogrammetry system (Object Modeler), mainly for digitizing small and medium-size objects and a custom modeling framework (Interactive Model Refinement and Rendering tool) in order to refine the digitized artifact. The 3D models are accompanied by images, texts, metadata information, sounds and movies (Fig. 4.5). These virtual reconstructions (3D models and accompanying data sets) are represented as eXtensible Markup Language (XML) based data to allow interoperable exchange between the museum and external heritage systems.

These virtual reconstructions are stored in a MySQL database system and managed through the use of a specially designed Content Management Application,

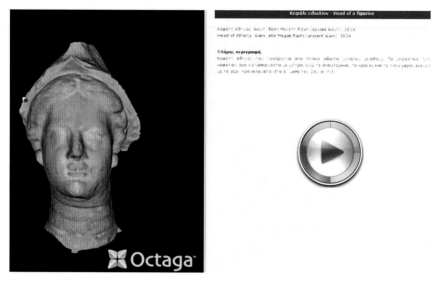

Fig. 4.4 Users can 'walk' freely in the virtual museum and interact with the artifacts. Once they select an artifact they can choose to zoom in, rotate it at the X-, Y-, Z-axis and read details in tags on the artifact itself

which also allows building and publishing virtual exhibitions on the Internet or in a museum kiosk system. The system is a complete tool that enables archiving of both content and context of museum objects. The described interactive techniques can transform the museum visitors 'from passive viewers and readers into active actors and players' [44].

4.3.1 The Virtual Reality Museum Technical Components

Two main components of the system are of interest for the evaluation: the Content Management Application (CMA) and Augmented Reality Interface (ARIF). CMA allows publishing of virtual museums to both Web and a specially designed application (ARIF) for switching between the Web and an AR system. The CMA application is implemented in Java, trademark of Oracle Corporation and it includes the database of the representations of cultural objects and their associated media objects, such as images, 3D models, texts, movies, sounds and relevant metadata. It enables user-friendly management of different types of data stored in the Virtual Museum database, through various managers, such as the *Cultural Object Manager* (deals with virtual representations of cultural artifacts), the *Presentation Manager* (manages virtual exhibitions with the help of templates) and the *Template Manager* (stores these visualization templates).

Fig. 4.5 The interface is ergonomically made so that it can tolerate errors. All icons, fonts and interactive objects are large and understandable

The ARIF component is a presentation or visualization framework that consists of three main sub-components:

- The *ARIF Exhibition Server*. Data stored in the Database is visualized on user interfaces via the ARIF Exhibition Server.
- The *ARIF Presentation Domains* with implemented web browser functionality, suited for web-based presentations.
- The *ARIF AR—Augmented reality functionality*. This sub-component provides an AR based virtual museum exhibition experience on a touch screen in the museum environment using table-top AR learning experiences, e.g., AR quizzes and on-line museum exhibitions.

4.3.2 The Virtual Reality Museum Cognitive Theory

It is difficult to reconcile inconsistent findings pertaining to the effect of playing cognitive training games on cognition [45–47], because the methodological differences between these studies are substantial. More research is required to elucidate what aspects of brain training games facilitate transfer to untrained cognitive abilities. Hence, the aim of our virtual museum was to test whether playing some simple memory exercises inside an ecologically valid 3D environment does transfer to different measures of executive functions in amnestic-type mild cognitive impairment (aMCI) older adults. Executive functions are a cognitive system that controls and manages other cognitive processes. For example, Updating is the ability to respond in a flexible and adaptive manner in order to keep up with the changes in the environment, e.g. during the period of road repairs you need to change your permanent route and use the new route until the repairs are finished.

According to the cognitive-enrichment hypothesis developed by Hertzog et al. [48], the trajectory of cognitive development across the life span is not fixed. Although the trajectory of cognitive development at normal seniors is largely determined by a lifetime of experiences and environmental influences, there is potential for discontinuity in the trajectory given a change in cognition-enriching behaviors. The cognitive-enrichment hypothesis is corroborated by ample evidence for plasticity, i.e., the potential for improvement of ability as a consequence of training [49] of everyday cognitive task-switching in the elderly population. There are some reports providing evidence of improvements in "Updating" as part of the bigger structure called "executive functions" [50–52]. There are also promising reports showcasing seniors "Shifting" ability improvement, as a mental process during which seniors redirect their focused attention from one channel of information to another as quickly as possible or change the course of their actions while maintaining accurate performance [53, 54]. Shifting can be initiated consciously or unconsciously by a stimulus in our surroundings or by habit. For example, while talking on the phone, we may have to switch to preventing a small grandchild from touching a sharp object. Many older people encounter shifting problems; they may find it difficult and frustrating to try to change their thinking, routines or actions. Those who do not train their shifting ability may have problems changing undesirable habits. Finally, Davidson et al. [55] and Karbach and Kray [56] reported improvement at "Inhibition", the ability to ignore irrelevant stimuli or suppress irrelevant reactions while performing a task. Inhibition includes the deliberate prevention of an act, behavior or response, when it is not desirable. At work, for example, we must sometimes ignore our co-workers' conversations and focus our attention on our own tasks. Training this ability helps us concentrate on relevant activities while ignoring disturbing stimuli. It will enable you, for example, to write a letter while the television is on. In addition, domains such as selective attention [57] and inductive reasoning [58] can be improved in older adults.

We now know that the virtue of a cognitive-training technique depends on the generalization or transfer of training to untrained tasks [59]. Different degrees of transfer have to be distinguished. The minimal degree of transfer that can occur is improvement within the same cognitive domain as subjected to training, assessed using different stimuli, and requiring a different response than the training task. This type of transfer is referred to as near transfer. Improvement of abilities in other cognitive domains than the cognitive domain subjected to training is referred to as far transfer.

Virtual Reality Museum exploration and interaction activities are considered to provide an ideal context for cognitive enrichment [60, 61]. The unique characteristics of virtual museums presumed to facilitate transfer are their motivating nature, frequent presentation of feedback, precise reinforcement schedules, and stimulus variability, analogues to the basic characteristics of good video games, which are designed to enhance learning through effective learning principles supported by research in the Learning Sciences [62]. As a result of their entertainment value, virtual museums maintain the motivation to engage in practice for much longer than monotonous laboratory tasks or traditional training programs. Frequent feedback supports motivation and is also important for conditioning the desired level of performance. When the difficulty level of the task is continuously adapted to the performance, players will constantly be challenged at the limits of their ability. It is in particular the phase of skill-acquisition that calls for cognitive control (CC), whereas continued performance at a mastered level is associated with cognitive load automation and release of CC resources [63, 64]. Furthermore, small increments of difficulty level maximize the proportion of successful experiences with the task. The stimulus variability also plays an important role in training CC, because it helps to generalize learnt cognitive skills to multiple stimulus contexts.

Transfer of virtual reality museum interventions to CC has, however, not been demonstrated consistently. Owen et al. [46], for instance, demonstrated that playing computerized cognitive training games like Nintendo's® Dr. Kawashima's Brain Training™ was not more beneficial for CC functions than answering general knowledge questions online. It is being assumed that because the sample of participants in Owen et al.'s study was very heterogeneous and included both young and old adults, it is possible that improvements of cognitive test performance were attenuated in young adults due to ceiling performance at pretest. This could have obscured possible transfer of training in the sub-sample of older adults. The notion that sample heterogeneity can confound the observed effect of virtual reality training substantially is corroborated by Feng et al. [65]. They found no effect of playing action virtual reality games on spatial attention in a sample of young adults. However, separate analysis of the effect in males and females revealed that females did actually benefit from playing. In addition, in the Owen et al. study the participant sample was heterogeneous with respect to training adherence, so participants who completed only two training sessions could have had a negative impact on aggregated training outcomes. Another aspect of Owen et al.'s study that makes the observed absence of transfer difficult to interpret is that transfer was assessed using a test battery comprising only four cognitive tests, three of which were measures of working memory capacity.

Ackerman et al. [45] demonstrated that sample heterogeneity cannot account for Owen et al.'s [46] findings. They found that playing cognitive training games (Nintendo® Wii™ Big Brain Academy™) does not benefit cognitive abilities to a greater extent than reading assignments do, in a homogeneous sample of healthy seniors on a relatively fixed and extensive training schedule. Moreover, a broader assessment of cognitive abilities of interest was made than in Owen et al.'s study. Still, Ackerman et al. focused predominantly on reasoning ability and perceptual processing speed, while a large share of the cognitive games under study taxed working memory updating and the large variety of the tasks probably stimulated participants' attention and task set shifting. Inclusion of transfer tasks, gauging working memory, updating and set shifting, in Ackerman et al.'s study could have led to different conclusions regarding transfer of playing cognitive training games.

Conversely, there is also some evidence against Owen et al.'s [46] and Ackerman et al.'s [45] pessimistic conclusions regarding the beneficial effects of playing virtual reality educational games on CC functions. Namely, Peretz et al. [47] found a larger improvement of visuospatial working memory, visuospatial learning, and focused attention after playing Cognifit Personal Coach® cognitive training games than after playing conventional 3D videogames that were matched for intensity, in a sample of older adults. Even though there is some theoretical overlap in the cognitive functions assessed by Peretz et al. and Owen et al. and Ackerman et al., the specific cognitive tests used to assess transfer in these studies was different. It is conceivable that some cognitive tests are more sensitive to transfer effects than others, which might explain the discrepant results of these studies.

Furthermore, playing 3D videogames not specifically designed for cognitive training can also improve CC functions in older adults. Basak et al. [66] demonstrated that playing a particular complex 3-D real-time strategy game (Rise of Nations) was associated with greater improvements of shifting, updating, and inductive reasoning than observed in the control condition. It must be noted that the control group in this study was a no-contact control group, so it is not certain to what extent the observed improvements in the videogame group are attributable to placebo-effects. Nevertheless, the improvements of CC in this study were larger than practice effects due to repeated exposure to the same cognitive test.

It has been argued that failures to demonstrate far transfer of playing cognitive training games in the population of older adults may be due to a general age-related decrease of the extent to which learning transfers to untrained abilities [45]. This assertion is supported by Ball et al.'s [57] finding that cognitive strategy training programs for improving memory, processing speed and reasoning, respectively, were associated with improvements within the trained cognitive domain but not with far transfer to untrained cognitive abilities of older adults. In contrast, however, far transfer of practicing basic cognitive tests has been reported repeatedly in the cognitive aging literature [56, 67–69]. Brain training games like Nintendo's® Dr. Kawashima's Brain Training™ share many task components of basic cognitive laboratory tasks and videogames have several additional characteristics facilitating transfer [61]. Therefore, it is reasonable to expect that transfer of computerized cognitive training games in the population of older adults is replicable.

4.3.3 The Virtual Reality Museum Cognitive Exercises

In general three tasks have been identified and developed within the scope of this study. While the complete cognitive stimulation is expected to encompass several tasks from each cognitive domain (memory, attention, executive functions), this study is aiming to evaluate the task domains of navigation, spatial orientation and spatial memory. Tasks were chosen specifically for patient with aMCI for whom it is essential to be spatially oriented in order to live independently.

Generalization of skills during cognitive stimulation towards daily-life settings has only received little support in the literature [70]. More specifically, task-focused training appears to show no transfer to situations outside of the training situation and the effectiveness of strategy training requires further evidence. While the external validity of training applications seems to be of central importance to the patients' success in their daily life, most traditional rehabilitation studies have not successfully demonstrated such transfer yet. Even though principles of context-sensitive rehabilitation have been mentioned in several literature reviews [71], context-sensitivity is often not associated with transfer to activities of daily life. This is because context-sensitive tasks are essentially based on the unique experiences that a patient has in his daily life. Hence, a transfer is often not necessary as training tasks are either identical to common daily chores or replicate them as closely as possible. Nonetheless, when traditional process-specific tasks are combined with individualized context, task generalization across similar daily activities seems to be of relevance.

The Virtual Reality Museum is designed to speed up auditory processing, improve working memory, improve the accuracy and the speed with which the brain processes speech information and reengage the neuromodulatory systems that gate learning and memory. To reverse cognitive disuse and drive brain plasticity, the program strongly engages the brain with demanding exercises and an adaptive and reward-based daily training schedule, consistent with the recommendations of Tucker-Drob [72]. This procedure is based on practices of context-sensitive rehabilitation suggested by Ylvisaker [73]. Cognitive exercises provided by it are divided into three interrelated categories, that, in aggregate, span the cognitive functions of seniors:

- Listen and Plan: Seniors follow instructions to locate and find items in an order. Instructions become more difficult (phonetically and syntactically) progressively (purpose: training on spatial navigation abilities and planning following complex instructions with continuous processed speech).
- Storyteller: Seniors hear segments of museum items stories and are asked to answer a set of questions concerning the details of the respective segment (purpose: training on story comprehension and memory).
- Exer-gaming: Seniors are asked to actually represent the "scene" depicted at the archeological artifacts or multimedia description, e.g. movement, dance, wedding (purpose: training on executive function and orientation/praxis).

This type of intervention was used in the recent study by Smith et al. [69], which was the first double-blind large-scale clinical trial that demonstrated marked improvement not only in the trained task, but also in several generalized measures of memory and perception of cognitive performance in everyday life, relative to an active control group that received a frequency and intensity-matched cognitive stimulation program.

The modularity of the cognitive tasks above also reflects the standards of current process-specific assessments [74]. The Virtual Reality Museum Listen and Plan exercise is a spatial navigation task that is implemented as close as possible to the actual Archeological Museum of Aiani, at Kozani, Greece from where we took the layout and archeological artifacts. Consequently, this virtual navigation task is also meaningful for clinical decision-making about real-world behavior as well.

Transfer was assessed by comparing performance on a battery of cognitive tests before and after the intervention. Taking into account that some cognitive tests may be more sensitive to transfer effects than others, several measures of updating, shifting, and inhibition were included in the test battery. Although it is assumed that training interventions boost functional or even plastic changes to the brain, neuronal correlates of the training induced changes in intervention studies were only examined in the last decade [75]. Knowledge about the intervention related neuronal and functional changes is additionally useful in order to understand the efficiency of the training and transfer effects to other tasks [76]. Therefore, in the present study we used event-related brain potentials (ERPs) derived from the electroencephalogram (EEG) in order to study more closely the neuronal processes which are affected by the training intervention.

4.4 Research Methodology

4.4.1 Design

Single-site randomized controlled double-blind trial.

4.4.2 Participants

One hundred and fourteen patients with MCI according to the revised Petersen criteria [77], aged between 65 and 88 years, were recruited to participate in the experimental study, which was conducted in Alzheimer Hellas day clinic Agios Ioannis at Thessaloniki, Greece between May 2011 and October 2012. The participants were randomly assigned to the training groups. We excluded subjects who met criteria for dementia (DSM-IV), AD (NINCDS-ADRDA), depressive episode (IDC-10), subjects with cerebrovascular disease (Hachinski scale score $\geq$4), and those with any other medical or psychiatric identifiable cause accounting for their complaints.

The neuropsychological battery used for the pre- and post- testing included tests for the assessment of memory (Rey Auditory Verbal Learning Test—RAVLT), language and semantic memory (15-items short-form of the Boston Naming Test, category fluency), praxis and visuospatial skills (Rey complex figure copy), attention and executive function (Symbol Digit Modalities Test, Trail Making part A and B, Stroop interference Test [78] and letter fluency). A cognitive domain was judged as impaired when subjects scored 1.5 SD below values for age and education matched controls in at least one test. According to the results of the neuropsychological exploration, subjects were classified as pure amnestic MCI (a-MCI), patients fulfilling Petersen's criteria for amnestic MCI, with memory being the only affected domain (see Table 4.1 for details).

Participants also received an Auditory ERP-recording completed using a *Nihon Kohden–Neuropack* M1 MEB-9200 evoked potential/EMG measuring system. Event-related-potentials (ERPs) are used as a noninvasive clinical marker for brain function in human patients (Fig. 4.6). Auditory ERPs are voltage changes specified to a physical or mental occurrence that can be recorded by EEG [79]. Different ERPs were used in order to pinpoint the functional processes which would be improved by the cognitive process training and which may be affected by retesting. The principal ERP components elicited after task-relevant visual stimuli are among others the N1, the anterior N2, the P2, and the P3b. In Fig. 4.7, an example of an Auditory ERP signal can be seen. The signal can be divided into two parts, a pre-stimuli section consisting of a baseline with no clear potentials and a post-stimuli section consisting of various potentials. The first positive potential is called P1, followed by a negative potential N1, then P2, N2, and so forth. The latency of these potentials is measured from onset of stimuli to the peak of the potential. Sometimes the peaks are named using the latency, e.g. if N1 occur at a latency of 40 ms it is named N40 or if P3 occur at a latency of 300 ms it is named P300. The baseline amplitude is the difference between the peak of a potential and the mean of the pre-stimulus baseline. The baseline measurement used to discriminate between the MCI amnestic patients and the controls in our study is shown in Fig. 4.6 [80].

4.4.3 Procedures

Thirty-nine of the participants represented a virtual reality museum cognitive training group—experimental group (remaining N = 32; 12 men, mean age: 70.5 years; range 65–82; seven drop-outs because of technical problems, illness, and tenancy changeover). The other participants formed an active control group (N = 39; 16 men, mean age: 69.7 years; range: 65–88; no drop-outs) and a non-contact control group (remaining N = 34; 13 men, mean age: 70.9 years; range: 65–87; two dropouts because of illness). The virtual reality museum cognitive training group was exposed to a multilayered cognitive training over a period of

Table 4.1 Demographic characteristics and cognitive status of the participant groups

Group	Cognitive training	Active control	Non-contact control	Statistical significance
Mean age	70.5 years (4.3)	69.7 years (4.5)	70.9 (4.4)	$F_{(2, 102)} = 1$, $P = 0.36$
MMSE score	26.8 (3.6)	26.2 (3.6)	26.2 (3.1)	$F_{(2, 102)} = 1.4$, $P = 0.24$
Stroop-test (color repetition)	73.4 (34.24)	74.4 (32.2)	70.6 (23.40)	
RAVLT-immediate recall	15.4 (4.3)	15.5 (4.6)	15.0 (3.1)	
RAVLT-delayed recall	1.6 (1.5)	1.7 (1.5)	2.2 (1.5)	
RAVLT-recognition	5.6 (2.2)	5.5 (2.2)	7.4 (1.9)	
BNT	10.42 (2.46)	10.60 (1.91)	11.22 (1.90)	
Category fluency	10.6 (3.98)	11.3 (3.1)	11.2 (4.3)	
Letter fluency	7.4 (3.54)	7.1 (2.6)	6.0 (3.4)	
Ray figure copy	34.6 (1.3)	32.7 (1.9)	28.9 (8.5)	
Ray figure immediate recall	11.9 (9.2)	11.4 (9.2)	7.0 (4.7)	
Ray figure delayed-recall	11.6 (9.4)	10.6 (9.1)	7.2 (4.6)	
Ray figure recognition	6.6 (2.9)	6.3 (2.5)	6.0 (1.4)	
Forward digit repetition	6.2 (1.1)	6.1 (1.1)	5.8 (1.1)	
Backward digit repetition	3.8 (0.8)	3.9 (0.7)	2.2 (1.3)	
Trail-making test B	193.9 ms (98.5)	179.0 (83.7)	188.8 ms (55.1)	
GDS	10.3 ± 2.5	11.3 ± 3.1	13.3 ± 2.5	

Standard deviations are given in parentheses behind the mean values. There were no significant group differences as is indicated by the statistical analysis (last column)

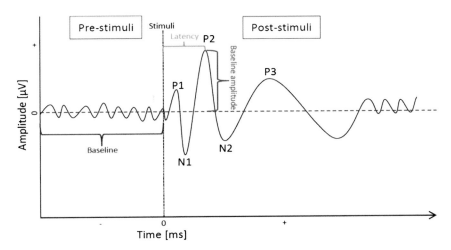

Fig. 4.6 Illustration of a possible Auditory ERP signal. On the X-axis the time is shown with 0 at the stimuli. The Y-axis is the amplitude with 0 at the baseline. In the pre-stimuli window a baseline is visible from which a horizontal average can be calculated

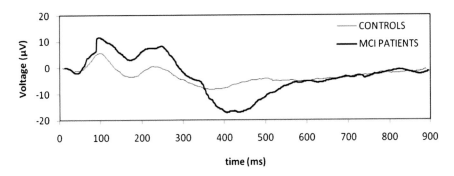

Fig. 4.7 Grand average baseline AERP waveforms for MCI amnestic patients at our study and comparison to baseline for controls

5 month. At the same time, the active control group is a sample of the MCI amnestic population from the Agios Ioannis day clinic in Thessaloniki that received a learning-based memory training approach in which participants used computers to make cognitive exercises, viewed DVD-based educational programs on history, art and literature or participated at puzzle solving exercises. The active control group was required to have high face validity and match the experimental group for daily and total training time, interesting audiovisual content, and computer use. Thus the AC cognitive training program employed a learning-based memory training approach in which participants used computers to view DVD-based educational programs on history, art and literature.

The participants in the virtual reality museum cognitive training and the active control group trained twice a week for 90 min across 5 months. The virtual reality museum cognitive training was conducted on a one-to-one basis while the active control trainings were conducted in small groups with not more than 12 participants by professional psychologists. Two extra sessions were offered at the end of the program for those participants who missed the regular sessions. The participants were not encouraged to train outside the training sessions.

4.4.4 Data Recordings and Analysis

4.4.4.1 Electrophysiological Recording

The Electroencephalogram (EEG) was recorded from 32 active electrodes positioned according to the extended 10–20 system (the electrodes mounted directly on the scalp included the following positions: C3, C4, CP3, CP4, CPz, Cz, F3, F4, F7, F8, FC3, FC4, FCz, Fp1, Fp2, Fpz, Fz, O1, O2, Oz, P3, P4, P7, P8, PO3, PO4, POz, Pz, T7, and T8). Electrodes A1 and A2 were placed at the left and right earlobes. The horizontal and vertical EOG was measured by electrodes placed at the outer canthi (LO1, LO2) and above and below both eyes (SO1, SO2, IO1, IO2). Electrode impedance was kept below 10 kΩ. The amplifier band pass was 0.01–140 Hz. EEG and EOG were sampled continuously with a rate of 2,048 Hz. Data was archived on a hard disk with triggers using post-session annotation.

Offline, the EEG was downscaled to a sampling rate of 500 Hz by using the software Neuroworkbench (Nihon-Kohden, Japan). The epochs were 1,200 ms long ranging from 100 ms before and 1,000 ms after stimulus onset. All epochs with EEG amplitudes of more than ±120 μV or with drifts of more than 150 μV within 300 ms were discarded. For all participants and conditions at mean 48 epochs (Min $=$ 17; Max $=$ 53; SD $=$ 7.3) of the epochs remained for averaging after artefact rejection and correction. The epochs were averaged according to the stimulus conditions (target trials versus non-target trials) and referenced to linked earlobes (excluding the EOG electrodes). For stimulus locked averages only correct epochs were used, excluding trials with false alarms or misses. A digital low-pass filter was set at 17 Hz in order to reduce oversampling.

4.4.4.2 Analysis

Statistical analyses were performed by means of repeated measures ANOVAs with Greenhouse-Geisser corrected degrees of freedom. In case of significant main effects (if the factor included more than two levels) or interactions, additional ANOVAs were applied for post hoc testing of contrasts and simple effects. For response times (RTs; correct commission trials) the ANOVA included the within factor time (session one, session n) and the between factor group (virtual reality museum cognitive training group, active control group, no-contact control group). Separate ANOVAs were carried out for false alarms and for misses, because they are different types of errors either demanding a response or not. Both analysis included the factors time and group.

The peak amplitude and latency of the N1 potential was measured at the two occipital electrodes O1 and O2 were the potential showed its maximum. The N2 was quantified as the mean amplitude in the time interval between 240 and 300 ms at the electrodes FCz, Cz and CPz where maximum amplitude resulted. A reliable measurement of the peak was not possible due to the overlapping P2, and P3b potentials. The P2 potential was quantified in amplitude and latency as the local maximum at the electrodes FCz, Cz and CPz in the search interval between 200 and 400 ms where it showed the highest peaks. The peak amplitude and latency of the P3b potential was measured as the local maximum at the electrodes Cz, CPz and Pz in the search interval between 400 and 700 ms where it showed the highest amplitudes.

Six separate ANOVAs were carried out for the peak amplitudes and latencies of the N1, P2 and the P3b, respectively, including the between subject factor group and the within subject factors session (session one, session two), stimulus type (target, non-target) and electrodes (O1 and O2 for the N1; FCz, Cz, and CPz for the P2 potential; Cz, CPz, and Pz for the P3b potential, resp.). An additional ANOVA was carried out for the N2 mean amplitudes including the between subject factor group and the within subject factors session, stimulus type, and electrodes (FCz, Cz, and CPz).

We also used sLORETA [81] in order to closer examine the underlying neuronal changes of the expected training effect of stimulus feature processing as reflected by the P2. We examined only the target condition because the training gains may especially help to improve target detection. The program sLORETA estimates the sources of activation on the basis of standardized current density at each of 6,239 voxels in the grey matter of the MNI-reference brain with a spatial resolution of 5 mm. The calculation is based upon a linear weighted sum of the scalp electric potentials with the assumption that neighboring voxels have a maximal similar electrical activity. The voxel-based sLORETA images were first computed for each individual averaged ERP in the target condition in the interval from 170 to 190 ms surrounding the P2 peak. Then, the differences of the sLORETA images between test sessions were statistically compared between groups using the sLORETA voxelwise randomization test (5,000 permutations) which is based on statistical nonparametric mapping (SnPM) and implemented in sLORETA. Two independent group tests were carried out for comparison of the three groups (cognitive training group versus no-contact control groups, and versus social control group). The tests were performed for an average of all time frames in the interval with the null hypothesis that (T1 group A–T2 group A) = (T1 group B–T2 group B). The tests were corrected for multiple comparisons [82].

4.5 Results

4.5.1 Neuropsychological Variables Outcome

In the virtual reality museum and active control aMCI group, there were significant differences between the delayed-recall scores on the RAVLT at baseline and those at both the 5-month follow-up (1.6 ± 1.5 vs. 4.4 ± 1.5, $p = 0.04$; 1.6 ± 1.5 vs. 4.6 ± 2.3, $p = 0.04$) (Table 4.2). The immediate recall scores on the Rey Osterrieth Complex Figure (11.9 ± 9.2 vs. 15.8 ± 9.4; $p = 0.04$), the Trail-Making B (193.9 ± 98.5 vs 104.1 ± 28.7; $p = 0.04$) and the MMSE (26.8 ± 3.6 vs. 28.2 ± 2.5; $p = 0.04$) were significantly improved only at the 5-month follow-up in the virtual reality museum aMCI group. There was a tendency toward improvement of the digit span forward scores (6.2 ± 1.1 vs. 7.8 ± 1.3; $p = 0.07$) at the follow-up of the virtual reality museum aMCI group and a general training-induced BNT scores improvement (10.6 ± 1.9 vs. 12.0 ± 2.0; $p = 0.07$) compared to the baseline scores in the virtual reality and the active control aMCI group (Table 4.2). The GDS score was also improved after cognitive training, but the difference did not reach statistical significance (10.3 ± 2.5 vs. 8.9 ± 1.7; $p = 0.23$). There were no significant differences between the baseline and follow-up scores in other outcome measures in the MCI wait-list control group.

4.5.2 Electrophysiological Measures Outcome

The P300 component latency and amplitude among the experimental groups (as detected on the Pz electrode) for the two conditions (target and no-target auditory stimuli) before and after training are summarized in Table 4.3. When the non-target stimulus was presented, the P300 latency following training was significantly shorter in both memory training groups (Table 4.3). The P300 amplitude was significantly higher after training on both groups. However, when the target stimuli was presented, the P300 latency following training was significantly shorter in both research groups; the Virtual Reality Museum latencies were significantly longer than those of the **Active Control**; and the amplitude was significantly lower for the **Active Control** than for the Virtual Reality Museum.

Our results are in line with previous training studies which also found evidence for improvements of specific cognitive functions after cognitive training in older participants (e.g. for working memory: [83], e.g., for dual task performance: [54]).

Performance improvements of older participants were also found for cognitive training of visual conjunction search in other training studies [84, 85]. These studies found evidence that seniors has learning skills just as good as the young ones to efficiently use feature information and selectively attend to those objects in the search array that share common features with the target. Our findings however go further, because we used for a first time a 3D Virtual Museum environment and

Table 4.2 Changes in outcome variables in the participants with aMCI

	Virtual museum aMCI group		Active control aMCI group		Normal control aMCI group	
	Baseline	Follow-up	Baseline	Follow-up	Baseline	After 20 weeks
RAVLT, immediate recall	15.4 ± 4.3	16.6 ± 5.1	15.5 ± 4.6	15.6 ± 4.1	15.0 ± 3.1	12.8 ± 5.9
RAVLT, delayed recall	1.6 ± 1.5	4.4 ± 1.5*	1.7 ± 1.5	4.6 ± 2.3*	2.2 ± 1.5	2.4 ± 2.6
RAVLT, recognition	5.6 ± 2.2	7.0 ± 1.9	5.5 ± 2.2	6.4 ± 2.3	7.4 ± 1.9	7.4 ± 0.9
ROCF copy	34.6 ± 1.3	36.0 ± 0.0	32.7 ± 1.9	34.2 ± 1.6	28.9 ± 8.5	26.2 ± 8.8
ROCF, immediate recall	11.9 ± 9.2	16.8 ± 9.4*	11.4 ± 9.2	11.0 ± 3.4	7.0 ± 4.7	9.3 ± 5.4
ROCF, delayed recall	11.6 ± 9.4	16.3 ± 8.9	10.6 ± 9.1	15.4 ± 8.1	7.2 ± 4.6	9.6 ± 5.6
ROCF, recognition	6.6 ± 2.9	7.0 ± 2.8	6.3 ± 2.5	7.4 ± 2.5	6.0 ± 1.4	5.0 ± 0.7
Digit span forward	6.2 ± 1.1	7.8 ± 1.3†	6.1 ± 1.1	7.2 ± 1.1	5.8 ± 1.1	6.4 ± 1.5
Digit span backward	3.8 ± 0.8	4.0 ± 1.6	3.9 ± 0.7	3.6 ± 0.9	2.2 ± 1.3	2.6 ± 0.5
Stroop, color reading	73.4 ± 35.2	86.6 ± 26.8	74.4 ± 32.2	80.2 ± 23.3	70.6 ± 23.4	59.8 ± 39.9
Category fluency	10.6 ± 2.2	13.4 ± 5.7	11.3 ± 3.1	13.2 ± 4.4	11.2 ± 4.3	11.6 ± 4.8
Letter fluency	7.4 ± 3.6	8.6 ± 4.8†	7.1 ± 2.6	7.6 ± 2.9	6.0 ± 3.4	6.0 ± 5.0
TRAIL-B	193.9 ± 98.5	104.10 ± 28.7*	179.0 ± 83.7	210.0 ± 62.6	188.8 ± 55.1	228.8 ± 75.0
BNT score	10.4 ± 1.9	15.4 ± 2.4†	10.6 ± 1.9	12.0 ± 2.0†	11.2 ± 1.9	10.0 ± 2.2
MMSE score	26.8 ± 3.6	28.2 ± 2.5*	26.2 ± 3.6	27.0 ± 2.6	26.2 ± 3.1	24.6 ± 4.6
GDS score	10.3 ± 2.5	8.9 ± 1.7	11.3 ± 3.1	9.9 ± 2.7	13.3 ± 2.5	14.9 ± 2.2

* $p < 0.05$, † $p = 0.07$

Table 4.3 Effect of cognitive training on P300 latency and amplitude (Pz electrode): mean (standard deviation)

Measures	Virtual museum			Active control			F		
	Before training	After training	T	Before training	After training	T	Maineffect: training	Maineffect: group	Interaction: training with group
Latency target stimuli	447.52 (84.0)	394.49 (60.53)	9.11***	399.47 (82.80)	365.41 (65.14)	6.64**	14.30*** (1.59)	3.91* (1.59)	1.89 (1.59)
Amplitude target stimuli	4.07 (2.65)	4.39 (1.71)	3.67*	4.73 (2.70)	5.11 (1.81)	4.12*	4.46* (1.59)	4.32* (1.59)	0.89 (1.59)
Latency non-taget stimuli	468.75 (108.40)	418.47 (79.22)	6.87**	437.39 (139.10)	395.34 (88.66)	4.21*	9.56** (1.59)	3.4* (1.59)	0.23 (1.59)
Amplitude non-target stimuli	4.18 (2.50)	4.42 (2.12)	2.21	4.30 (2.76)	4.58 (1.93)	1.94	3.18 (1.59)	1.06 (1.59)	0.45 (1.59)

$p < 0.05$; ** $p < 0.01$; *** $p < 0.001$

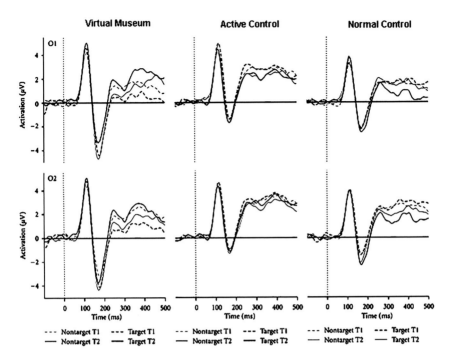

Fig. 4.8 Stimulus-locked event-related potentials at the occipital electrodes *O1* and *O2* separately for target and non-target trials, for the first (T1) and the second test session (T2) as well as for the Virtual Museum cognitive training group, the Active Control group and the normal control group

showed the neuronal correlates of the functional processes suggesting improvements via the VR training as follows:

- In the Virtual Museum cognitive training occipotal N1 enhancement was evident post-training compared to pre-training for non-target stimuli. This suggests that the participants developed mechanisms for enhanced attention of arrays, which were not immediately recognized as targets, that is, the non-targets (Fig. 4.8).
- The frontal N2 enhancement was also evident post-training compared to pre-training for non-target stimuli. However, as this effect failed to reach significance, it can only be speculated that also the subsequent processing or even inhibition of the non-target stimuli improved after cognitive training. Based on the enhanced attention in non-target trials in the Virtual Museum cognitive training group as was reflected in the N1 amplitude, one may expect also a decrease in the false alarm rate (Fig. 4.8).

The N2 (see Fig. 4.8) showed a maximum at the electrodes FCz (1.2 μV) and Cz (1.4 μV) and was less negative at CPz (2.1 μV; main effect of f electrodes:

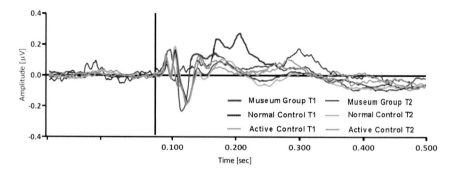

Fig. 4.9 Changes in P300 amplitudes in the Museum Group (Pz average) compared to the other groups pre- (T1) and post- (T2) training

$F(2,204) = 30.7$, $P < 0.001$). The tree-way-interaction of the factors session x stimulus type x group reached also a significance ($F(2,102) = 3.01$; $P < 0.003$).

The increased amplitude of the P300 in target trials may suggest that feature based stimulus processing was significantly improved in our older participants after only the Virtual Reality Museum cognitive training (Fig. 4.9). Consequently, the improved discrimination of stimulus features in target-present trials should decrease the likelihood of missed targets and increase the likelihood of target detection. This effect on performance data was evident in our cognitive training group post-training compared to the pre-training session and also when compared to the control groups.

The sLORETA analysis of the P300 amplitude differences between test sessions elucidates the neuronal basis of the training gain. Specifically, activation in the lingual and parahippocampalgyri was increased only in the cognitive training group and not in the two control groups. Most importantly, the increased P300 amplitude together with the significant changes in brain activation show that the cognitive training caused a change in brain processes on a functional level in a near transfer task of visual search. Both regions are anatomically and functionally connected [86] and are discussed as being sensitive for global visual feature processing [87], as well as the global processing of spatial layout [88] and surface properties like color and texture of scenes and objects in visual arrays [86]. For our training group we found that the cognitive process training improved the textual and spatial processing of visual arrays in general. The use various kinds of visual material like pictures, objects, and text pages that were used in various tasks in the training sessions did improve one basic cognitive process of global processing of visual arrays. The present results also suggest the P300 potential of the ERP as a possible marker for the improvement of this cognitive process (Fig. 4.10).

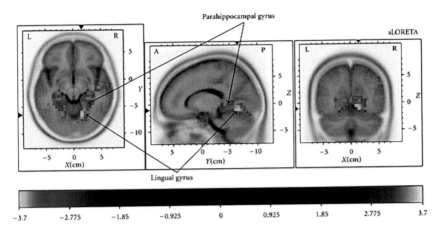

Fig. 4.10 Graphical representation of the sLORETA results comparing the differences of the target-P300. The *blue colour* indicates local maxima of lower activation in the first compared to the second test session for the cognitive training group in the right lingual and parahippocampal gyri, which may explain the amplitude difference of the P300 between sessions in the tested interval surrounding the P300 peak

4.6 Discussion

In the present study we were able to distinguish the functional processes which were sensitive to the training intervention from retest effects. The effect of test sessions on the topography of the P300 applies to all groups. We assume that the P300 may reflect memory-based stimulus processing. Thus, whereas attentional processing of target-absent trials (N1 results) and feature-based stimulus processing of target-present trials (P3 results) were only modulated by the cognitive training intervention, the improvement of stimulus categorization, which is based on memory representations (P300), was sensitive to retesting. In our study, the amplitude of the P300 component increased and latency shortened significantly following training in both experimental groups. This adds to the evidence that the P3b contains a component related to response selection or execution [89, 90]. The idea of a functional compromise associated with MCI is not new, and previous studies have reported a higher degree of functional impairment in MCI subjects when compared with matched healthy subjects [91–99].

To a limited extent, the present findings support Basak et al.'s [66] finding that inhibition can be improved by playing videogames and Schmiedek [100] demonstration that functional impairment can be improved by practicing basic cognitive tasks. The results from the present study suggest that modest improvements of the functional ability, processing speed and memory can also be achieved by means of playing virtual reality cognitive training games. A similar partially positive result of 3D games for cognition-enriching everyday activities and processing speed was reported by Nouchi et al. [101].

Not all 3D virtual reality environments however are created equal [60] and given an individual's stage of cognitive development, one environment can be more beneficial for cognitive functions than the other. For example, the cognitive training games used in the Virtual Reality Museum were very similar to those used in an actual educational museum visit [57]. Preliminary evidence for far transfer of the cognitive training was found in the present study using the neural correlates. The different extent of transfer in our study may be explained by the additional focus of the aMCI group to use specific strategies to perform the training tasks.

All our data support the *a priori* intuitive notion that highly cognitive-dependent skills are more likely to be affected as a consequence of the Virtual Reality Museum cognitive training, and that aMCI subjects show significant improvement in these functional domains. On the other hand, it is noteworthy that differences between groups were not restricted to the neuropsychological variables or the neural correlates, but also to behavioral areas as well, such as depression and motivation, although this change was not significant. As suggested by Green and Bavelier [61], motivation is a key condition for transfer to occur. The engaging nature of the virtual reality museum used in the present study could thus have facilitated transfer of training. It is clear that more research in this direction is required. Nevertheless, it can be concluded that our findings support the notion of plasticity in the neural system underlying virtual reality cognitive training and point to a relationship between the more ecological validity of Virtual Reality Museum and enhancement of specific cognitive skills.

4.7 Conclusion

The results from our study suggest that older adults do not need to be technologically savvy to benefit from virtual reality training. Almost none of the aMCI participants in the reviewed studies had prior experience with the technologies (i.e., video games, computers) used in the intervention study and yet they were still able to benefit from these novel approaches. Previous research has shown participants' prior use of computers was not significantly associated with acquisition of computer skills during training sessions, suggesting older adults can benefit from novel technologies [102].

Despite common misconceptions older adults do enjoy learning to use new technology, perceptions of the computerized training programs were positive for the older adults who completed computerized training [103]. In spite of many older adults reporting anxiety about using unfamiliar technology at the beginning of training, most reported high levels of satisfaction after training was completed. Some patients also stated they could use their new video game skills to connect more with their grandchildren, like we have seen many times in the literature [104]; whereas others were very willing to learn to use video games and believed they could be a positive form of mental exercise [105].

In conclusion, the present study lends modest support to the notion that playing virtual reality cognitive training games improves untrained cognitive functions in aMCI. Since these functions facilitate adaptive behavior in various contexts, improved cognitive processing can be expected to help older adults to overcome cognitive challenges in their daily routines. Virtual Reality provides an entertaining and thus motivating tool for improving cognitive and executive functions. The Virtual Reality Museum doesn't require physical well-being and mobility of the participant as much as physical exercise interventions; although these seem to be more effective in buffering decline of executive function [106]. Additionally, the virtual reality museum is not expensive to administer as compared to interventions supervised by a therapist. Virtual Reality cognitive rehabilitation, such as the process-specific RehaCom tasks used by Weiand [107] appear to be successful at keeping patients motivated for continuous training even after the supervised sessions at the clinic have finished. As such, the process-specific training seems to be a good choice for long-term self-guided exercises. The present study suggests that the Virtual Reality Museum should not be dismissed as a cognitive training tool.

Even within the homogeneous sample of older adults that participated in the present study, some participants benefited more from playing the virtual reality museum than others. A variety of factors may be responsible for individual differences in sensitivity to cognitive training. For instance, recent findings from our lab indicate that inter-individual genetic variability modulates transfer of training to untrained tasks [108]. Therefore, caution concerning the interpolation of aggregate data to individuals is advised, and individual differences in cognitive training outcomes are an important topic to be addressed in future studies. Geusgens et al. [109] reviewed 41 studies specifically looking for transfer effects during cognitive rehabilitation. They only included studies that trained compensation strategies as opposed to cognitive skills training. Out of the 41 reviewed studies, 36 were able to demonstrate some form of transfer. However, only 22 studies actually evaluated transfer to daily-life activities while the others looked at either simulated lab-based activities or activities that were very similar to the previously trained ones. Out of these 22 studies, 18 were able to show transfer of learned abilities, but only six included statistical evidence for their results. Furthermore, the sample sizes of most studies were very small or based on single-case designs. Consequently, no clear-cut conclusions for or against strategy training transfer to daily activities can be drawn.

The artwork of the virtual reality museum we presented here was maybe not nearly as advanced and capturing as commercial off-the-shelf games, which could create even higher levels of realism. Modern game engines already provide the technology to develop environments that can be easily recognized by users and allow for high visual quality. Transparency and "realism" in a broader sense can relate to plausibility and place illusions which are described by Slater [110]. Plausibility illusion refers to the fact that the user believes the virtual scenario is actually occurring. It is caused by events and the scenario relating directly to the user (e.g. virtual character talking to user). For example, a cognitive task that is embedded in a user-relevant scenario directly relates to the therapy goal of the patient and represents a desired outcome of the patient's rehabilitation (e.g. virtual kitchen with cooking tasks relates to the scenario that the patient aims to engage in independently at home). This stands in contrast to the abstract nature of traditional neuropsychological tests which may have little in common with real-world scenarios (e.g. using abstract objects for mental rotation). Scenarios of high realism are believed to be of advantage when patients deny their cognitive deficits. The realism of a task can potentially lead patients to compare their performance with common standards and past experiences and make them realize that their cognitive abilities may not match their subjective perception. This is the basis for patients actively engaging in cognitive training and making progress throughout their cognitive rehabilitation. As the growing number of serious games suggests, engaging game-like training content appears to be a method of choice to prevent frustration and boredom of users.

It is important to note that inconsistencies may be due to several factors not related to the actual training program itself, including different cognitive outcome measures and modifications of the training program. The electrophysiological data helped to elucidate the functional processes which were sensitive to the training intervention and, on the other hand, to retest effects due to task repetition. Additionally, the mediating neuronal basis of the training gain was identified, thus, underlining the efficiency of the training to induce functional changes in the brain. More specifically, the cognitive training especially improved the global feature processing of visual arrays which may explain the improvement in target detection within a given time window in the near transfer task of visual conjunction search. These results cannot be explained by test repetition or by the mere social interaction of the training intervention, suggesting that a multilayered formal cognitive training is sufficient to facilitate neuronal plasticity in older age.

Our study bears several shortcomings which may give directions for further studies. First of all, the cognitive training was multidimensional and aimed mainly at enhancing basic and executive functions tested by a number of our tasks in order to improve daily life activities. As the training was domain unspecific, it is not possible to show divergent results in two or more tasks in the effects of the training procedure. Further studies which aim to evaluate broad cognitive trainings should bear in mind (1) to use more than one transfer task which assess the same cognitive function in order to show convergent effects of the training and/or, (2) to use transfer tasks assessing cognitive functions which were not intended to be improved by the training in order to show divergent effects. An additional shortcoming of the present study is the fact that the virtual museum interface system was outdated compared to the more intuitive solution provided recently by Microsoft Kinect[1] for full-body tracking [111]. The virtual reality museum group received basic PC-practice which may have made them more experienced with computer technology than the other groups. However, although modern interaction devices such as the Microsoft KINECT 3D sensor for natural gesture interaction is

---

[1] Microsoft Kinect—www.kinectforwindows.org/www.xbox.com/kinect.

more senior-friendly [112], our study interaction with the PC was reduced to a minimum and the manual responses were collected with special response buttons and not with a computer keyboard or a mouse. Therefore, we do think that a more advanced interaction for the cognitive training group may elicit even more transfer effects. Further training studies should try to exclude any confounding effect of the training procedure on the evaluation of the training effects.

Older adults are the now fastest growing segment of Internet users [113]. According to a 2010 Pew Internet and American Life survey, 78 % of adults aged 50–64 years and 42 % of adults older than 65 years of age use the Internet. This is a sharp increase from 2000 when only 50 % of adults 50–64 years and 15 % of adults older than 65 years of age used the Internet [114]. As ownership of personal computers continues to grow and older adults have access to the Internet [115], cognitive training programs need to take fuller advantage of these outlets to improve cognitive function and delay cognitive decline in later life.

References

1. Virvou, M., Katsionis, G.: On the usability and likeability of virtual reality games for education: the case of VR-ENGAGE. Comput. Educ. **50**, 154–178 (2008)
2. Lewis, T.M., Aggarwal, R., Rajaretnam, N., Grantcharov, T.P., Darzi, A.: Training in surgical oncology: the role of virtual reality simulation. Surg. Oncol. **20**, 134–139 (2011)
3. Rizzo, A., Kim, G.J.: A SWOT analysis of the field of virtual reality rehabilitation and therapy. Presence **14**(2), 119–146 (2005)
4. Rizzo, A., Schultheis, M., Kerns, K.A., Mateer, C.: Analysis of assets for virtual reality applications in neuropsychology. Neuropsychol. Rehabil. **14**(1/2), 207–239 (2004)
5. Laczo, J., Vlcek, K., Vyhnalek, M., Vajnerova, O., Ort, M., Holmerova, I., Tolar, M., Andel, R., Bojar, M., Hort, J.: Spatial navigation testing discriminates two types of amnestic mild cognitive impairment. Behav. Brain Res. **202**, 252–259 (2009)
6. Petersen, R.C., Smith, G.E., Waring, S.C., Ivnik, R.J., Tangalos, E.G., Kokmen, E.: Mild cognitive impairment: clinical characterization and outcome. Arch. Neurol. **56**(3), 303–308 (1999)
7. Gauthier, S., Reisberg, B., Zaudig, M., Petersen, R.C., Ritchie, K., Broich, K., Belleville, S., et al.: Mild cognitive impairment. Lancet **367**(9518), 1262–1270 (2006)
8. Petersen, R.C.: Mild cognitive impairment as a diagnostic entity. J. Intern. Med. **256**(3), 183–194 (2004)
9. Winblad, B., Palmer, K., Kivipelto, M., Jelic, V., Fratiglioni, L., Wahlund, L.-O., Nordberg, A., et al.: Mild cognitive impairment–beyond controversies, towards a consensus: report of the International Working Group on Mild Cognitive Impairment. J. Intern. Med. **256**(3), 240–246 (2004)
10. Eschweiler, G.W., Leyhe, T., Klöppel, S., Hüll, M.: New developments in the diagnosis of dementia. Dtsch. Ärzteblatt Int. **107**(39), 677–683 (2010)
11. Hanfelt, J.J., Wuu, J., Sollinger, A.B., Greenaway, M.C., Lah, J.J., Levey, A.I., Goldstein, F.C.: An exploration of subgroups of mild cognitive impairment based on cognitive, neuropsychiatric and functional features: analysis of data from the National Alzheimer's Coordinating Center. Am. J. Geriatr. Psychiatry: Official J. Am. Assoc. Geriatr. Psychiatry **19**(11), 940–950 (2011)

12. Petersen, R.C., Roberts, R.O., Knopman, D.S., Boeve, B.F., Geda, Y.E., Ivnik, R.J., Smith, G.E., et al.: Mild cognitive impairment: ten years later. Arch. Neurol. **66**(12), 1447–1455 (2009)
13. Gold, D.A.: An examination of instrumental activities of daily living assessment in older adults and mild cognitive impairment. J. Clin. Exp. Neuropsychol. 37–41 (2012)
14. Perneczky, R., Pohl, C., Sorg, C., Hartmann, J., Komossa, K., Alexopoulos, P., Wagenpfeil, S., et al.: Complex activities of daily living in mild cognitive impairment: conceptual and diagnostic issues. Age Ageing **35**(3), 240–245 (2006)
15. Guerreiro, R., Wojtas, A., Bras, J., Carrasquillo, M., Rogaeva, E., Majounie, E., Cruchaga, C., et al. : TREM2 Variants in Alzheimer's Disease. N. Engl. J. Med. (2012) (121114171407007)
16. Patel, B.B., Holland, N.W.: Mild cognitive impairment: hope for stability, plan for progression. Clevel. Clin. J. Med. **79**(12), 857–864 (2012)
17. Albert, M.S., DeKosky, S.T., Dickson, D., Dubois, B., Feldman, H.H., Fox, N.C., Gamst, A., et al.: The diagnosis of mild cognitive impairment due to Alzheimer's disease: recommendations from the National Institute on Aging-Alzheimer's Association workgroups on diagnostic guidelines for Alzheimer's disease. Alzheimer's & Dement. J. Alzheimer's Assoc. **7**(3), 270–279 (2011)
18. Wang, H.-M., Yang, C.-M., Kuo, W.-C., Huang, C.-C., Kuo, H.-C.: Use of a modified spatial-context memory test to detect amnestic mild cognitive impairment. PLoS ONE **8**(2), e57030 (2013)
19. Pai, M.C., Jacobs, W.J.: Int. J. Geriatr. Psychiatry **19**, 250–255 (2004)
20. Cherrier, M.M., Mendez, M., Perryman, K.: Neuropsychiatry Neuropsychol. Behav. Neurol. **14**, 159–168 (2001)
21. Kalova, E., Vlcek, K., Jarolimova, E., Bures, J.: Behav. Brain Res. **159**, 175–186 (2005)
22. Kessels, R.P., Feijen, J., Postma, A.: Dement. Geriatr. Cogn. Disord. **20**, 184–191 (2005)
23. Mapstone, M., Steffenella, T.M., Duffy, C.J.: A visuospatial variant of mild cognitive impairment: getting lost between aging and AD. Neurology **60**(5), 802–808 (2003)
24. Hort, J., Laczó, M., Vyhnálek, M., Bojar, J., Bureš, J., Vlček, K.: Spatial navigation deficit in amnestic mild cognitive impairment. Proc. Natl. Acad. Sci. **104**(10), 4042–4047 (2007)
25. Rey, B., Alcañiz, M.: Research in Neuroscience and Virtual Reality. InTech (2010)
26. Cornwell, B.R., Johnson, L.L., Holroyd, T., Carver, F.W., Grillon, C.: Human hippocampal and parahippocampal theta during goal-directed spatial navigation predicts performance on a virtual Morris water maze. J. Neurosci.: Official J. Soc. Neurosci. **28**(23), 5983–5990 (2008)
27. Harvey, C.D., Collman, F., Dombeck, D.A., Tank, D.W.: Intracellular dynamics of hippocampal place cells during virtual navigation. Nature **461**(7266), 941–946 (2009)
28. Plancher, G., Tirard, A., Gyselinck, V., Nicolas, S., Piolino, P.: Using virtual reality to characterize episodic memory profiles in amnestic mild cognitive impairment and Alzheimer's disease: influence of active and passive encoding. Neuropsychologia **50**(5), 592–602 (2012)
29. Slater, M., Antley, A., Davison, A., Swapp, D., Guger, C., Barker, C., Pistrang, N., et al.: A virtual reprise of the Stanley Milgram obedience experiments. PLoS ONE **1**, e39 (2006)
30. Waller, D., Richardson, A.R.: Correcting distance estimates by interacting with immersive virtual environments: effects of task and available sensory information. J. Exp. Psychol. Appl. **14**(1), 61–72 (2008). doi:10.1037/1076-898X.14.1.61
31. Klauer, K.C., Zhao, Z.: Double dissociations in visual and spatial short-term memory. J. Exp. Psychol. Gen. **133**(3), 355–381 (2004)
32. Lithfous, S., Dufour, A., Després, O.: Spatial navigation in normal aging and the prodromal stage of Alzheimer's disease: insights from imaging and behavioral studies. Ageing Res. Rev. **12**(1), 201–213 (2012)
33. O'Keefe, J., Dostrovsky, J.: Brain Res. **34**, 171–175 (1971)
34. Morris, R.G., Garrud, P., Rawlins, J.N., O'Keefe, J.: Nature **297**, 681–683 (1982)

35. Astur, R.S., Taylor, L.B., Mamelak, A.N., Philpott, L., Sutherland, R.J.: Behav. Brain Res. **132**, 77–84 (2002)
36. Feigenbaum, J.D., Morris, R.G.: Neuropsychology **18**, 462–472 (2004)
37. Maguire, E.A., Burgess, N., Donnett, J.G., Frackowiak, R.S., Frith, C.D., O'Keefe, J.: Science **280**, 921–924 (1998)
38. Abrahams, S., Pickering, A., Polkey, C.E., Morris, R.G.: Neuropsychologia **35**, 11–24 (1997)
39. De Lillo, C., James, F.C.: Spatial working memory for clustered and linear configurations of sites in a virtual reality foraging task. Cogn. Process. **13**(Suppl 1), S243–S246 (2012)
40. Koenig, S., Crucian, S., Dalrymple-Alford, J., Dünser, A.: Assessing navigation in real and virtual environments: a validation study. Int. J. Disabil. Hum. Dev. **10**(4), 325–330 (2010)
41. Caffo, A., De Caro, M., Picucci, L., Notarnicola, A., Settanni, A., Livrea, P., Lancioni, G., Bosco, A.: Reorientation deficits are associated with amnestic mild cognitive impairment. Am. J. Alzheimer's Dis. Dement. **27**(5), SAGE Aug 1 (2012)
42. Liarokapis, F., Anderson, E.: Using augmented reality as a medium to assist teaching in higher education. To appear in Eurographics 2010, Education Program, Norrkfping, Sweden, 4–7 May 2010
43. Tsolaki, M., Kounti, F., Agogiatou, C., Poptsi, E., Bakoglidou, E., Zafeiropoulou, M., Soumbourou, A., et al.: Effectiveness of nonpharmacological approaches in patients with mild cognitive impairment. Neuro-degenerative Dis. **8**(3), 138–145 (2011)
44. Tsatali, M., Tarnanas, I., Malegiannaki, A., Tsolaki, M.: Does cognitive training with the use of a virtual museum improve neuropsychological performance in aMCI? In: 22nd Alzheimer Europe Conference in Vienna, Assistive Technologies P3, 121–134 (2012)
45. Ackerman, P.L., Kanfer, R., Calderwood, C.: Use it or lose it? WII brain exercise practice and reading for domain knowledge. Psychol. Aging **25**, 753–766 (2010)
46. Owen, A.M., Hampshire, A., Grahn, J.A., Stenton, R., Dajani, S., Burns, A.S., Howard, R.J., Ballard, C.G.: Putting brain training to the test. Nature **465**, 775–778 (2010)
47. Peretz, C., Korczyn, A.D., Shatil, E., Aharonson, V., Birnboim, S., Giladi, N.: Computer-based, personalized cognitive training versus classical computer games: a randomized double-blind prospective trial of cognitive stimulation. Neuroepidemiology **36**, 91–99 (2011)
48. Hertzog, C., Kramer, A.F., Wilson, R.S., Lindenberger, U.: Enrichment effects on adult cognitive development: can the functional capacity of older adults be preserved and enhanced? Psychol. Sci. Public Interest **9**, 1–65 (2009)
49. Denney, N.W.: A model of cognitive-development across the life-span. Dev. Rev. **4**, 171–191 (1984)
50. Baron, A., Mattila, W.R.: Response slowing of older adults: effects of time-limit contingencies on single- and dual-task performances. Psychol. Aging **4**, 66–72 (1989)
51. Buschkuehl, M., Jaeggi, S.M., Hutchison, S., Perrig-Chiello, P., Dapp, C., Muller, M., Breil, F., Hoppeler, H., Perrig, W.J.: Impact of working memory training on memory performance in old–old adults. Psychol. Aging **23**, 743–753 (2008)
52. Dahlin, E., Neely, A.S., Larsson, A., Backman, L., Nyberg, L.: Transfer of learning after updating training mediated by the striatum. Science **320**, 1510–1512 (2008)
53. Sammer, G., Reuter, I., Hullmann, K., Kaps, M., Vaitl, D.: Training of executive functions in Parkinson's disease. J. Neurol. Sci. **248**, 115–119 (2006)
54. Bherer, L., Kramer, A.F., Peterson, M.S., Colcombe, S., Erickson, K., Becic, E.: Transfer effects in task-set cost and dual-task cost after dual-task training in older and younger adults: further evidence for cognitive plasticity in attentional control in late adulthood. Exp. Aging Res. **34**, 188–219 (2008)
55. Davidson, D.J., Zacks, R.T., Williams, C.C.: Stroop interference, practice, and aging. Neuropsychol. Dev. Cogn. B AgingNeuropsychol. Cogn. **10**, 85–98 (2003)
56. Karbach, J., Kray, J.: How useful is executive control training? Age differences in near and far transfer of task-switching training. Dev. Sci. **12**, 978–990 (2009)

57. Ball, K., Berch, D.B., Helmers, K.F., Jobe, J.B., Leveck, M.D., Marsiske, M., Morris, J.N., Rebok, G.W., Smith, D.M., Tennstedt, S.L., Unverzagt, F.W., Willis, S.L.: Effects of cognitive training interventions with older adults: a randomized controlled trial. JAMA **288**, 2271–2281 (2002)

58. Schmiedek, F., Lovden, M., Lindenberger, U.: Hundred days of cognitive training enhance broad cognitive abilities in adulthood: findings from the COGITO study. Front. Aging Neurosci. **2**, 27 (2010)

59. Klingberg, T.: Training and plasticity of working memory. Trends Cogn. Sci. **14**, 317–324 (2010)

60. Achtman, R.L., Green, C.S., Bavelier, D.: Video games as a tool to train visual skills. Restor. Neurol. Neurosci. **26**, 435–446 (2008)

61. Green, C.S., Bavelier, D.: Exercising your brain: a review of human brain plasticity and training-induced learning. Psychol. Aging **23**, 692–701 (2008)

62. Gee, J.P.: What Video Games Have to Teach Us about Learning and Literacy. Palgrave Macmillan, New York, NY (2007)

63. Shiffrin, R.M., Schneider, W.: Controlled and automatic human information processing: II. Perceptual learning, automatic attending and a general theory. Psychol. Rev. **84**, 127–190 (1977)

64. Logan, G.D.: Automaticity, resources, and memory: theoretical controversies and practical implications. Hum. Factors **30**, 583–598 (1988)

65. Feng, J., Spence, I., Pratt, J.: Playing an action video game reduces gender differences in spatial cognition. Psychol. Sci. **18**, 850–855 (2007)

66. Basak, C., Boot, W.R., Voss, M.W., Kramer, A.F.: Can training in a real-time strategy video game attenuate cognitive decline in older adults? Psychol. Aging **23**(4), 756–777 (2008)

67. Mahncke, H.W., Connor, B.B., Appelman, J., Ahsanuddin, O.N., Hardy, J.L., Wood, R.A., Joyce, N.M., Boniske, T., Atkins, S.M., Merzenich, M.M.: Memory enhancement in healthy older adults using a brain plasticity-based training program: a randomized, controlled study. Proc. Natl. Acad. Sci. U.S.A. **103**, 12523–12528 (2006)

68. Uchida, S., Kawashima, R.: Reading and solving arithmetic problems improves cognitive functions of normal aged people: a randomized controlled study. Age (Dordr.) **30**, 21–29 (2008)

69. Smith, G.E., Housen, P., Yaffe, K., Ruff, R., Kennison, R.F., Mahncke, H.W., Zelinski, E.M.: A cognitive training program based on principles of brain plasticity: results from the improvement in Memory with Plasticity-based Adaptive Cognitive Training (IMPACT) study. J. Am. Geriatr. Soc. **57**, 594–603 (2009)

70. Cicerone, K.D., Dahlberg, C., Kalmar, K., Langenbahn, D.M., Malec, J.F., Bergquist, T.F., Felicetti, T., Giacino, J.T., Harley, P.J., Harrington, D.E., Herzog, J., Kneipp, S., Laatsch, L., Morse, P.A.: Evidence-based cognitive rehabilitation: recommendations for clinical practice. Arch. Phys. Med. Rehabil. **81**, 1596–1615 (2000)

71. Cicerone, K.D., Dahlberg, C., Malec, J.F., Langenbahn, D.M., Felicetti, T., Kneipp, S., Ellmo, W., Kalmar, K., Giacino, J.T., Harlez, P., Laatsch, L., Morse, P., Catanese, J.: Evidence-based cognitive rehabilitation: updated review of the literature from 1998 through 2002. Arch. Phys. Med. Rehabil. **86**, 1596–1615 (2005)

72. Tucker-Drob, E.M.: Neurocognitive functions and everyday functions change together in old age. Neuropsychology **25**, 368–377 (2011)

73. Ylvisaker, M., Turkstra, L.S., Coelho, C.: Behavioral and social interventions for individuals with traumatic brain injury: a summary of the research with clinical implications. Semin. Speech Lang. **26**(4), 256–257 (2005)

74. Sohlberg, M.M., Mateer, C.A. (eds.): Cognitive rehabilitation: an integrative neuropsychological approach. Guilford Press, New York (2001)

75. Mozolic, J.L., Hayasaka, S., Laurienti, P.J.: A cognitive training intervention increases resting cerebral blood flow in healthy older adults. Front. Hum. Neurosci. **4** (2010)

76. Lustig, C., Shah, P., Seidler, R., Reuter-Lorenz, P.A.: Aging, training, and the brain: a review and future directions. Neuropsychol. Rev. **19**(4), 504–522 (2009)

77. Petersen, R.C., Parisi, J.E., Dickson, D.W., et al.: Neuropathologic features of amnestic mild cognitive impairment. Arch. Neurol. **63**(5), 665–672 (2006)
78. Stroop, J.R.: Studies of interference in serial verbal reactions. J. Exp. Psychol. **18**, 643–662 (1935)
79. Papaliagkas, V., Kimiskidis, V., Tsolaki, M., Anogianakis, G.: Usefulness of event-related potentials in the assessment of mild cognitive impairment. BMC Neurosci. **9**, 107 (2008)
80. Kimiskidis, V., Papaliagkas, V.: Event-related potentials for the diagnosis of mild cognitive impairment and Alzheimer's disease. Expert Opin. Med. Diagn. **6**(1), 15–26 (2012)
81. Pascual-Marqui, D.: Standardized low-resolution brain electromagnetic tomography (sLORETA): technical details. Methods Find. Exp. Clin. Pharmacol. D **24**, 5–12 (2002)
82. Holmes, A.P., Blair, R.C., Watson, J.D.G., Ford, I.: Nonparametric analysis of statistic images from functional mapping experiments. J. Cereb. Blood Flow Metab. **16**(1), 7–22 (1996)
83. Li, S.C., Schmiedek, F., Huxhold, O., Röcke, C., Smith, J., Lindenberger, U.: Working memory plasticity in old age: practice gain, transfer, and maintenance. Psychol. Aging **23**(4), 731–742 (2008)
84. Dennis, W., Scialfa, C.T., Ho, G.: Age differences in feature selection in triple conjunction search. J. Gerontol. B **59**(4), P191–P198 (2004)
85. Ho, G., Scialfa, C.T.: Age, skill transfer, and conjunction search. J. Gerontol. B **57**(3), P277–P287 (2002)
86. Cant, J.S., Goodale, M.A.: Attention to form or surface properties modulates different regions of human occipitotemporal cortex. Cereb. Cortex **17**(3), 713–731 (2007)
87. Mechelh, A., Humphreys, G.W., Mayall, K., Olson, A., Price, C.J.: Differential effects of word length and visual contrast in the fusiform and lingual gyri during reading. Proc. Royal Soc. B **267**(1455), 1909–1913 (2000)
88. Epstein., R.A.: Parahippocampal and retrosplenial contributions to human spatial navigation. Trends Cogn. Sci. **12**(10), 88–396 (2008)
89. Falkenstein, M., Hohnsbein, J., Hoormann, J.: Effects of choice complexity on different subcomponents of the late positive complex of the event-related potential. Electroencephalogr. Clin. Neurophysiol. **92**(2), 148–160 (1994a)
90. Falkenstein, M., Hohnsbein, J., Hoormann, J.: Time pressure effects on late components of the event-related potential (ERP). J. Psychophysiol. **8**(1), 22–30 (1994b)
91. Tam, C.W., Lam, L.C., Chiu, H.F., Lui, V.W.: Characteristic profiles of instrumental activities of daily living in Chinese older persons with mild cognitive impairment. Am. J. Alzheimers Dis. Dement. **22**, 211–217 (2007)
92. Pereira, F.S., Yassuda, M.S., Oliveira, A.M., Forlenza, O.V.: Executive dysfunction correlates with impaired functional status in older adults with varying degrees of cognitive impairment. Int. Psychogeriatr. **20**, 1104–1115 (2008)
93. Ahn, I.S., et al.: Impairment of instrumental activities of daily living in patients with mild cognitive impairment. Psychiatry Inv. **6**, 180–184 (2009)
94. Burton, C.L., Strauss, E., Bunce, D., Hunter, M.A., Hultsch, D.F.: Functional abilities in older adults with mild cognitive impairment. Gerontology **55**, 570–581 (2009)
95. Schmitter-Edgecombe, M., Woo, E., Greeley, D.R.: Characterizing multiple memory deficits and their relation to everyday functioning in individuals with mild cognitive impairment. Neuropsychology **23**, 168–177 (2009)
96. Aretouli, E., Brandt, J.: Everyday functioning in mild cognitive impairment and its relationship with executive cognition. Int. J. Geriatr. Psychiatry **25**, 224–233 (2010)
97. Bangen, K.J., et al.: Complex activities of daily living vary by mild cognitive impairment subtype. J. Int. Neuropsychol. Soc. **16**, 630–639 (2010)
98. Teng, E., Becker, B.W., Woo, E., Cummings, J.L., Lu, P.H.: Subtle deficits in instrumental activities of daily living in subtypes of mild cognitive impairment. Dement. Geriatr. Cogn. Disord. **30**, 189–197 (2010)

 99. Teng, E., Becker, B.W., Woo, E., Knopman, D.S., Cummings, J.L., Lu, P.H.: Utility of the functional activities questionnaire for distinguishing mild cognitive impairment from very mild Alzheimer disease. Alzheimer Dis. Assoc. Disord. **24**, 348–353 (2010)
100. Schmiedek, F., Bauer, C., Lovden, M., Brose, A., Lindenberger, U.: Cognitive enrichment in old age: web-based training programs. GeroPsych **23**(2), 59–67 (2010)
101. Nouchi, R., Taki, Y., Takeuchi, H., Hashizume, H., Akitsuki, Y., Shigemune, Y., Sekiguchi, A., Kotozaki, Y., Tsukiura, T., Yomogida, Y., Kawashima, R.: Brain training game improves executive functions and processing speed in the elderly. PLoS ONE **7**, e29676 (2012)
102. Saczynski, J.S., Rebok, G.W., Whitfield, K.E., Plude, D.J.: Effectiveness of CD-ROM memory training as a function of within-session autonomy. Int. J. Cogn. Technol. **9**(1), 25–33 (2004)
103. Lee, B., Chen, Y., Hewitt, L.: Age differences in constraints encountered by seniors in their use of computers and the internet. Comput. Hum. Behav. **27**(3), 1231–1237 (2011)
104. Torres, A.: Cognitive effects of video games on older people. ICDVRAT **19**, 191–198 (2008)
105. Belchior, P.D.C. Cognitive training with video games to improve driving skills and driving safety among older adults [dissertation]. ProQuest Information and Learning (2008)
106. Colcombe, S., Kramer, A.F.: Fitness effects on the cognitive function of older adults: a meta-analytic study. Psychol. Sci. **14**, 125–130 (2003)
107. Weiand, C.: Neuropsychologische Behandlungsmethoden im Vergleich—Eine randomisierte klinische Studie [Comparing neuropsychological treatment methods—A randomized clinical trial]. PhD dissertation, University of Konstanz, Konstanz (2006)
108. Colzato, L.S., Van Muijden, J., Band, G.P.H., Hommel, B.: Genetic modulation of training and transfer in older adults: BDNF Val66Met polymorphism is associated with wider useful field of view. Front. Psychol. **2**, 199 (2011)
109. Geusgens, C., Winkens, I., van Heugten, C., Jolles, J., van den Heuvel, W.: The occurence and measurement of transfer in cognitive rehabilitation: a critical review. J. Rehabil. Med. **39**(6), 425–439 (2007)
110. Slater, M.: Place illusion and plausibility can lead to realistic behaviour in immersive virtual environments. Philos. Trans. R. Soc. B: Biol. Sci. **364**, 3549–3557 (2009)
111. Lange, B., Rizzo, A., Chang, C.-Y., Suma, E., Bolas, M.: Markerless Full Body Tracking: Depth-Sensing Technology within virtual environments. Paper presented at the Interservice/Industry Training, Simulation, and Education Conference (I/ITSEC), Orlando (2011)
112. Nebelrath, R., Lu, C., Schulz, C.H., Frey, J., Alexandersson, J.: A gesture based system for context-sensitive interaction with smart homes. In: Wichert, R., Eberhardt, B. (eds.), 4. AALKongress, pp. 209–222 (2011)
113. Hart, T.A., Chaparro, B.S., Halcomb, C.G.: Evaluating websites for older adults: adherence to 'senior-friendly' guidelines and end-user performance. Behav. Info. Technol. **27**(3), 191–199 (2008)
114. Pew Internet and American Life Project: Changes in internet use by age, 2000–2010. Accessed 24 Jul 2011 (2010)
115. Gamberini, L., Alcaniz, M., Barresi, G., Fabregat, M., Ibanez, F., et al.: Cognition, technology and games for the elderly: an introduction to ELDERGAMES project. PsychNology J **4**(3), 285–308 (2006)

Chapter 5
Adaptive Cognitive Rehabilitation

Inge Linda Wilms

Abstract Since the emergence of computers, it has been recognized that they might play a role in brain rehabilitation efforts. However, the complexity of cognitive skills combined with the challenges of translating into software the skilled, therapeutic activities relating to rehabilitation of brain injury have prevented real breakthroughs in the area of relearning and retraining of cognitive abilities. The real potential of advanced technology has so far not been unleashed in current implementations of training resulting in lack of long-term efficacy and generalization of training. This paper will attempt to provide an overview of some the challenges facing anyone entering the exiting field of training for cognitive recovery and propose a way forward for future cognitive rehabilitation using advanced computer technology.

Keywords Cognitive rehabilitation · Adaptive training · Artificial intelligence · Experience-based plasticity · Computer-based training

5.1 Introduction

Cognitive rehabilitation research provides a fascinating insight into the inner workings of the bits and pieces in the brain that make up the cognitive facilities and skills. Rehabilitation of brain injury is the art of combining the hopes and

I. L. Wilms (✉)
Brain Rehabilitation using Advanced Technology Laboratory (BRATLab),
Department of Psychology, University of Copenhagen, Oester Farimagsgade 2 A,
1353 Copenhagen K, Denmark
e-mail: inge.wilms@psy.ku.dk

I. L. Wilms
Center for Rehabilitation of Brain Injury, Amagerfaelledvej 56 A,
2300 Copenhagen S, Denmark

A. L. Brooks et al. (eds.), *Technologies of Inclusive Well-Being*,
Studies in Computational Intelligence 536, DOI: 10.1007/978-3-642-45432-5_5,
© Springer-Verlag Berlin Heidelberg 2014

wishes of the patient with a realistic and achievable plan for recovery based on the combined knowledge, experience and observations from a host of highly skilled clinicians. To create the best plan for recovery, clinicians must be able to draw from a large box of tools. Training for recovery of cognitive skills is only one way to assist the patient back to a reasonable quality of daily living. Never the less, research in neuroplasticity and how the brain reacts to stimuli may offer new and exciting methods in the future.

5.2 The Challenges

Anyone entering the exiting field of cognitive recovery training is facing three major challenges.

The first is that the tools for assessing the scope of the brain injury are not designed with therapy and training in mind. They are primarily tools for localizing and diagnosing the extent of damage. This means that they rarely provide data which directly can be used to determine what to train, how to train and the progress to be expected. The diagnostics tools are also very diverse in terms of interpretation and often lack sensitivity to general training-induced changes.

The second challenge is that no two injuries produce the same type of problems in the patients. Although the result of injury may be categorized within certain cognitive boundaries like speech, attention or memory, the damage done is individual and so is the subsequent behavioural consequences and effect. A function is rarely totally destroyed but may function erratically, slowly or perfectly well despite damage. Add to this that cognitive ability may fluctuate during the day or week depending on the general state of the patient. This influences the observation and interpretation of the severity and ability of the patients in terms of training.

The third challenge is that the learning process needed to reacquiring a cognitive skill may not be similar to the learning process of acquiring it initially. You rarely achieve rehabilitative improvement of memory and attention simply by just being exposed to problems requiring memory and attention. How to engage the recovery processes of the brain in a manner tailored to overcome the individual destruction is far from fully understood let alone documented as guidelines to training.

5.2.1 Fundamental Learning and Adaptation

> When an axon of cell A is near enough to excite a cell B and repeatedly or persistently takes part in firing it, some growth process or metabolic change takes place in one or both cells such that A's efficiency, as one of the cells firing B, is increased (p. 62) [1].

The experience-based plasticity of the brain has been defined as the ability of the nervous system to respond to intrinsic or extrinsic stimuli through a reorganization of its internal structure [2] and is primarily believed to be the result of long-term synaptic and axonal changes in the neural substrate [3]. This reorganization may be observed in various way, e.g. as regional changes in weight and volume of the neural substrate subserving a function [e.g. 4] or as localized changes in metabolism on fMRI imagery [e.g. 5]. So in order to understand how an injured brain may recover, it is important to touch upon how brains learn and adjust in general.

In 1985, Rumelhart and McClelland [6] proposed that the data carrying structures in the brain were organized in neural networks in which memory was stored across a landscape of interconnected neurons, each contributing to the storage and retrieval through weighted modulation. Learning was defined as a basic weight adjustment in the network based on the statistical propensity between input and output stimuli through feed forward and feedback realignment. By adjusting individual weights for triggering activity in the neuron, a group of neurons would be able to express complex behaviour beyond the ability of the individual parts of the network. In other words, our skills and knowledge which may be destroyed in brain injury is stored in a manner not easily reproduced. It is a product of our individual experiences through life.

Figure 5.1 illustrates how the healthy brain may organize and strengthen connectivity and activation of individual elements of the network through feedback mechanisms. It is a simplified model of how repeated training of a skill may induce changes that improves the effectiveness and fluency of said skill. The model depicts an untrained network (1), and how it changes in response to focused, repeated activity (2). Increased and constant activity will produce stronger and faster connections between nodes in the network (3) and reduce the dependency on lesser pathways (4) which will then be reduced (5). It suggests that what would normally be considered the hardware layer—and as such fixed—is in itself an adaptable entity influenced by continued training and stimuli.

Although simplified, the model above provides a representation of the current knowledge and principles of experience-based plasticity. It summarizes the following points relevant to relearning an injured skill. Firstly, that the internal organization needed for interaction in the neural substrate is formed or honed in response to activity and experience; secondly, that the organization of the neural substrate may be accommodated internally in a manner which may differ from individual to individual although the surface behaviour may seem similar. Considering that the learning and honing of skills take an individual course for each of us, no two people are likely to achieve a skill in precisely the same manner. There are however, elements that all training needs to adhere to in order to have an effect at the level of plasticity [8].

Two elements of training which are of particular interest in relation to computer-based training in general and game-based training in particular are the adaptive level of difficulty and feedback. In the next part, examples of research will be presented that point to interesting aspects of these two elements in relation to computer-based training.

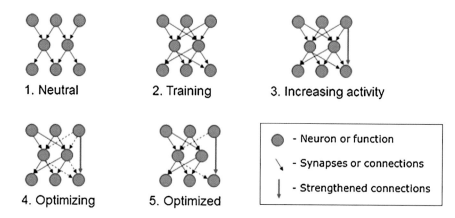

1. Neutral   2. Training   3. Increasing activity

4. Optimizing   5. Optimized

- Neuron or function

- Synapses or connections

- Strengthened connections

Fig. 5.1 The taxonomy of skill learning at the neural level. As training increases, the connectivity between active areas is increased and optimized by constant stimulation. Adapted from Robertson and Murre [7]

5.2.2 Brain Injury: The Damage Done

The most common causes of injury are ischemic attacks, haemorrhage and head trauma, but also illness and anoxia may cause lasting non-progressive injury to the brain [9]. At the neurophysiological level, the initial destruction from obstruction of blood flow or disease may cause damage and loss of neural substrate serving as basis for cognitive as well as motor functions. This includes destruction of neurons and synaptic connections as well as axonal pathways between more distant areas of the brain. This in turn may cause further disruption due to imbalance in signals caused by lack of inhibitory or excitatory signals from the destroyed areas [2]. Lack of inhibitory signals from extinct or damaged areas may cause overexcitement of other areas resulting in erratic firing or response to stimuli. Destruction or reduction in neural pathways may cause asynchronous data processing resulting in slow or delayed processing of incoming stimuli. Circuitry unaffected by the physical injury itself may be affected by the erratic feedback signals [10].

More succinctly though, internal constructs may be damaged by injury and cause strange problems like the attention deficit neglect, which is the inability to attend to stimuli presented to the left side of space. The brain keeps internal maps of the topography of the body and the surrounding world in order to determine the spatial coordinates of objects and stimuli [11]. This internal representation may also be affected by injury causing invalid translation and response to feedback stimuli and consequently invalid learning and adaptation [12, 13]. After the injury, the adaptive learning mechanisms of the brain will continue responding to stimuli even though they may be considered erratic responses to initiated action. This faulty adaptation may lead to a state called learned non-use, where parts of the brain and the subsequent motor control become dormant due to initial decrease in motor feedback during the initial phases of brain injury [14–16].

Figure 5.2 depicts two types of injury to the neural network from Fig. 5.1. In the first case (6.a), the network pathways between two areas are severed, leaving only small and decrepit pathways for the signals between two components of the network. In the second case (6.b), the pathways are more or less intact but the foundation of the function has been diminished or destroyed. The point here is that the intricate neural network established through experience and training, is ripped apart in brain injury and the foundation for execution of a skill is impaired with bits and pieces still responding to signals and stimuli. This foundation for relearning and rehabilitation is dramatically different from the foundation present when learning a new skill in an uninjured brain.

5.2.3 Recovering from Brain Injury Through Training

Experience-based plasticity offers hope for future rehabilitation training as a mean for recovery after brain injury. The fact that the plastic mechanisms of the brain induce change as a result of experience and activity has fuelled extensive research into understanding the nature of these mechanisms and the conditions for their control and harnessing [2, 7, 8, 17]. In rehabilitation research, knowledge about experience-based plasticity has slowly but fundamentally changed the perception that injury to the neural substrate of functions of the brain would result in final and permanent impairment. Increasing amount of evidence supports that experience-based plasticity may be a major contributor to the recovery from acquired brain injury at several different levels. One is the reactivation of neural substrate rendered dormant due to "learned non-use" [14, 15, 18–20]. Taub et al. [21] had observed that a temporary disruption or depression of motor activity due to injury would reduce the use and function of upper extremities after recovery. They demonstrated that subsequent brief and intensive training forcing the use of the affected limb would indeed improve voluntary control and function. Similar effects have been demonstrated in rehabilitation of language deficits in aphasia [20] and further studies have indicated that maladaptive plasticity may be responsible for learned non-use and that the effects can be reversed through training [22–25]. Experience-based plasticity has also been demonstrated in the reorganization of the internal topological maps which models our body in relation to the world around us and allows a correct interpretation of the origin of sensation or calculation of the position of objects [11, 12, 26–28].

There is the apparent paradox that the destruction of the neural foundation for a skill or function does not permanently damage the ability to express the skill at the surface level. In the case of 6.b in Fig. 5.2, one might ask what hope is there for recovery when the neural foundation has been destroyed? The REF (Reorganization of Elementary Functions) model attempts to bridge the apparent paradox that destruction of the neural substrate in an area known to subserve a specific function may not result in total inability to express the function [29–31]. In this model, the observed expression of a function may be accomplished through

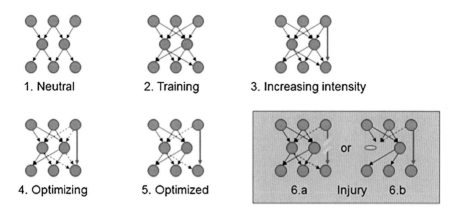

Fig. 5.2 Graphical example of types of injury to the foundation of the neural network from Fig. 5.1. Adapted from Robertson and Murre [7]

activation of different combinations (Algorithmic Strategies or AS) of elementary subfunctions. Training is required to establish new AS combinations of elementary subfunctions and in this way, training shapes, develops and to a certain extent also limits the skilled ability. At surface level, improvements to a specific skill may be observed, but internally, the observed results are now mediated through the activation of novel combinations of subfunctions. In other words, experience-based plasticity can be observed both at a cellular level as well as at a higher functional level.

However, the task of actually determining the training required to alleviate the effects of injury is a daunting task, particularly when determination depends only on data about the injury location and the expression of the functional impairment as measured by the diagnostic tools. In addition, recovery from brain injury may be defined differently depending on perspective. If skills are considered to be tools needed to solve a task then, on the surface level, recovery from injury is the reinstatement of the ability to execute the now impaired task. Viewed in this context, vocal speech is a tool for communication. Since, communication can be achieved through other means than vocal speech e.g. using writing or artificial speech generation, recovery in this sense might be achieved by training other ways to communicate. If, on the other hand, the production of speech is considered to be a task, vocal speech recovery would be understood as the reestablishment of the ability to speak. The training in this case would be aimed at recovering the sub-skills needed in the production of speech.

So planning a path for recovery cannot depend solely on knowledge of how a healthy brain executes a task, but must rely also on knowledge of the internal or external resources available after injury and how best to shape the training to support the mechanisms of experience-based plasticity.

5.3 Advanced Technology and Cognitive Rehabilitation

Well then, is technology the answer? To some extend yes, but bear in mind that technology also introduces new challenges in the translation of known therapy into computer-based versions as well as in the development of new tools. Seemingly equal types of training may in fact differ in subtle but important ways which will be demonstrated a little later but first a brief history of the use of computers in rehabilitation training.

5.3.1 A Brief History

In 1968, *computer-assisted instruction* was being promoted at all major universities in the US with the specific goal of achieving individualized instructional education of students [32]. The idea was to let the student progress through a particular subject at an individual pace and route catering to the differences in ability and motivation. The expectation at the time was that the rapid technological development would take care of issues like the exorbitant cost of hardware, lack of random access storage devices, and limited access to computer mainframes. Time proved them right on these accounts, but Atkinson and Wilson raised another and more profound concern—the lack of theoretical background and common framework of metrics for the definition of learning goals and the subsequent evaluation of achievements. They correctly pointed out that an evaluation of a computer-assisted instructional system is partly an evaluation of the software and equipment being used but, more importantly, it is an evaluation of the software designer and how the stated learning goals have been translated and transformed into an instructional system based on computer technology [32, p. 76].This point has been raised again with standard computer-based training [33–35] and in the use of virtual reality [36–39]. In other words, the development of a successful computer-based training system requires a system developer that understands how to combine the training and learning strategies with the potential of advanced technology. Without fully understanding the basic elements needed in training, the introduction of advanced technology is just adding further complexity into the rehabilitation equation.

Computer technology found its way into cognitive rehabilitation training in the early 1980' with the advent of mini- and microcomputers. Even though the cost of equipment was exorbitant compared to today, it was recognized that technology could play a vital role in both the assessment of deficits, as a cognitive prosthetics, and as a way to offer cost effective treatment to more patients by improving the intensity, contents, and delivery of training [e.g. 40–43]. The reports from the 1980s and early 1990s demonstrated huge enthusiasm in the application of computer training within areas like aphasia [e.g. 40, 41, 44], attention deficits such as neglect [e.g. 45–47], and memory deficits [e.g. 48]. The reasons for introducing computer technology in training then were basically the same as today—to increase therapeutic efficiency by providing easy and inexpensive access to therapy, to improve the content and quality, and to increase the intensity of therapy [39, 40, 49, 50].

5.3.2 Assessment

The benefits of computer usage was fairly quickly realized in the area of assessment, where the use of technology introduced improved monitoring and recording of response time, accuracy and behaviour during assessment. Old and favoured paper-and-pencil tests like line-bisection and cancellation tests in neglect have been converted to computer-based versions which have led to increased sensitive to the symptoms of neglect [51–55], as well as expansions of said tests to include observations and assessment of aberrant behaviour [54, 56–61], and the detection and separation of multiple disorders [62]. New assessment techniques have also been introduced e.g. with the use of virtual reality that, in addition to the recording and monitoring benefits, provide assessment opportunities situated in more realistic environments [50, 63]. Last but not least, new meta test systems are being introduced, which are able to select the best assessment tools based on the actual performance of the patient during testing [64].

5.3.3 Treatment

Not surprisingly however, efficient use of computer technology in treatment has turned out to be much more difficult than other areas of use. It was established fairly quickly that computer-based mindless drill-training, in which the patients were made to repeat the same tasks over and over again, had little or no effect at all [44]. It was also observed that although the patients did improve on the task being trained, often no generalization or cross-over effect to activities of daily living was found [e.g. 65–67]. An early review in 1997 concluded that the effect of the training resulted from the language content of computer-based training and **not** from the mere use of the computer per se [68]. Another early review at the time, covering a wider range of cognitive treatments based on computer technology, concluded that efficacy of computer-based treatment did not differ from normal treatment [69]. Others became so desolate by the lack of results in computer-based treatment that they strongly cautioned the use of computer training pending further research [33].

By the end of the 1990s the initial excitement of using computers in cognitive rehabilitation treatment was replaced by resignation to the fact that computers in themselves did not solve the fundamental problems of cognitive treatment. As in 1968, researchers thirty years later began to realize that using computer technology to facilitate learning required standardized measurements of efficacy; treatment that was solidly based on learning and practice; better design and testing of user interfaces, and a deeper understanding of the mechanisms of the deficit being treated [34, 42].

5.3.4 Computer-Based Rehabilitation Training Today

Today, there seems to be two main approaches to the use of technology in experience-based treatment. The first is to try to discover and understand every aspect of a deficit through detailed assessment. Specific solutions are then developed to train these aspects of the impairment. Prism Adaptation Therapy is one such example [70]. The second approach is to develop therapy systems that simulate reality hoping that interaction with the systems will stimulate the brain to regain lost ability, accepting that it may be well into the future before all aspects of brain injury are even uncovered let alone understood. With the introduction of virtual reality applications, it has become possible to emulate real world situations in a safe and controlled environment. It has so far been proven successful in teaching for instance neglect patients how to navigate safely in traffic [65, 71], in training children with ADHD [72], and in desensitizing soldiers with PTSD [73].

Games technology is an area that slowly has begun to interest rehabilitation researchers. Action computer games have elements that can keep gamers occupied with seemingly boring and trivial activities for hours on end, in the attempt to achieve rewards and skills. Most often, they contain elements of practice with increased levels of difficulty and they are set in environments that require activation of different cognitive skills, are motivating, and to some extent even addictive. More importantly, action games may model challenges, speed, progression and even general content to the particular preference of the gamer either stated explicitly through games parameters or implicitly by measuring the ability of the gamer [74–77]. Many of these elements in games resemble to some extent those found to induce experience-based plasticity [8] in computer-based therapy.

5.3.5 The Challenge of Individuality in Injury
and Treatment

Even though advanced technology has been available for some time and keeps inspiring to new and more exiting features of training and therapy, the fundamental issue of what type of training has an effect at which point in time is still unresolved. Training a new skill, cognitive or physical, require that the trainee progress through levels of training with increased difficulty and challenge [e.g. 20, 78, 79]. In order to progress steadily, the trainee has to practise to achieve ease and precision in the execution of the skill [79, 80]. The same is basically true for many of our cognitive skills like speech, writing, problem solving etc. Although the complex neural foundation of cognitive skills is probably there from birth, many cognitive skills are honed during childhood and adolescence. The first challenge in learning a skill using computer technology is therefore to recognize the logical sequence of steps needed to acquire competency in general. Expertise within a skill is not achieved through cognitive reflection alone, if at all. The ability to execute a skill may require consistent and frequent use and practice not to deteriorate, but even a rusty skill can normally be recovered with a little practice.

However, a failing skill due to brain injury is not the same as the lack of a skill never learned. Reclaiming a skill, impaired or destroyed by brain injury, may be more challenging for a number of reasons. They all have to be considered when attempting to reconstruct the impaired ability, be it using computer-based or face-to-face therapy sessions. Firstly, there may still be traces of knowledge left or more precisely areas of neurons and connections that respond erratically to stimuli causing maladaptive learning. As mentioned earlier in the chapter, the failing ability may be due to destruction of tissue both full and partial, destroyed or diminished network connectivity, asynchronous processing caused by slowed metabolisms, or even the lack of inhibitory or excitatory signals from other areas of the brain. The current state of diagnostics and assessment tools rarely, if ever, offer full details of what may be the underlying cause of the functional deficit. Secondly, no two injuries are the same and all patients are different. Motivation and goals for the individual patient may vary greatly dependent on the patient's ability, attitude, and even awareness of injury. As a consequence, what may be extremely difficult to one patient to master and achieve may be easy for another patient even though they have been diagnosed with the same impairment. Forcing patients to go through a fixed set of steps in training may in fact be counterproductive as the training may reinforce aberrant behaviour of the neural substrate. Thirdly, brain injury may cause fluctuation in performance due to fatigue and enhanced sensitivity to lack of sleep and changes in the environment. What may seem easy in one session may be very difficult in the next even for the same patient. It is here the challenge but also one of the benefits of technology in rehabilitation training presents itself.

5.3.6 The Same is not the Same

To create computer-based training systems, you have to define specifically what you think will have an effect based on rehabilitation models and theory and on observations from clinical practice. The detailed test and investigation required when building a training system will often reveal that what seems to be the same behaviour at a surface level may not always be so. Any implementation of computer technology is in essence a translation of ideas or practice into an automated environment. The designer of the system depends on rigorous definitions of goals and predefined targets to be able to implement and test the functionality of the system. This conversion of ideas into a computer-based, stringent reality may reveal information on what actually affects the brain and what does not, as exemplified by the following study that was done as an attempt to create a computer-based version of prism-based training for the attention deficit neglect [81].

5.3.7 The Prism Example

For more than a century, prism adaptation has been used to study experience-based plasticity and in particular, how the visuomotor system adapts to the visual distortion created by the prisms. Stratton [82] was the first to test if the angle of the retinal projection of the visual image was a determinant for the subsequent perception, by rotating visual input 180°. He and others after him found that the brain will adapt to the distortion over time enabling the exposed subject to navigate and perceive the world as before.

In 1998, Rossetti et al. published [83] a seminal study which demonstrated that exposure to prism adaptation might alleviate some of the symptoms related to egocentric visual neglect in patients, regardless of the severity of neglect. Prism Adaptation Therapy (PAT) has since become one of the most promising therapies in the treatment of egocentric visual neglect [70, 84–86].

In PAT, patients typically sit in front of a box which hides the movement of the arm. The therapist is positioned at the opposite end and states which of three targets a patient should point at. A temporary discrepancy between the internal representation and the actual position and extension of the body is created by letting visual input pass through prism goggles. The prism goggles cause a distorted projection usually 10° rightward (Fig. 5.3).

The typical training consists of three steps. The first step is an initial measure of the proprioceptive accuracy of the subject (without goggles), usually established by letting the subjects pointing at targets. The movement of their arm and hand is disguised underneath a non-transparent barrier (blinded) [70, 88].

The second step is to expose the subjects to the visual distortion induced by the prism goggles. During exposure, subjects repeat the target pointing task, still without being able to see their arm movement. However, they are provided with feedback on pointing precision by seeing their fingertip in relation to the target, which allows them to adjust to the exposure distortion trial by trial. This step consists of 90 pointing trial.

The third step is basically similar to the first step. Visual input is restored to normal by removing the prism goggles and the blinded pointing accuracy is re-measured. In healthy subjects, exposure to prism goggles produces an adaptation effect—the after-effect—that can be observed as a left-ward deviation in pointing accuracy once the prism goggles has been removed. The size of the after-effect correlates with the degree of distortion induced by prism goggles, the larger the deviation, the larger the size [89].

Fig. 5.3 A prismatic right-
shift causes targets in actual
position *A* to appear to be at
location *B* [87]

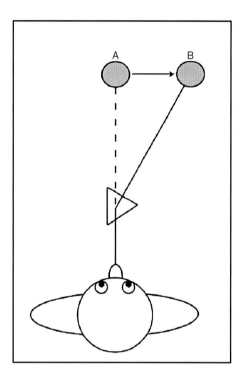

In an attempt to reduce workload on the therapists and provide a platform for patients to train at home, a computer-based version of the equipment used for Prism Adaptation Therapy was created by the author. The pointing targets were displayed on a touchscreen with the lower part disguised by a barrier, which hid the movement of the arm and hand. Subjects would receive feedback on pointing precision on the touchscreen in the shape of an "X", which would be placed next to the target in a distance equal to the distance from target to actual pointing position underneath the box.

A test was conducted to verify that the computer-based version would produce after-effects similar to those produced by the standard therapist sessions of PAT. The results were very surprising. In fact, the visual aftereffect produced in the computer session was approximately half the size of the aftereffect produced by the standard session. A series of studies were conducted to determine the cause and it turned out that only seeing the fingertip as feedback produced the same after-effect [81] in both sessions. Seeing "X" or images of fingers on the touchscreen did not. So an apparently similar type of implementation of PAT produced vastly different results. Without the additional studies, the conclusion might very well have been that computer training does not work, but instead it turned out that the effect of PAT was related to the type of feedback provided.

5.4 Artificial Intelligence and Rehabilitation

The use of technology may provide important information and input to understanding the brain. As the previous example demonstrated, a seemingly minor change may turn up new data on the components of visuomotor adaptation. The next part details how artificial intelligence may be used to set the level of difficulty at an individual basis.

5.4.1 Adjusting Level of Difficulty in Cognitive Rehabilitation

When training a patient, a skilled therapist is able to constantly monitor and modify the training activity to match the patient's mood, skills and learning rate. If a particular area seems difficult to master, the therapist can simplify the training, choose another approach or postpone the task until later. Constraint-induced language therapy is an example of this type of training [20, 90], where a small group of patients are playing a card game, which requires various forms of communication that can be adjusted by the therapist according to the skills and capability of the individual player. As the patients become more competent, they must use more sophisticated verbal skills in communication to win the game.

Moving up one step, the therapist can pre-plan a certain set of tasks that the patient need to accomplish in a predefined sequence. Usually progression from one task to the next is determined by a percentage of correct answers. Basic computer assistance can be introduced as in Mortley et al. [91], where the computer is monitoring the patient's performance and providing the therapist with data necessary to determine how to progress. The tasks can also be pre-programmed by the therapist as in the "Afasi-assistant" system [92], where the patient is guided by the computer through the pre-programmed set of computer-based training tasks, aimed at improving anomia step by step, based on a specific level of patient performance.

More automatic progression is accomplished by pre-programming (scripting) a set of paths which may be activated when reaching a pre-programmed level of competency as in computerized aphasia training [93–96]. Another example is the CogMed[1] system aimed at improving short-term memory. Here, the user is progressing through levels of increasingly more difficult tasks, but is required to master one before advancing to the next. CogMed has the added feature that difficulty may automatically regress, if the user repeatedly fails a task at a higher level [97–99]. A similar method of progression was administered in perimetry treatment of visual impairments [100].

---

[1] CogMed is a computer-based training system published by Pearson Education Inc.

Cueing is another way of increasing or reducing difficulty. By letting the patient get access to more or less assistive information during training, a task can become easier or harder without breaking the general progress of the patient. Examples are found in word mobilization, where the display of pictures of objects or written words may cue verbal pronunciation [93, 94, 101–105] or attention training where attention may be directed towards a particular item or spot using sound or light after a period of time [e.g. 106, 107].

A recent development within computerized assessment is to let the scores from one test be the input for selection of further tests in which testing parameters and elements are set to match the performance and skills of the individual patient [64, 108]. Since computerized tests can be scored instantly, the scores themselves along with the detected characteristics of behaviour during assessment may be feed directly into the assessment system and serve as selection criteria for the next test to follow. The potential to feed these results into training systems which then tailor a plan based on the assessment scores may be a viable way for future research into technological advanced training.

Although the controls asserted above, at surface level, may be termed advanced control of the progression and delivery of training, most of them still only partially address the more fundamental challenges in the delivery of treatment—the diversity and uniqueness in the impact of brain injury and the fluctuating performance of the patient. The path to achieving a skill may vary considerably from patient to patient and computerized training needs to be flexible and modular since even small variations in training may influence difficulty for the individual patient. One approach to this challenge has been demonstrated in one of author's studies [77]. In this study, the question of difficulty was approached from a different angle using user modelling and artificial intelligence to determine and control difficulty and progression. A fundamental set of parameters each controlled a particular aspect of difficulty. In combination, the value of the parameters defined the task difficulty as a continuum from easy to difficult based on the real-time characteristics of an individual patient. Artificial intelligence algorithms were used to monitor progress and adjust the value of each parameter accordingly to match the progress and state of the patient.

Furthermore, to verify the potentials of this approach a patient was subjected to three weeks of visual attention training in which difficulty was controlled by three parameters: number of items on the screen, length of a word displayed inside each item and finally the variety of letters used to compose the words. The AI engine controlling the parameters was fairly simple, but the result was a fairly complex set of combinations only possible to control using online assessment and adaptation of the parameters. The study demonstrated that the AI engine was able to construct a level of difficulty that challenged the patient and changed as the patient improved. In this study, the reaction time of the patient was fed into the algorithm as feedback, but from the perspective of the AI algorithms, it might as well have been error rate or galvanic skin response, EEG input or any type of feedback indicating treatment progress. Quantitative measurement that changes in response to training might be used as valid input to the AI algorithms as long as they fit within a set range of acceptable responses. Even this range may initially be calculated using the AI algorithms and in this manner, the use of AI would accommodate for the fact that learning and ability do not always follow a straight progressive line but may occur in jumps [7] and even temporarily relapse [77].

5.5 Concluding Comments

The use of computer technology in rehabilitation of patients with brain injury is not a new trend. Computer technology is used pervasively within all major aspects of cognitive rehabilitation research: Mapping, analysis, diagnostics, prosthetics and therapy.

However, in the field of cognitive rehabilitation training, the use of technology is still at a very basic level mostly being computer versions of existing training. Only few attempts have been made to combine the continuous mapping of the learning ability of the patients with adaptive software and much more research and collaboration across fields of expertise is needed to move forward. It is the hope of the author that by creating systems where test and training are fully integrated, it will be possible to maintain individual user profiles that reflect the current state of progress as input to adaptable training software systems.

References

1. Hebb, D.O.: The Organization of Behaviour: A Neuropsychological Theory. Wiley, New York (1949)
2. Cramer, S.C., et al.: Harnessing neuroplasticity for clinical applications. Brain **134**, 1591–1609 (2011)
3. Abbott, L.F., Nelson, S.B.: Synaptic plasticity: taming the beast. Nat. Neurosci. **3**(11), 1178–1183 (2000)
4. Rosenzweig, M.R., Bennett, E.L.: Psychobiology of plasticity: effects of training and experience on brain and behavior. Behav. Brain Res. **78**(1), 57–65 (1996)
5. Thimm, M., et al.: Impact of alertness training on spatial neglect: a behavioural and fMRI study. Neuropsychologia **44**(7), 1230–1246 (2006)
6. McClelland, J.L., Rumelhart, D.E.: Distributed memory and the representation of general and specific information. J. Exp. Psychol. Gen. **114**(2), 159–188 (1985)
7. Robertson, I.H., Murre, J.M.J.: Rehabilitation of brain damage: brain plasticity and principles of guided recovery. Psychol. Bull. **125**(5), 544–575 (1999)
8. Kleim, J.A., Jones, T.A.: Principles of experience-dependent neural plasticity: Implications for rehabilitation after brain damage. J. Speech Lang. Hear. Res. **51**(1), S225–S239 (2008)
9. Sundhedsstyrelsen.: Hjerneskaderehabilitering-en medicinsk teknologivurdering in Medicinsk Teknologivurdering. In Hørder, M., Beck, M., Andersen, S.E. (eds.). Sundhedsstyrelsen: København (2011)
10. Redding, G.M., Wallace, B.: Prism adaptation and unilateral neglect: review and analysis. Neuropsychologia **44**(1), 1–20 (2006)
11. Redding, G.M., Rossetti, Y., Wallace, B.: Applications of prism adaptation: a tutorial in theory and method. Neurosci. Biobehav. Rev. **29**(3), 431–444 (2005)

12. Ramachandran, V.S., Hirstein, W.: The perception of phantom limbs. The DO Hebb lecture. Brain **121**(9), 1603–1630 (1998)
13. Ramachandran, V.S., Rogers-Ramachandran, D.: Synaesthesia in phantom limbs induced with mirrors. In: Proceedings of Biological Sciences, vol. 263(1369), pp. 377–386 (1996)
14. Pulvermüller, F., Berthier, M.L.: Aphasia therapy on a neuroscience basis. Aphasiology **22**(6), 563–599 (2008)
15. Taub, E.: Harnessing brain plasticity through behavioral techniques to produce new treatments in neurorehabilitation. Am. Psychol. **59**(8), 692–704 (2004)
16. Taub, E., Uswatte, G.: Constraint-induced movement therapy: answers and questions after two decades of research. NeuroRehabilitation **21**(2), 93–95 (2006)
17. Duffau, H.: Brain plasticity: from pathophysiological mechanisms to therapeutic applications. J. Clin. Neurosci. **13**(9), 885–897 (2006)
18. Meinzer, M., Breitenstein, C.: Functional image studies of treatment-induced recovery in chronic aphasia. Aphasiology **22**(12), 1251–1268 (2008)
19. Meinzer, M., et al.: Functional re-recruitment of dysfunctional brain areas predicts language recovery in chronic aphasia. Neuroimage **39**(4), 2038–2046 (2008)
20. Pulvermüller, F., et al.: Constraint-induced therapy of chronic aphasia after stroke. Stroke **2001**(32), 1621–1626 (2001)
21. Taub, E., Uswatte, G., Pidikiti, R.: Constraint-induced movement therapy: a new family of techniques with broad application to physical rehabilitation—a clinical review. J. Rehabil. Res. Dev. **36**(3), 237–251 (1999)
22. Breier, J.I., et al.: Changes in language-specific brain activation after therapy for aphasia using magnetoencephalography: a case study. Neurocase **13**(3), 169–177 (2007)
23. Maher, L.M., et al.: A pilot study of use-dependent learning in the context of constraint induced language therapy. J. Int. Neuropsychol. Soc. **12**(6), 843–852 (2006)
24. Sterr, A., Saunders, A.: CI therapy distribution: theory, evidence and practice. NeuroRehabilitation **21**(2), 97–105 (2006)
25. Nudo, R.J., Plautz, E.J., Frost, S.B.: Role of adaptive plasticity in recovery of function after damage to motor cortex. Muscle Nerve **24**(8), 1000–1019 (2001)
26. Fernandez-Ruiz, J., et al.: Rapid topographical plasticity of the visuomotor spatial transformation. J. Neurosci. **26**(7), 1986–1990 (2006)
27. Gauthier, L.V., et al.: Remodeling the brain—plastic structural brain changes produced by different motor therapies after stroke. Stroke **39**(5), 1520–1525 (2008)
28. Ward, N.S.: Neural plasticity and recovery of function. In: Laureys S. (ed.) Boundaries of Consciousness: Neurobiology and Neuropathology, pp. 527–535. Elsevier Science Bv, Amsterdam (2005)
29. Mogensen, J., Malá, H.: Post traumatic functional recovery and reorganization in animal models: a theoretical and methodological challenge. Scand. J. Psychol. **50**(6), 561–573 (2009)
30. Mogensen, J.: Reorganization of the injured brain: Implications for studies of the neural substrate of cognition. Front. Psychol. 2:7 (2011). doi:10.3389/fpsyg.2011.00007
31. Mogensen, J.: Almost unlimited potentials of a limited neural plasticity. J. Conscious. Stud. **7**(8), 13–45 (2011)
32. Atkinson, R.C., Wilson, H.A.: Computer-assisted instruction. Science **162**(3849), 73–77 (1968)
33. Robertson, I.: Does computerized cognitive rehabilitation work? A review. Aphasiology **4**(4), 381–405 (1990)
34. Robertson, I.H.: Setting goals for cognitive rehabilitation. Curr. Opin. Neurol. **12**(6), 703–708 (1999)
35. Ting, D.S.J., et al.: Visual neglect following stroke: current concepts and future focus. Surv. Ophthalmol. **56**(2), 114–134 (2011)
36. Myers, R.L., Laenger, C.J.: Virtual reality in rehabilitation. Disabil. Rehabil. **20**(3), 111–112 (1998)

37. Rizzo, A.A., Buckwalter, J.G., Neumann, U.: Virtual reality and cognitive rehabilitation: a brief review of the future. J. Head Trauma Rehabil. **12**(6), 1–15 (1997)
38. Rose, F.D., Brooks, B.M., Rizzo, A.A.: Virtual reality in brain damage rehabilitation: review. Cyberpsychol. Behav. **8**(3), 241–262 (2005). discussion 263–71
39. Tsirlin, I., et al.: Uses of virtual reality for diagnosis, rehabilitation and study of unilateral spatial neglect: review and analysis. CyberPsychol. Behav. **12**(2), 175–181 (2009)
40. Katz, R.C., Nagy, V.T.: A computerized treatment system for chronic aphasic patients. In: Clinical Aphasiology Conference, BRK Publishers, Oshkosh (1982)
41. Mills, R.H.: Microcomputerized auditory comprehension training. In: 12th Clinical Aphasiology Conference, BRK Publishers, Oshkosh, 6–10 June 1982
42. Loverso, F.L., et al.: The application of microcomputers for the treatment of aphasic adults. In: Clinical Aphasiology Conference, Ashland, BRK Publishers (1985)
43. Dick, R.J., et al.: Programmable visual display for diagnosing, assessing and rehabilitating unilateral neglect. Med. Biol. Eng. Comput. **25**(1), 109–111 (1987)
44. Katz, R.C., Nagy, V.T. A computerized approach for improving word recognition in chronic aphasic patients. In: Clinical Aphasiology Conference, BRK Publishers, Phoenix (1983)
45. Bergego, C., et al.: Rehabilitation of unilateral neglect: A controlled multiple-baseline-across-subjects trial using computerised training procedures. Neuropsychol. Rehabil. **7**(4), 279–293 (1997)
46. Robertson, I., Gray, J., McKenzie, S.: Microcomputer-based cognitive rehabilitation of visual neglect: three multiple-baseline single-case studies. Brain Inj. **2**(2), 151–163 (1988)
47. Robertson, I.H., et al.: microcomputer-based rehabilitation for unilateral left visual neglect: a randomized controlled trial. Arch. Phys. Med. Rehabil. **71**(9), 663–668 (1990)
48. Middleton, D.K., Lambert, M.J., Seggar, L.B.: Neuropsychological rehabilitation: microcomputer-assisted treatment of brain-injured adults. Percept. Mot. Skills **72**(2), 527–530 (1991)
49. Katz, R.C.: Application of computers to the treatment of US veterans with aphasia. Aphasiology **23**(9), 1116–1126 (2009)
50. Rizzo, A.A., et al.: Analysis of assets for virtual reality applications in neuropsychology. Neuropsychol. Rehabil. **14**(1–2), 207–239 (2004)
51. Anton, H.A., et al.: Visual neglect and extinction: a new test. Arch. Phys. Med. Rehabil. **69**(12), 1013–1016 (1988)
52. Liang, Y.Q., et al.: A computer-based quantitative assessment of visuo-spatial neglect using regression and data transformation. Pattern Anal. Appl. **13**(4), 409–422 (2010)
53. Potter, J., et al.: Computer recording of standard tests of visual neglect in stroke patients. Clin. Rehabil. **14**(4), 441–446 (2000)
54. Rabuffetti, M., et al.: Touch-screen system for assessing visuo-motor exploratory skills in neuropsychological disorders of spatial cognition. Med. Biol. Eng. Comput. **40**(6), 675–686 (2002)
55. Rengachary, J., et al.: Is the posner reaction time test more accurate than clinical tests in detecting left neglect in acute and chronic stroke? Arch. Phys. Med. Rehabil. **90**(12), 2081–2088 (2009)
56. Baheux, K., et al.: Virtual reality pencil and paper tests for neglect: a protocol. CyberPsychol. Behav. **9**(2), 192–195 (2006)
57. Baheux, K., et al.: Diagnosis and rehabilitation of hemispatial neglect patients with virtual reality technology. Technol. Health Care **13**(4), 245–260 (2005)
58. Broeren, J., et al.: Neglect assessment as an application of virtual reality. Acta Neurol. Scand. **116**(3), 157–163 (2007)
59. Donnelly, N., et al.: Developing algorithms to enhance the sensitivity of cancellation tests of visuospatial neglect. Behav. Res. Methods Instrum. Comput. **31**(4), 668–673 (1999)
60. Guest, R., et al.: Using image analysis techniques to analyze figurecopying performance of patients with visuospatial neglect and control groups. Behav. Res. Methods Instrum. Comput. **36**(2), 347–354 (2004)

61. Guest, R.M., Fairhurst, M.C., Potter, J.M.: Diagnosis of visuo-spatial neglect using dynamic sequence features from a cancellation task. Pattern Anal. Appl. **5**(3), 261–270 (2002)
62. Beis, J.M., Andre, J.M., Saguez, A.: Detection of visual field deficits and visual neglect with computerized light emitting diodes. Arch. Phys. Med. Rehabil. **75**(6), 711–714 (1994)
63. Fordell, H., et al.: A virtual reality test battery for assessment and screening of spatial neglect. Acta Neurol. Scand. **123**(3), 167–174 (2011)
64. Gur, R.C., et al.: A cognitive neuroscience-based computerized battery for efficient measurement of individual differences: Standardization and initial construct validation. J. Neurosci. Methods **187**(2), 254–262 (2010)
65. Katz, N., et al.: Interactive virtual environment training for safe street crossing of right hemisphere stroke patients with unilateral spatial neglect. Disabil. Rehabil. **27**(20), 1235–1243 (2005)
66. McCall, D., et al.: The utility of computerized visual communication for improving natural language in chronic global aphasia: Implications for approaches to treatment in global aphasia. Aphasiology **14**(8), 795–826 (2000)
67. Ramsberger, G.: Achieving conversational success in aphasia by focusing on non-linguistic cognitive skills: a potentially promising new approach. Aphasiology **19**(10/11), 1066–1073 (2005)
68. Katz, R.C., Wetz, R.T.: The efficacy of computer-provided reading treatment for chronic aphasia adults. J. Speech Lang. Hear. Res. **40**(3), 493–507 (1997)
69. Chen, S.H.A., et al.: The effectiveness of computer-assisted cognitive rehabilitation for persons with traumatic brain injury. Brain Inj. **11**(3), 197–209 (1997)
70. Frassinetti, F., et al.: Long-lasting amelioration of visuospatial neglect by prism adaptation. Brain **125**, 608–623 (2002)
71. Kim, J., et al.: Virtual environment training system for rehabilitation of stroke patients with unilateral neglect: crossing the virtual street. Cyberpsychol. Behav. **10**(1), 7–15 (2007)
72. Rizzo, A.A., et al.: A virtual reality scenario for all seasons: The virtual classroom. CNS Spectr. **11**(1), 35–44 (2006)
73. Rizzo, A., et al.: Virtual reality exposure therapy for combat-related PTSD. In: Shiromani, P.J., Keane, T.M., LeDoux, J.E. (eds.) Post-Traumatic Stress Disorder, pp. 375–399. Humana Press, New York (2009)
74. Charles, D., Black, M.: Dynamic player modeling: a framework for player-centered digital games. In: Proceedings. of the International Conference on Computer Games: Artificial Intelligence, Design and Education (2004)
75. Charles, D., et al.: Player-centred game design: Player modelling and adaptive digital games. In: Proceedings of the Digital Games Research Conference (2005) (Citeseer)
76. Spronck, P.H.M.: Adaptive game AI. Universitaire Pers Maastricht, Maastricht (2005)
77. Wilms, I.: Using artificial intelligence to control and adapt level of difficulty in computer-based, cognitive therapy—an explorative study. J. Cybertherapy Rehabil. **4**(3), 387–396 (2011)
78. Rasmussen, J.: Skills, rules, and knowledge; signals, signs, and symbols, and other distinctions in human performance models. IEEE Trans. Syst. Man Cybern. **13**(3), 257–266 (1983)
79. Kolb, D.A.: Experimental Learning: Experience as the Source of Learning and Development, vol. 256, Financial Times/Prentice Hall, New Jersey (1983)
80. Lave, J., Wenger, E.: Situated Learning: Legitimate Peripheral Participation (Learning in doing: social, cognitive and computational perspectives). Cambridge University Press, Oxford (1991)
81. Wilms, I., Malá, H.: Indirect versus direct feedback in computer-based prism adaptation therapy. Neuropsychol. Rehabil. **20**(6), 830–853 (2010)
82. Stratton, G.M.: Some preliminary experiments on vision without inversion of the retinal image. Psychol. Rev. **3**(6), 611–617 (1896)
83. Rossetti, Y., et al.: Prism adaptation to a rightward optical deviation rehabilitates left hemispatial neglect. Nature **395**(6698), 166–169 (1998)

84. Serino, A., et al.: Effectiveness of prism adaptation in neglect rehabilitation a controlled trial study. Stroke **40**(4), 1392–1398 (2009)
85. Serino, A., et al.: Neglect treatment by prism adaptation: what recovers and for how long. Neuropsychol. Rehabil. **17**(6), 657–687 (2007)
86. Vangkilde, S., Habekost, T.: Finding wally: prism adaptation improves visual search in chronic neglect. Neuropsychologia **48**(7), 1994–2004 (2010)
87. Vangkilde, S.: Prisme-baseret rehabilitering af visuel neglekt. Dissertation, University of Copenhagen, Copenhagen (2006)
88. Redding, G.M., Wallace, B.: Components of prism adaptation in terminal and concurrent exposure: organization of the eye-hand coordination loop. Percept. Psychophys. **44**(1), 59–68 (1988)
89. Fernández-Ruiz, J., Díaz, R.: Prism adaptation and aftereffect: specifying the properties of a procedural memory system. Learn. Mem. **6**(1), 47–53 (1999)
90. Meinzer, M., et al.: Intensive language training enhances brain plasticity in chronic aphasia. BMC Biol. **2**(20), 1–9 (2004)
91. Mortley, J., Enderby, P., Petheram, B.: Using a computer to improve functional writing in a patient with severe dysgraphia. Aphasiology **15**(5), 443–461 (2001)
92. Pedersen, P.M., Vinter, K., Olsen, T.S.: Improvement of oral naming by unsupervised computerised rehabilitation. Aphasiology **15**(2), 151–169 (2001)
93. Katz, R.C., Nagy, V.T.: An intelligent computer-based spelling task for chronic aphasic patients. In: Clinical Aphasiology Conference, BRK Publisher, Seabrook Island (1984)
94. Katz, R.C., Nagy, V.T.: A self-modifying computerized reading program for severely impaired aphasic adults. In: Clinical Aphasiology Conference, BRK Publishers, Ashland (1985)
95. Katz, R.C., Wertz, R.T.: Hierarchical computerized language treatment for aphasic adults. Int. J. Rehabil. Res. **12**(4), 454–455 (1989)
96. Wertz, R.T., Katz, R.C.: Outcomes of computer-provided treatment for aphasia. Aphasiology **18**(3), 229–244 (2004)
97. Klingberg, T.: Computerized training of working memory in children with ADHD. Eur. Neuropsychopharmacol. **17**, S192–S193 (2007)
98. Lundqvist, A., et al.: Computerized training of working memory in a group of patients suffering from acquired brain injury. Brain Inj. **24**(10), 1173–1183 (2010)
99. Westerberg, H., et al.: Computerized working memory training after stroke: a pilot study. Brain Inj. **21**(1), 21–29 (2007)
100. Schmielau, F., Wong Jr, E.K.: Recovery of visual fields in brain-lesioned patients by reaction perimetry treatment. J. Neuroeng. Rehabil. **4**, 1–16 (2007)
101. Abel, S., et al.: Decreasing and increasing cues in naming therapy for aphasia. Aphasiology **19**(9), 831–848 (2005)
102. Breitenstein, C., et al.: Five days versus a lifetime: Intense associative vocabulary training generates lexically integrated words. Restor. Neurol. Neurosci **25**(5–6), 493–500 (2007)
103. Fink, R.B., et al.: Computer-assisted treatment of word retrieval deficits in aphasia. Aphasiology **19**(10–11), 943–954 (2005)
104. Kim, K., et al.: A virtual reality assessment and training system for unilateral neglect. CyberPsychol. Behav. **7**, 742–749 (2004)
105. Ramsberger, G., Marie, B.: Self-administered cued naming therapy: a single-participant investigation of a computer-based therapy program replicated in four cases. Am. J. Speech Lang. Pathol. **16**(4), 343–358 (2007)
106. Myers, R.L., Bierig, T.A.: Virtual reality and left hemineglect: a technology for assessment and therapy. CyberPsychol. Behav. **3**(3), 465–468 (2000)
107. Robertson, I.H., et al.: Phasic alerting of neglect patients overcomes their spatial deficit in visual awareness. Nature **395**(6698), 169–172 (1998)
108. Donovan, N.J., et al.: Conceptualizing functional cognition in traumatic brain injury rehabilitation. Brain Inj. **25**(4), 348–364 (2011)

Chapter 6
A Body of Evidence: Avatars and the Generative Nature of Bodily Perception

Mark Palmer, Ailie Turton, Sharon Grieve, Tim Moss and Jenny Lewis

Abstract Complex Regional Pain Syndrome (CRPS) causes dramatic changes in perceptions of the body that are difficult to communicate. This chapter discusses a proof of concept study, evaluating an interactive application designed to assist patients in describing their perception of their body. This was tested with a group of CRPS patients admitted to a 2 week inpatient rehabilitation program who used the application in a consultation with a research nurse. The chapter draws upon audio recordings and a structured questionnaire designed to capture the experience of using the tool. Participants' reports of the positive impact of being able to see what they had previously only been able to feel (and often unable to describe) are examined. This is considered alongside studies within cognitive neuroscience examining phenomena such as the Rubber Hand Illusion and considers what this might mean for the representation of our body within virtual spaces.

Keywords Complex regional pain syndrome · Rubber hand illusion · Bodily perception · Avatar · Phenomenology · Emergent

M. Palmer (✉) · A. Turton · T. Moss
University of the West of England (UWE) Bristol, Coldharbour Lane,
Frenchay, Bristol, BS16 1QY, UK
e-mail: Mark.Palmer@uwe.ac.uk

A. Turton
e-mail: Ailie.Turton@uwe.ac.uk

T. Moss
e-mail: Tim.Moss@uwe.ac.uk

S. Grieve · J. Lewis
Royal National Hospital for Rheumatic Diseases NHS Foundation Trust,
Upper Borough Walls, Bath BA1 1RL, UK
e-mail: sharon.grieve@rnhrd.nhs.uk

J. Lewis
e-mail: jenny.lewis@rnhrd.nhs.uk

A. L. Brooks et al. (eds.), *Technologies of Inclusive Well-Being*,
Studies in Computational Intelligence 536, DOI: 10.1007/978-3-642-45432-5_6,
© Springer-Verlag Berlin Heidelberg 2014

6.1 Introduction

In this chapter we introduce work that has been conducted to develop an 'avatar' based tool to allow patients experiencing Complex Regional Pain Syndrome (CRPS) to describe their perceptions of their body. We will first describe the nature of the condition and the ways that it problematizes the communication of altered body perception for patients.

Having been introduced to the condition we will then draw upon a number of studies that support the idea that our bodily perception is an emergent phenomenon that is susceptible to alteration. We will first examine research within cognitive neuroscience that draws attention to the plasticity of bodily and spatial perception. This will initially focus upon work addressing the rubber hand illusion [1–6] and then consider studies investigating whether this is a phenomenon that can also be discovered within the representation of the body in virtual worlds.

We will then consider the role that the somatosensory system plays within the 'appearance' of the body. This will also open upon a philosophical consideration of the role that touch plays within this, drawing upon the phenomenologies of Husserl and Merleau-Ponty. Our consideration of the phenomenology of Merleau-Ponty will also 'touch upon' the issue of space and this will be considered alongside the work of Berti and Frassinetti's [7] examining way that tool use can remap the perception of space

The consideration of these issues will then open upon the ways in which we might be able to rethink the notion that sensation is mapped onto the body, but rather it is the means by which it is given; drawing upon Damascio's consideration of Spinoza. This will also be addressed through McCabe's research [8] investigating the ways in which the depiction of body movement can affect the symptoms of patients with fibromyalgia and consider whether the disturbance/alteration of these processes might begin to provide a theoretical framework to understand conditions such as CRPS.

Having examined the ways in which notions of bodily perception might be rethought we will return to the development of the body perception tool. The data generated through interviews with CRPS patients using the application in a clinical setting then provides the opportunity to further examine our understanding of body perception. Drawing on these results we will consider the scope the tool has to help patients 'make sense' of their condition and consider its potential for further development.

6.2 Complex Regional Pain Syndrome

Until recently Phantom limb syndrome was thought to be an unusual phenomenon experienced by a limited number of people who had lost a limb. It's now widely recognized as being a common experience of amputees [9]. One reason for this was that amputees did not complain of these strange sensations, fearing that they might not be believed or be labeled as mentally affected. A similar set of circumstances appear to have affected our perception of Complex Regional Pain Syndrome (CRPS). CRPS is a chronic pain condition without known cause usually affecting a single limb. In addition to sensorimotor and autonomic changes suffers also describe changes in body perception [10, 11] such that their perception of their CRPS affected limb is different from its objective appearance. For example, a person with CRPS may perceive a dramatic enlargement of sections of their limb, or sections of limb as missing; as a result people with CRPS have reported they have found it hard to talk about this experience of altered body perception.

First identified during the American civil war by Dr. Silas Weir Mitchell, CRPS was named *Causalgia*. The term was derived from the Greek for pain and heat which are symptoms typical of the syndrome. Associated with the body's extremities it can affect whole or even multiple limbs. CRPS is now defined as either CRPS-I or II dependent on the presence (II) or absence (I) of identifiable nerve damage. There are no objective tests to diagnose this condition, so diagnosis relies on signs and symptoms that meet a diagnostic checklist [12]. The nature of the pain suffered is common to both forms, although in CRPS II the pain suffered is disproportionate to the injury sustained. The condition is acknowledged to be a multi-system syndrome that involves aberrant changes in vasomotor function, inflammatory mechanisms and cortical processing [13]. These changes are probably triggered initially by a peripheral insult but it quickly evolves into a centrally driven condition for which there is currently no cure. For the vast majority, symptoms resolve in a matter of months and by 1 year 80–85 % of patients will be asymptomatic [14]. However the remaining 15–20 % continue to experience disabling symptoms and are commonly left with long-term functional impairments as well as the psychosocial impact of chronic pain [15].

As noted above, a common factor in CRPS is chronic pain associated with heat, this is often reported as an extreme burning sensation but it can also involve contradictory experiences of hot and cold sensation in the affected region. This is accompanied by extreme sensitivity and painful reactions to, and aversion from, everyday sensations such as the touch of clothing [16]. Other symptoms that may be experienced include the perceived (and sometimes actual) swelling of limbs and the sense that the affected region does not belong to the patient; in some cases this can lead to a desire to have the limb amputated. Perhaps one of the most perplexing aspects of these symptoms for patients and clinicians is that they can vary through the course of a day.

As a result of these unexplained experiences there has been a tendency to believe that there is a psychosomatic basis to CRPS. However Beerthuizen et al. [17] conducted a systematic review of research dealing with the relationship between CRPS1 and a range of psychological factors (covering issues such as mood, stress reactions, and personality traits) believed to have an impact on the onset and maintenance of CRPS1 and discovered evidence that contradicted such a conclusion. They examined nine studies dealing with somatization, of these four showed that CRPS1 patients demonstrated no differences in the level of somatization demonstrated by their control groups and three evidenced lower levels. When considered alongside studies examining a range of other traits their conclusion was that there was 'no relationship between psychological factors and CRPS1'. In fact the frustration of trying to understand and convey these symptoms can be more of an issue; one patient involved with our study noted that 'I really thought I was losing it.'

If the range of experiences described are not the result of psychosomatic tendencies how might we begin to understand these symptoms?

6.3 The Rubber-Hand Illusion

Although the causes of CRPS are as yet unknown, cognitive neuroscience has shown that body perception can be affected by what has become known as the rubber hand illusion (RHI).The illusion was first reported by Botvinick and Cohen [1] who described the illusion and its capacity to lead participants to perceive that a rubber hand belonged to their body. This was generated by placing the subject's hand on a table in front of them whilst hiding it from their view behind a standing screen. A life size rubber hand was then placed in front of the subject and they were asked to look at the rubber hand whilst it and their own hand were simultaneously brushed (Fig. 6.1). All of the participants (10) who took part in the study reported at some stage that they felt the sensation where they saw the rubber hand and many reported that they felt that the rubber hand was their own. The illusion has become the basis of claims made by many and in particular artists, that vision forms a primary role in the formation of body image [18]; however this isn't surprising given that the title of the report was *Rubber hands 'feel' touch that eyes see.*

Botvinick and Cohen also conducted tests which exposed participants to the illusion for a prolonged period of up to half an hour. Participants were then asked to move their 'free' hand under the table along a straight edge until it aligned with the perceived position of the hand subject to the illusion. It was found that the illusion caused a displacement in proprioceptive perception dependent on the reported duration of the illusion (subsequent studies refer to the proprioceptive drift between the perceived and actual position of the participant's hand). Botvinick and Cohen proposed that the illusion was generated as a result of the 'spurious reconciliation of visual and tactile inputs' leading to a distortion of a sense of position, in other words in reconciling these 'inputs' vision took precedence over proprioception.

Tsakiris and Haggard [3] sought to establish whether the RHI was the result of a bottom–up process of sensory integration or whether a bodily schema influenced its creation. Their tests involved synchronous and asynchronous stimulation of single and multiple fingers to establish whether only the stimulated fingers were susceptible to the illusion or if adjacent fingers were also affected. In addition they conducted tests utilising congruent and incongruent positioning of real and rubber hands; the incongruent position involving a −90° turn of the rubber hand so its

Fig. 6.1 The rubber hand
illusion; the participant's
hands are un-shaded and the
rubber hand *gray*

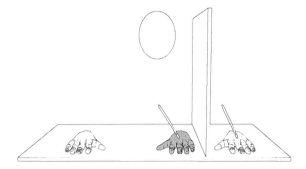

fingertips pointed towards the hidden hand (see Fig. 6.2). Tests with a neutral object replacing the rubber hand were also conducted allowing the 'stimulation' of an object that did not represent a part of the body.

Their outcomes demonstrated a significant difference in the proprioceptive drift that occurred between congruent and incongruent positions, the illusion being far stronger when stimulation was synchronous. When a neutral object was used a proprioceptive drift *away* from the object occurred which was described as a 'perceptual repulsion'. This lead Tsakiris and Haggard to propose that such repulsion occurred when self-attribution did not occur. The results obtained through the use of a neutral object were also contrary to the findings of Armel and Ramachandran's [2] whose results suggested that a neutral object (which in their tests was a table top) could become incorporated into the body schema.

In order to investigate whether the illusion was a bottom up phenomena or subject to the influence of a pre-existing bodily schema, tests were conducted where the index and little fingers were stimulated and participants were asked to identify where they perceived their middle finger to be. The outcome was that subjects perceived the location of their middle finger to be in a position relative to the fingers receiving the stimulation. Tsakiris and Haggard believed that, contrary to Armel and Ramachandran's [2] proposal that the RHI resulted from the construction of the body image from concurrent visual and tactile sensation, the fact that un-stimulated fingers 'followed' fingers receiving stimulation demonstrated the illusion resulted from the integration of visual input with a pre-existing representation of the body. As a result they concluded the bottom up combination of visual and tactile data was a necessary, but not sufficient condition in the creation of the illusion.

Whilst Tsakiris and Haggard believed they had demonstrated that a pre-existing bodily schema impacted upon the creation of the RHI, Ehrsson et al. [4] felt it was *...clearly important to find out whether an illusory feeling of ownership can be induced in the absence of visual input...* This was investigated by blindfolding participants and instead of brushing the hand, they moved the subjects own hand to touch the rubber hand whilst simultaneously touching the participants remaining hand in the 'same' place (Fig. 6.3).

Fig. 6.2 Tsakiris and
Haggard's use of incongruous
hand positions

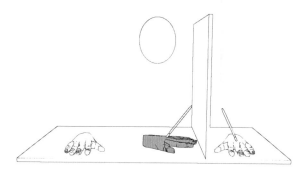

Fig. 6.3 Ehrsson, Holmes
and Passingham the
experimenter (*light gray*)
moved the subjects hand to
touch the rubber hand (*dark
gray*) whilst simultaneously
stimulating the same point
on the subject's hand

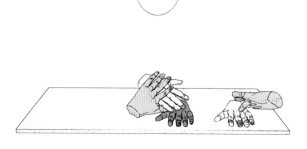

Using this method it was discovered that the illusion could be generated using synchronous touching *without* the need for a visual representation. Further tests asked subjects to locate the index finger on the hand subject to the illusion. This revealed an error in locating the finger, the scale of which appeared to be dependent upon the veracity of the illusion. What lies in common between Tsakiris and Ehrsson's research is that temporally correlated stimulation was a vital factor in the formation of the illusion. Furthermore fMRI scans conducted during one series of tests by Ehrsson highlighted the involvement of the premotor activity in the reconciliation of these sensations; they also noted that activity in that region increased with the participant's rating of the strength of the illusion. Given that Ehrsson et al. created the illusion without the input of vision the notion that a correlation between visual and tactile data is a *necessary* condition for the illusion can no longer be maintained. In conclusion this demonstrates that the RHI can be induced via congruent tactile stimulation alone and visual feedback is not required.

Further experiments relating to the RHI have sought to establish whether its affect could also be applied using virtual environments. Ehrsson worked with Slater, Perez-Marcos and Sanchez-Vives [5] to explore whether the illusion could be generated using computer graphics and a virtual limb. To examine this participants stood with their right arm resting hidden on a pedestal whilst they were provided with a stereoscopic 3D view of a virtual arm (extended in front of them) projected onto a screen (Fig. 6.4). A 6° of freedom tracking system[1] was used on the stereo glasses worn by the participants so that the projection would appropriately update with changes in the position and orientation of the participants head. A 'wand' was used to touch the subject's arm and generate the data to position the image of a 'ball' used to touch the arm in virtual space. As has been the method in prior studies of the RHI the stimulation of the participants' real and virtual arms was conducted synchronously and asynchronously. The tests also included rotating the virtual arm whilst measuring the EMG data coming from the subject's arm to examine whether muscle activity resulted from seeing movement applied to the virtual arm.

The outcome of these tests was that participants' sense of possession of the virtual arm was broadly similar to those of the rubber hand in the RHI. Those experiencing synchronous stimulation reported a far higher level of illusion and proprioceptive drift was only demonstrated when stimulation was synchronous. There was a positive correlation between the reported strength of the illusion and the number of EMG activity onsets when the virtual arm was rotated; demonstrating a connection between the depiction of movement on screen, ownership of the limb and premotor activity. As a result of these outcomes it appears that the sense of possession observed in the RHI can also be experienced through a virtual object.

Whilst Slater et al. focused on substituting a virtual limb for the rubber hand used in the RHI, Yuan and Steed [6] utilised a virtual body; seeking to explore whether immersive virtual reality (IVR) is capable of generating the illusion.

> Our claim is that an illusion very similar to the rubber hand illusion is "automatically" induced by active use of the virtual body in an IVR.

Yuan and Steed's protocol involved the use of a head mounted display which provided the user a first person perspective and occluded their view of the physical environment. They hypothesised that if a proprioceptive match between the virtual and actual body was achieved the subject would associate the virtual body with their own. Perceiving a weakness of prior studies to have been the passivity of participants their tests were devised so that users were asked to undertake a series of tasks with their right hand using a wand to perform tasks such as placing 'balls' through 'holes' in a 'table' within the IVR. Participants were instructed to keep their left hand still during the tests to enable the placement of sensors to detect the participant's galvanic skin response (GSR) and avoid variations in the data due to movement. Combined with the fact that participants were seated at a table, this enabled Yuan and Steed to use the wand's position to judge the position of the user's arm through the use of inverse kinematics and 'map' the position of the avatar's arm to the user.

Participants' sense of ownership of the avatar within IVR was explored using questions broadly similar to those used by Botvinick and Cohen. Yuan and Steed asked participants if the virtual arm felt as if it was their own, whilst other questions

---

[1] Allowing tracking of position and orientation.

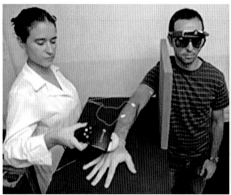

Fig. 6.4 Slater, Perez-Marcos, Ehrsson, Sanchez-Vives: towards a digital body: the virtual arm illusion (images courtesy of Mel Slater)

reflected the more active nature of the tests and addressed issues such as the sensation of holding virtual objects. An additional indicator utilised were the GSR readings used to detect the subject's stress levels when a virtual object fell towards their active arm. This was based Armel and Ramachandran's [2] examination of ownership through the stress reactions (indicated through GSR) when a finger on the rubber hand was bent back to what would be a painful position on a real hand.

Yuan and Steed reported that participants experienced a similar sense of ownership of the virtual body as was experienced of the rubber hand in the RHI. Similarly ownership was only demonstrated when the participant's interactions were represented through the use of the avatar and not the neutral 'object' used in some tests. The GSR demonstrated that threat to the virtual body produced a positive response, although this was not as strong as the reactions reported by Armel and Ramachandran (although it is worth noting that their tests represented a far more direct 'assault' upon the body) some participants moved their arm out of the way of the falling 'object'. As a result of these tests Yuan and Steed claimed that they had 'shown that an "IVR arm ownership illusion" exists'.

Although not a part of these studies it is worth noting that users of IVR systems such as Char Davies' installations Osmose (1995) and Ephémère (1998) have anecdotally reported high degrees of immersion. What is worth noting about these IVR environments is that Davies removed (or more to the point did not add) a visual representation of the body (the user's physical body being occluded through the use of a head -mounted display). However this does not mean that the body was not present. This is because Davies used its movement to enable navigation through the virtual environment. When we walk we lean in our direction of travel to initiate a step, Davies tracked the lean of the torso so that this could be used to generate a forward momentum/glide through the virtual environment (although a crude analogy, for those familiar with games one might consider the torso to be akin to a'thumb' or joystick generating movement through the world).

Although the/a body was not rendered visible within the virtual world the simultaneity of the user's bodily movements and the movement of their viewpoint within the IVR generated a convincing sense of presence. In fact as a user of Osmose and Ephémère, Palmer also reported that when passing though virtual objects he experienced a physical sensation as these 'passed through' him.

Although there is still much to discover about the RHI its existence begins to illustrate a number of interesting points. Perhaps the most significant of these is that perception of the body is malleable. This appears to originate in the reconciliation of sensation and may even operate at the level of one type of sensation such as touch. As discussed above vision is not a necessary condition of the illusion, indeed it appears that incongruous visual representation of limbs can *disrupt* the illusion.

Given that a sense of bodily presence can be achieved without the visual representation of the body Armel and Ramachandran's results (of possession arising from the brushing of a table top) might arise from of a lack of representation (the table top being a background object) rather than the table being a 'substitute' for the arm. If this were the case it may be that incongruent representation of a limb (Tsakiris and Haggard, Yuan and Steed) has a greater impact upon a sense of possession than its visual absence. In fact further studies conducted by Tsakiris et al. [19] addressing the RHI discovered that

> A conflict between visual and tactile percepts does not induce the illusion. The present study showed that right frontal cortex monitors the perception of body-related sensory signals when conflicts arises, blocking the attribution of the rubber hand to one's own body.

If it is the case that incongruent representation blocks the attribution of the rubber hand as a part of the body there appears to be an underlying mechanism into which the visual plays.

6.4 The Somatosensory System

The Somatosensory system incorporates the receptors and processing that provide our sense of touch, temperature, body position (proprioception) and pain (nociception). McCabe et al. [8] have proposed that disruptions within the somatosensory system can begin to account for the symptoms described within CRPS. They note how body image is formed through the integration and processing of multimodal sensory perceptions that involve the peripheral and central nervous system (CNS). However as we have seen through the RHI this is not a unidirectional system, whereby the CNS simply responds to a peripheral stimulation. If this were the case the proprioceptive drift that results occurs as a consequence of the illusion would not occur; because sensation would simply be experienced in the place it was received. Given that the CNS plays a key role in the generation of bodily perception, it is possible to see how disruption within the CNS may affect bodily perception within CRPS.

This has some far reaching consequences for what we might understand bodily perception to be. The body is such a constant within our perception that it appears to anchor all other sensation. We take the body as a given around which the differentiation between self and other is based and around which our perceptions are formed. The phenomenologist Edmund Husserl expresses this when he states that

> The Body is in the first place the medium of all perception...the zero point of orientation... each thing that appears has eo ipso and orientating relation to the body [20].

In his consideration of the body as the bearer of sensation Husserl begins from the starting point of the body as an object with parts that can be perceived *just like other things* [20, p. 152]. However such a consideration is restricted to visual appearances and it is with reference to the sensation of touch that it differs. Although Husserl recognises the importance of kinaesthetic sensation, touch has a particular importance since it is constitutive of both the perception of objects and the body. The way that we visually perceive our body is not the same as it is perceived through touch, that is something *touching which is touched* [20, p. 155]. In fact Husserl goes so far as to state that *A subject whose only sense was the sense of vision could not at all have an appearing body* [20, p. 158].

It is interesting that the dual nature of this sensation appears to anchor sensation to the world; indeed this might be the 'mechanism' that draws together the hands in the experiments of Ehrsson, Holmes and Passingham; the somatosensory system pulling together the *hand touching* and the one *which is touched* into the same space. In fact the anchoring of sensation that occurs through touch means that it is quite literally the touch stone of perception; however the danger is that having begun to see its affect upon perception, we can miss other aspects of what it means to be an embodied subject. In the introduction to Neural Signatures of Body Ownership Tsakiris et al. [19] quote James [21] who stated that *When I decide to write I do not need to look for my hand in the same way that I have to look for a pen or a piece of paper, for the simple reason my hand is "always there"*, indeed the sensation of the body as "always there" appears to underpin Tsakiris' notion of a pre-existing bodily schema which plays into the creation of the RHI. But does the "always there" nature of the body in perception mean that we have a tendency towards perceiving it (and as Husserl's analysis began with) as being *like other things*? Within the phenomenology of Maurice Merleau-Ponty we begin to see a more nuanced understanding of the relationship *between* senses and what this may imply for the perception of the body.

> If my arm is resting on the table I should never think of saying that it is beside the ash-tray in the way the ash-tray is beside the telephone. The outline of my body is a frontier across which ordinary spatial relations do not cross. This is because its parts are interrelated in a peculiar way: they are not spread out side by side, but envelope each other [22].

If as Merleau-Ponty suggests, the body's parts envelop each other in such a way that the body is an outline across which ordinary spatial relationships do not cross, the way in which a bodily schema (if one exists) might sit in the world may not be a straight forward spatial mapping. Indeed it appears that such an 'enveloping' of sensation might be the means by which localized sensation 'meets' in the RHI causing proprioceptive drift. Whilst the body might be *the medium of all perception* and *the zero point of orientation* it does not mean that we should not consider what it means if the enveloping of sensation is the nature of sensation 'before' it becomes 'rendered' spatially and gives us the body as it is perceived. Needless to say this opens up the possibility that the body is subject to change through the underlying relationships of sensation itself.

What is interesting is that, in contrast to Merleau-Ponty's assertion that the *body is a frontier across which ordinary spatial relations do not cross*, within CRPS we discover patients who can experience a limb to be foreign to their body and who possess 'a poor awareness of its location in space' [8]. In fact when asked to move an affected limb we have observed that suffers of CRPS often do not move the limb as any other part of their body but move it from the unaffected region of their body, as if the limb were an *object* within their personal space (and therefore affectively 'besides' rather than enveloped within their body). In this regard we discover circumstances within which ordinary spatial relationships *have* entered the body. Arguably if subjects 'reject' or block the formation of the RHI as a result of incongruent sensation (or the use of a natural object) if a consequence of CRPS is that sensation of the affected limb feels markedly different from its visual appearance this might account for the lack of ownership felt by patients. As we shall see when we examine the reactions of patients who have used the body image tool, many have stated that although they know how their limb feels, they have never had the opportunity to see it before.

Rather than possessing the body as a 'given' to which sensation 'adheres' it appears that bodily perception emerges from the relationship between sensation and the CNS. If sensation is not something appended to the body, 'external' sensation and proprioceptive perception will envelope each other as they are processed by the CNS generating our body perception and sense of space. Such a process would account for the possibility of those suffering from CRPS feeling that a limb does not belong to them. Although Tsakiris and Haggard have suggested that the RHI is subject to a bodily schema, such a schema might not be top down and predicative, indeed this is an oddly 'anthropomorphic' view of this mechanism. If the body-image is the result of a series of underlying relationships the apparent presence of a bodily schema might simply be an attractor within the system. Given this, shifts in the relationships that lead to the generation of body perception can have significant outcomes.

6.5 The Body and Space

As a result of our analysis of the RHI we have suggested that there is a constitutive relationship between touching and the touched which is generative in nature. This means we should attempt to understand this relationship in terms of its production rather than the imposition of a model of the body upon sensation. However whilst we might want to avoid the imposition of a (normalizing human) form upon perception, it is also the case that perception arises from a particularly human perspective and this should feature within our consideration. In other words, although there are 'elements' that are generative, as a generative process there are elements of what it means to have a body enfolded within this. In this sense the body is never merely *the zero point of orientation*, a mathematical point from which measurements are made of an otherwise homogenous space; rather it is the origin of our interactions with the world. If we consider the relationship between the body and space, space is always loaded with a particular kind of potential. If we return to Mearleau-Ponty's phenomenology he notes that

> The example of instrumentalists shows even better how habit has its abode neither in thought nor in the objective body, but in the body as mediator of a world. It is known that an experienced organist is capable of playing an organ which he does not know...

> He does not learn objective spatial positions for each stop and pedal, nor does he commit them to 'memory'. During the rehearsal, as during the performance, the stops, pedals and manuals are given to him as nothing more than possibilities of achieving certain emotional or musical values, and their positions are simply the places through which this value appears in the world [22, p. 145].

We need not be an accomplished performer to experience this either; if we change the place where we store regularly used items, such as the cutlery drawer in a kitchen, we often find ourselves going to the wrong drawer for weeks on end because this action is not a part of conscious thought; for instance we might be in the habit of making tea and fetching a spoon, we do not even think of cutlery as such but we are thinking of tea. This continues until our conscious mind begins to spot the error 'in progress' and effectively begins to re-write our habits. In fact these habitual actions can incorporate more than just our bodies. For instance if we observe sports persons who use objects such as bats, sticks, rackets or swords their movement towards the world incorporates the measure of those objects. For instance, a sportsperson does not merely reach for a ball, see that they cannot reach it and then lean to shorten the distance; on the contrary the appropriate lean is already incorporated into the reaching action because their racket is included in their bodily space. We can observe the disquieting effect that a change to these objects can make to our measure of the world such as when we switch from driving our regular car to a hire vehicle and endeavor to park it. To this extent Merleau-Ponty notes that it is possible for an object to become incorporated into the body.

> ... those actions in which I habitually engage incorporate their instruments into themselves and make them play a part in the original structure of my own body [22, p. 104].

In *When Far Becomes Near: Remapping of Space by Tool Use* Berti and Frassinetti [7] provides an insight into this phenomena. Neuroscience has observed that there might be a distinction between either the processes or regions of the brain involved with the handling of near (peripersonal) and far (extrapersonal) spaces; this is an understanding of space based upon human action and the spaces within which particular actions are required. Peripersonal space is defined as the space within which things are in hand reaching distance and extrapersonal space being that which is beyond our reach and requires some form of locomotion to be able to grasp an object. What interested Berti and Frassinetti were a number of cases where those who had suffered strokes showed dissociation between near and far spaces manifesting spatial neglect in only one or the other of those spaces [23–25].

A task used to establish the degree of neglect suffered by patients involves subjects bisecting a line in near and far spaces. Berti and Frassinetti reported the case of patient P.P., who having suffered a stroke, showed dissociation between near and far spaces where neglect occurred in a line-bisection test in near space but where this could be performed without impediment in extra personal space using a light pen. They predicted that if the bisection of the line in 'far' space were to be executed with a tool such as stick, the extended representation of the body should now include that 'far' space as 'near'. The outcome of this would then be that the same neglect that had occurred in near space would now be seen in 'far' space. The result was as predicted and when a stick was used instead of a light pen the line was misperceived in the same way as it was when within near space; far had indeed become near.

As a result of this, if we examine the issue in terms of the perception of bodily space, rather than the 'habitual' use of a tool being a requirement to achieve this, as was suggested by Merleau-Ponty, it seems their simple use is enough to incorporate them into the space defined by the body. In fact Berti and Frassinetti define the coding of near and far spaces as *a dynamic process*. Although in this instance we have not been directly considering bodily perception, the rapid changes that affect the mapping of bodily space suggests a dynamic process that does not (and perhaps even cannot) rely on a predefined model.

6.6 Changes in Body Perception

If we are considering the impact of the somatosensory system we should also be aware that it is known to be involved in more than the perception of what we might consider to be our own sensations. When viewing the experience of others and imagining activities of various sorts, the somatosensory and premotor cortices becomes active. Antonio Damascio describes the empathetic sensation of pain that we can experience under these circumstances as the "as-if-body-loop" mechanism, within which our empathy 'involves an internal brain simulation that consists of a rapid modification of ongoing body maps' [26].

This poses a dilemma for how we might think of these sensations, however

> The result of the direct simulation of body states in body-sensing regions is no different from that of filtering of signals hailing from the body…. What one feels then is based on "false" construction, not on the "real" body state.

Damascio's use of scare quotes is appropriate because the sensation is real, what differs is that in that empathizing with somebody who has (for example) grazed their knees, although we've physically not received that graze, we experience the sensation of a graze. In itself this might lead us to speculate whether this might underpin the EMG results seen by Ehrsson, Slater et al. or might even be a factor within the RHI itself; however whatever the mechanisms that underlie the RHI or CRPS, it is clear that the sensations of CRPS patients are real even though they apparently possess a 'normal' limb.

In considering these issues we must also be careful about how we consider the notion of a map. If we were to draw upon Tsakiris and Haggard the tendency might be to consider the map and its modification to be akin to a top down and predictive schema of the body. In itself this is a problematic notion for where would such a schema originate? If it were 'hard wired' into perception it would itself originate from that structure. If it were an a priori facet of perception one would have to question the reliability of such a schema unless it were based on the experience of individuals, and if it were based on experience such a schema could not be top down. Logically the most reliable reference for the body is a referral to, and enfolding of, bodily experience itself.

Damascio refers to a neural mapping of the body in terms of Spinoza's philosophy. Spinoza's thinking was markedly different to Descartes, Descartes proposed [27] that there are two substances, thought and extension. If we were to accept this we then encounter the problem of the mind/body divide; indeed if one considers the notion of a predictive bodily schema we have already implicitly assumed that such a divide exists. This is because a model would have to exist elsewhere so that sensation can be checked against it. In contrast, Spinoza proposed a single substance [28] and within this ontology, thought and extension are attributes of a single substance, expressing its essence in different ways. Damascio sees this as being akin to his proposal that 'mental processes are grounded in the brain's mapping of the body' [29], a mapping which is the result of sensory activity. In many ways we might even view this mapping *as* this sensory activity. In referring to Spinoza, Damascio draws upon the Scholium of Proposition 7, Part II of the Ethics which notes that 'Extension and the idea of the mode are one and the same thing, expressed in two ways.' Damascio states that he believes 'the foundational images in the stream of mind are images of some kind of body event, whether the event happens in the depth of the body or in some specialized sensory device near its periphery' [29, p. 197]. As a result of this we might find it to be less problematic if we were to consider the notion of 'mapping' as the *expression* of these events.

Given these factors it increasingly appears to be the case that our body perception results from a dynamic process. Although we might not be able to intervene in the underlying 'mechanisms' it seems logical that, much in the same way the RHI affected body perception, intervention in other processes might affect its expression. A condition that has similarities with CRPS is Fibromyalgia (FMS). The symptoms of FMS include widespread pain, hypersensitivity to sensory stimuli, phantom swelling of limbs and reduced sensitivity to the position of limbs and motor abnormalities such as tremors or slowness in movement.

In order to examine whether a dysfunction in the interaction between motor and sensory systems might be involved in symptoms experienced in FMS, McCabe et al. [8] conducted a series of tests using a mirror/whiteboard that created varying degrees of sensory conflict during congruent/incongruent limb movements. It itself this is interesting because of the differentiation that we have previously noted in experiments concerning congruent and incongruent limb positions and the RHI. In part this is because we have speculated over the acceptance or rejection of a limb based around these factors; however it should be noted that even if the limb is rejected this does not affect the sensation experienced within the limb. If the symptoms experienced in conditions such as FMS are the result of such sensory conflict, manipulating the perception of such movements could further impact upon the condition.

The results of these tests were that 89.7 % (26 out of 29) of patients with FMS involved in the tests reported changes in perception compared with 48 % of a healthy control group. The sensations experienced included...

> ...disorientation, pain, perceived changes in temperature, limb weight or body image. Subjects described how these symptoms were similar to those they experienced in a "flare" of their FMS. This led us to conclude that some sensory disturbances in FMS may be perpetuated by a mismatch between motor output and sensory feedback.

Whilst it appears that this mismatch plays a role within FMS, healthy participants also reported (albeit at a lower incidence) changes in perception. Rather than being a phenomena exclusively linked to the pathology of FMS, the results of these tests indicate that the underlying 'structure' of sensation is such that changes in the relationships between sensations within the somatosensory system can also affect body perception in healthy subjects. The anomalies that appear to exist within FMS therefore seem to make those who suffer from the condition all the more vulnerable to new anomalies. Once again this points towards what appears to be the emergent nature of body image. Given the close integration between (the enveloping) of bodily and spatial perception, it appears that conflicting sensation or faulty sensory integration might explain the experience of FMS and CRPS patients. But given that these experiences are not accompanied by physiological symptoms, how can conditions such as these be diagnosed?

6.7 A Need to Communicate Painful Contradictions

The relative lack of knowledge concerning CRPS can affect the time taken to make a diagnosis. As was noted by one patient in our study

> The thing I found difficult was getting this far, my GP knew nothing, so the diagnosis took forever...

This can often be exacerbated because those suffering from CRPS often question the nature of their experience. The contradictory nature of sensation and physical appearance was a constant theme when discussing the condition with patients

> The right side of my whole body actually feels quite normal, there no problem with that I don't have any difference in perception to what I see with that...

However these are not 'minor' contradictions, the disparity between experience and appearance is often marked

> I know there are fingers there and I even move them, I can't see fingers when I try closing my eyes to see it, I don't see anything, I just see a big blob...

Patients can also reject their limb. Even though patients recognize their limb doesn't appear to be physically different their experience of it is such that they shun it.

> There is a sense now of repulsion, I think is the word, I don't like looking at it.

Following the diagnosis of CRPS, a method currently used to assess the condition is the use of self-portrait sketches or drawings made by clinicians. This process can often be revealing for those suffering from CRPS because they have not fully engaged with the nature of their sensation. It seems that they often try to keep these contradictions at a distance rather than 'inhabit' the sensations that result from CRPS. One patient noted that

> ...it's quite new to me because I hadn't really thought about this until I came in here.

A contributory factor might well be that some patients experience an increase in the levels of pain if they focus on those sensations. Nevertheless it was also the case that if patients had considered the contradictory nature of their sensations it was something that could trouble them...

> ...it's a very strange thing... I really thought I was losing it

Although the use of drawings enables patients to describe the nature of the sensations, they have a number of limitations such as the ability of patients or staff to visually render these sensations. As a result of this it was decided to explore the potential of new media to provide a tool that more readily enabled patients to describe the nature of their sensation. The initial inspiration to explore this came from Alexa Wright's *After Image* project (1997) which dealt with the experience of amputees and phantom limbs. *After Image* was one of the first projects to be funded by the Welcome Trust's SciArt scheme. Wright interviewed and photographed eight people with amputated limbs. The resulting work included a 'straight' portrait of the subject alongside a digitally manipulated portrait depicting the experience of the phantom limbs based on Wright's interpretation of their experiences. These images were exhibited with an accompanying text, derived from the interview, in which each person describes their experience of their limb. Whilst Wright's work used Photoshop to manipulate photographic images of amputees, such an approach wasn't appropriate for a tool that could be easily used within a clinical setting. In addition the time intensive nature of such a process was not one that would allow patients to have an active involvement in the process. In addition the spatial anomalies experienced by CRPS patients also suggested that the development of a 3D tool would be more appropriate.

A specification for the tool was established using data from a previous exploratory study of body perception [11], and consultation with a person with CRPS.

It was determined the tool should allow the manipulation of the scale, position and surface texture of body segments whilst also allowing for these to be 'absent'/removed if desired. Scaling should include the ability to lengthen and shorten limb segments, as well as making them thicker and thinner. Body segments should be able to be moved and the extent of this should be such that they could be separated from each other and be placed in anatomically impossible positions. Colors and textures should be provided such that these might represent feelings of burning, cold, rough, smooth and lack of substance. Finally the tool should allow the viewing of the model or 'avatar' from different perspectives: front, back, left side, right side, through 360°.

Initial funding provided 2 weeks that allowed the creation of a prototype that fulfilled these criteria. The research was approved by the Local NHS Research Ethics Committee and the tool was tested with consenting patients admitted to an inpatient CRPS rehabilitation program. Participants were recruited from a tertiary referral service for those with CRPS based in the South West of England. Inclusion criteria were: a clinical diagnosis of CRPS [12], admission to the inpatient multi-disciplinary CRPS rehabilitation program at the Royal National Hospital for Rheumatic Diseases in Bath and the ability to understand and express themselves in English. Patients fitting the criteria were given the study information booklet by a member of the clinical team and asked to contact the research nurse if they were interested in participating. Informed consent was obtained from all participants prior to participation and is securely archived according to local NHS procedures. These initial tests led to a number of perceived opportunities to improve the system and this has led to a second set of tests which are still on-going at the time of writing.

Ten participants used the first version of the application in a single consultation with the research nurse. The nurse showed the participant the application, its capacity to alter the length, thickness and position of limb segments and the material and texture choices available for applying to the avatar's body parts. It was also stressed that any illustrative meaning attributed to the colors and textures were for the participant to decide. Once the scope of the software was demonstrated by the nurse she then operated it in response to instructions from the participant to achieve a representation according to their instructions. For example if a participant wished a limb (or portion thereof) to be altered in its length she would ask how long or short they wanted the limb segment to be; asking when to stop whilst increasing or decreasing it's length. Participants were asked to confirm they were satisfied with the accuracy of scaling after each manipulation.

Audio recordings were also made of the participants using the application to allowing interpretation of the images created and immediate reactions to the tool to be captured. Immediately after using the tool participants were asked to complete a structured questionnaire administered face to face with the research nurse. This had open questions to ascertain their views and experience of using the tool. The questionnaire was later modified to include a rating out of ten to determine how good a representation participants' thought the created image was.

In response to the question 'Did you find using the body perception application an acceptable way to communicate how you view or feel about your limb or body parts?'

All participants reported that the tool was a good method for communicating their body perception; both for themselves and for helping the clinician to understand patients' body perceptions. Of the seven participants asked to provide a rating out of ten with regards their satisfaction with the images they created: Three gave a rating of 7, one 8, one 9 and one 10. All participants were unanimous in the view that using the application was better than the standard interview addressing body perception undertaken earlier in their admission. They believed that the application was much more adaptable than a clinician's sketch and found it easy to use in consultation with the research nurse.

As a part of this process participants were also asked whether using the tool caused increased pain and distress. In most cases this was not an issue but in a number of instances increased pain was experienced, but what was interesting was that alongside this other benefits were expressed.

> No, I don't think I've got a bad feeling from doing this, it's not a bad feeling it's just to me looking at that.... puts it into perspective what I've got its just I don't know how to explain it, it looks in human form exactly how I feel and I've never had that...

6.8 An Erie Realisation and a Form of Acceptance

This moment of recognition was one of the interesting outcomes arising from the interviews. Quite often patients would express surprise at how they were able to depict their affected region using the system.

> It makes you see how distorted your (long pause), vision of your own body is, of your limb especially...

> It helps us to see... this is how much distortion you're seeing your body as

If one were to try and devise a metaphor for what many experienced, a crude comparison might be made with the sort of recognition adults express when they meet a sibling from whom they had been separated at birth. In a sense given that patients hadn't had prior opportunity to 'see' their affected limb, the tool provided the first opportunity for such recognition

> It wasn't that I disliked using it, it's just... for me as I say to visualise that how I feel I felt a bit emotional, but the more I'm looking at it, it's only because I'm sitting here thinking that is exactly how in my mind's eye what I look like so it was a bit of a shock I suppose.

And

> Patient: Because in your head I haven't said this word but I've felt this, you feel freakish, so you look at that and you think yeah that is how I feel

> Interviewer: So you're able to identify with the image on the screen?
> Patient: Yeah

What proved to be of interest were the feelings that were expressed alongside those moments of recognition. The patient who had previously commented *I don't like looking at it* when talking of their limb described their experience of the Body Image Tool and the image that they created in a different way...

> Patient: Seeing something and knowing that it's *your hand* is errm how can I put that into words, its erm, I don't know it I suppose accepting now that it's there, it's happened, I've got it..
> Interviewer: Does this help you accept it?
> Patient: *Yeah*, because you can see it...

There appears to be a degree of an increased level of acceptance of the limb having been able to 'see' the limb; however this would require further investigation to substantiate and would need to considered over and against other factors such as the nature of the pain suffered by patients. Nevertheless if Tsakiris and Haggard were correct in their notion that the perceptual drift *away* from neutral objects were the result of a *lack* of self-attribution the moment at which patients were able to recognise their limb may have led to an increased level of acceptance. However what also needs to be noted is that the nature of the sensation experienced by patients does appear to have been affected by this moment of recognition and in one instance it appeared to produce a flair up of the pain associated with the condition.

> Interviewer: Is it hard looking at it on the screen, does it cause you more discomfort?
> Patient: It does, I noticed its feeling quite sweaty and burning, whether that's just because it is, whether I'm looking at this... I think it is to do with this

Having provided a means to manipulate the scale and position of limbs, the means to communicate sensation was through the use of different materials and textures that allowed users to apply color and imagery such as ice and flames to regions of the body. It was the descriptive potential of such methods which we then prompted to examine.

6.9 The Nature of Pain

As was succinctly stated by one patient...

> You can't see pain

But then again the tool appeared to be enabling patients to see their limbs. The tool we had developed tended to focus upon the perception of the scale and position of limbs, whereas pain was being addressed (albeit region by region) by the application of a single color or texture.

> My hand feels as if it's absolutely on fire and then if somebody touches it, it feels cold, and this pins and needles and I don't know how to represent that...

In fact these contradictions appeared to be a characteristic feature of the pain experienced by many of the patients we spoke to.

> These fingers here are numb, I suppose in a way they should be transparent really shouldn't they... 'cause these three fingers here are numb, I feel I could just stab them, but it's still got that hypersensitivity in it.

As a part of our interview procedure patients were asked if there was anything that they felt was missing from the system.

> The electric shocks that go up and down it are obviously not there, so I don't know how you could represent those sort of things,...

Alongside this the sensation of 'pins and needles' described by one patient was, to varying degrees, also reported by others.

> Ok so the skin surface doesn't feel any different here, but when I've got my eyes closed it's very, it almost feels as if it's... if I say it's not there you know when you've sat on your foot and its gone to sleep, its that sort of feeling so you sort of know its there but if somebody said where is it its quite difficult to say it's just there...

As a result of this we embarked upon the development of techniques that would allow patients to approach a visual description of these sensations. In considering this, the ability to suitably depict contradictory sensations was an important factor; both sensations would need to be adequately represented. For instance if one sensation were 'black' and another 'white' we would want to retain the characteristics of each without reducing sensation to a 'grey'. Prior consideration had been given to the possibility of using particle effects to describe burning sensations and given that the sensation of 'electric shocks' was something we now wished to depict their use seemed apt.

But how should particle effects be applied to the 'avatar'? As noted the technique used in the first version of the tool had simply applied 'sensation' as a texture or color (Figs. 6.5, 6.6).

If particle effects were to be used to help evoke contradictory sensation the options available were to position them either internally or externally to the affected region. In the first iteration of the tool users had been given the option of using translucent textures, however only one patient had used this. External application of the effect might also allow a clearer view of the underlying texture. However there appeared to be distinct disadvantages to such an approach.

If these were applied externally the use of 'flame' or 'shock' effects might lead to comparisons with comic book character such as the 'human torch' or the various forms of 'force attacks' seen used by any number of superheroes. As such these seemed akin to an outward projection of such sensation. If we were then to compare imagery of this sort with an artist such as Hieronymus Bosch we can begin to see a means by which sensation is depicted as something inflicted upon the body through its internalization. If one considers the notion of shooting sensation or shocks these are things that we would also be likely to consider to be internal.

Fig. 6.5 Image generated using the first version of the Body Image Tool which provided the hand as a single entity

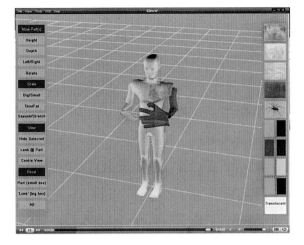

Fig. 6.6 In version one of the tool the users were restricted to blocking out regions using a single materials or texture

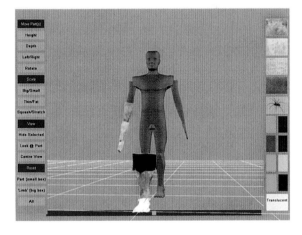

As a result of these considerations particle effects were implemented as an internal feature of the avatar. 'Flames' and ice colored 'shock' were provided along with spikes that pierced limbs. Further improvements were also made to the system based on user comments. Whereas the avatar originally only allowed the manipulation of the hand as a whole, the ability to manipulate fingers and their parts was introduced. A system was also introduced to measure the point at which patients felt their limb no longer belonged to them as well as a system to record this data.

Although at the time of writing we are only approximately half way through the trials of the second iteration of the tool and our protocols do not allow for the recording of interviews '(contrary to the first set of trials) there are some results that are worth noting. Out of the ten patients seen to date not one hasn't made use of the ability to use translucency and particle effects or spikes to describe their sensation, even if did not appear to be contradictory in nature (Figs. 6.7, 6.8, 6.9, 6.10).

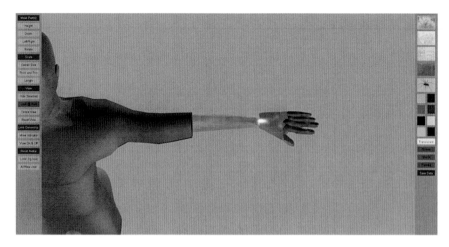

Fig. 6.7 Version two of the tool allowed the combination of particle effects and textures in the depiction of affected areas

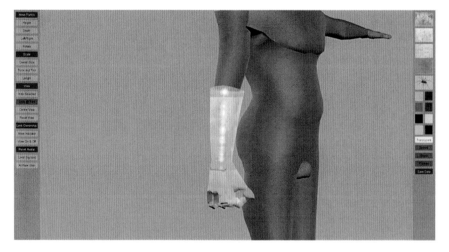

Fig. 6.8 The breaking down of the hand into smaller parts provided users a means to articulate the hand, here depicting it as clenched, as well as providing the opportunity for more discrete areas of sensation within it

What has emerged is a desire to more finely articulate the regions of sensation rather than simply containing them to the geometry that make up the body parts. A finer articulation of the kinds sensation depicted has also emerged, the sensation of compression and weight having been noted as something that would be useful.

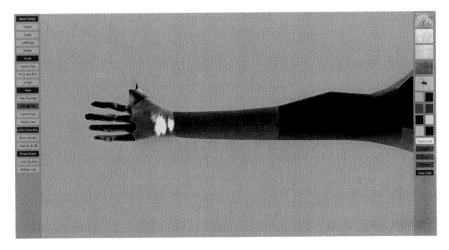

Fig. 6.9 Along with discrete areas of sensation the breaking up of the hand into smaller components also allowed users to remove part of their hand from its representation

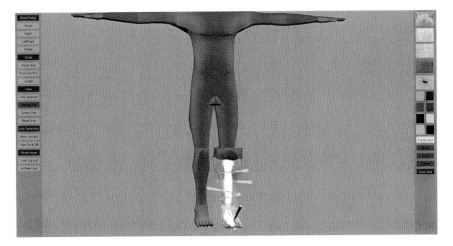

Fig. 6.10 Primarily designed to deal with contradictory sensations the use of particle and textures to this end was less common although still a feature within the tools use

6.10 Discussion

By far the lengthiest part of this chapter has been the consideration of the RHI. However the point of this has been to highlight the fact that body perception may not be the given we expect it to be; indeed the underlying mechanisms of the somatosensory system may be responsible for the experience of the body as an emergent phenomenon. Within such a schema it appears that vision plays a role and we have seen within Tsakiris and Haggard's work that vision seems to affect the acceptance of the RHI dependent upon whether we perceive sensation to be congruent or incongruent. However we have also seen that the RHI can be achieved without the involvement of visual perception as demonstrated by Ehrsson, Holmes and Passingham. As such the notion that vision is an a priori component of the RHI is refuted by such results; nevertheless it has also been demonstrated that vision can appear to have a pre-eminent and active role in the formation and denial of the illusion. Alongside this we have also seen that the utilisation of a whiteboard/mirror that utilises congruent and incongruent movements can disturb the perception of the body for both suffers of FMS and the control group; McCabe et al. suggesting that an underlying disturbance of the somatosensory system may be responsible for this.

Having worked with patients suffering from CRPS we have seen that the contradictory nature of the sensation they experience presents problems in the communication of their condition; indeed given that surprise was often expressed by patients at being able to see their affected limbs depicted, the act of literally making sense of their sensation prior to this appears to have been problematic. As such the tool seems to have provided a means to achieve this in such a way that some level of acceptance was gained; however further studies should be conducted to understand this better. However, if some degree of acceptance was achieved this has been at a level where the underlying mechanisms that have led to the condition have been left unaffected.

In terms of the body image tool there are still many improvements that could be made allowing a more granular depiction of sensation. Whilst at present all users stated that the system was easy to use, the danger is that the introduction of greater levels of granularity might lose the simplicity of the interface. Other challenges also remain such as being able to depict the sensation of compression or weight, although the latter might be addressed by the incorporation of some physics and inverse kinematics.

Beyond this it remains the case that if the representation of the body and its movements are capable of affecting perception there is the potential for using them in contexts beyond the description of body perception. For instance if we consider the movements that children might make when playing the part of a giant, these are often far more ponderous and stiff than those they might make depicting a mouse. If we were to track participants and represent their body and movements differently would this begin to affect their perception of their own body and movements? With the introduction of interfaces such as Microsoft's Kinect we may be moving into an arena where the ways in which users are depicted will affect more than just play. We are now opening upon an arena where there are great opportunities, but also one where we need to be cognizant of the potential affects that such interventions might have upon users.

References

1. Botvinick, M., Cohen, J.: Rubber hands 'feel' touch that eyes see. Nature **391**, 756 (1998)
2. Armel, K.C., Ramachandran, V.S: Projecting sensations to external objects: Evidence from skin conductance response. In: Proceedings of the Royal Society of London: Biological, **270**, 1499–1506 (2003)
3. Tsakiris, M., Haggard, P.: The rubber hand illusion revisited: Visuotactile integration and self-attribution. J. Exp. Psychol. Hum. Percept. Perform. **31**(1), 80–91 (2005)
4. Ehrsson, H., Holmes, N.P., Passingham, R.E.: Touching a rubber hand: Feeling of body ownership is associated with activity in multisensory brain areas. J. Neurosci. **25**(45), 10564–10573 (2005)
5. Slater, M., Perez-Marcos, D., Henrik Ehrsson, H., Sanchez-Vives M.V.: Towards a digital body: The virtual arm illusion. Fron. Hum. Neurosci. vol. 2, Article 6 (2008)
6. Yuan, Y., Steed, A.: Is the rubber hand illusion induced by immersive virtual reality? IEEE Virtual Reality **2010**, 95–102 (2010)
7. Berti, A., Frassinetti, F.: When near becomes far: Remapping of space by tool use. J. Cogn. Neurosci. **12**(3), 415–420 (2000)
8. McCabe, C.S., Cohen, H., Hall, J., Lewis, J., Rodham, K., Harris, N.: Somatosensory conflicts in complex regional pain syndrome type 1 and fibromyalgia syndrome. Curr. Rheumatol. Rep. **11**, 461–465 (2009)
9. Fraser, C.M., Halligan, P.W., Robertson, I.H., Kirker, S.G.B.: Characterising phantom limb phenomena in upper limb amputees. Prosthet. Orthot. Int. **25**, 235–242 (2001)
10. Moseley, L.: Distorted body image in complex regional pain syndrome. Neurology **65**, 773 (2005)
11. Lewis, J.S., Kersten, P., McCabe, C.S., McPherson, K., Blake, D.R.: Body perception disturbance: A contribution to pain in complex regional pain syndrome. Pain **133**(1–3), 111–119 (2007)
12. Harden, R.N., Bruehl, S., Perez, R.S., et al.: Validation of proposed diagnostic criteria (the "Budapest Criteria") for complex regional pain syndrome. Pain **150**(2), 268–274 (2010)
13. Marinus, J., Moseley, G.L., Birklein, F., et al.: Clinical features and pathophysiology of complex regional pain syndrome. Lancet Neurol **10**(7), 637–648 (2011)
14. Field, J., Warwick, D., Bannister, G.C.: Features of algodystrophy ten years after Colles' fracture. J. Hand. Surg. [Br.] **17B**(3), 318–320 (1992)
15. Geertzen, J.H.B., Dijkstra, P.U., van Sonderon, E.L., et al.: Relationship between impairment, disability and handicap in reflex sympathetic dystrophy patients; a long-term follow-up study. Clin. Rehabil. **12**, 402–412 (1998)
16. McCabe, C.S., Blake, D.R.: An embarrassment of pain perceptions? towards an understanding of and explanation for the clinical presentation of CRPS type 1. Rheumatology **47**, 1612–1616 (2008)
17. Beerthuizen, A., van't Spijker, A., Huygen, F., Klein, J., de Wit, R.: Is there an association between psychological factors and the complex regional pain syndrome type 1 (CRPS1) in adults? A systematic review. Pain **145**, 52–59 (2009)
18. Winkler, T.: Flying, spinning, and breaking apart: live video processing and the altered self. http://isea2011.sabanciuniv.edu/paper/flying-spinning-and-breaking-apart-live-video-processing-and-altered-self (2011)
19. Tsakiris, M., Hesse, M.D., Boy, C., Haggard, P., Fink, G.R.: Neural signatures of body ownership: A sensory network for bodily self-consciousness. Cereb. Cortex **17**, 2235–2244 (2007)
20. Husserl, E.: Ideas pertaining to a pure phenomenology and to a phenomenological philosophy book 2 (Rojcewic, R., Schuwer, A,. Trans.). Dordrecht: Kluwer (Original work published posthumously 1952) p. 61 (1980)
21. James, W.: The principles of psychology. Harvard University Press, Cambridge (1890)

22. Merleau-Ponty, M.: *Phenomenology of perception* (Smith, C,. Trans.). London: Routledge. (Original work published 1945) p. 98 (1962)
23. Halligan, P., Marshall, J.M.: Left neglect for near but not for far space in man. Nature **350**, 498–500 (1991)
24. Cowey, A., Small, M., Ellis, S.: Left visuo-spatial neglect can be worse in far than near space. Neuropsychologia **32**, 1059–1066 (1994)
25. Vuilleumier, P., Valenza, N., Mayer, E., Reverdin, A., Landis, T.: Near and far visual space in unilateral neglect. Ann. Neurol. **43**, 406–410 (1998)
26. Damascio, A.: Looking for Spinoza, p. 115. Hienemann, London (2003)
27. Descartes, R,.: Meditations on First Philosophy (J Cottingham Trans.) Cambridge: Cambridge University Press (Original work published 1641) (1996)
28. Pinoza, B.: *The Ethics*(S Shirley, Trans.). Indianapolis: Hackett (Original work published 1677) (1992)
29. Damascio, A.: Looking for Spinoza, p. 115. Hienemann, London (2003)

Chapter 7
Virtual Teacher and Classroom for Assessment of Neurodevelopmental Disorders

Thomas D. Parsons

Abstract Differential diagnosis and treatment of neurodevelopmental disorders that impact the brain's frontostriatal system require assessments that can differentiate the overlapping symptoms. Previous research has most often relied on paper-and-pencil as well as computerized psychometric tests of executive functions. Although these approaches provide highly systematic control and delivery of performance challenges, they have also been criticized as limited in the area of ecological validity. A possible answer to the problems of ecological validity in assessment of executive functioning in HFA children is to immerse the child in a virtual classroom environment where s/he interacts with a virtual human teacher.

Keywords Autism · Attention deficit hyperactivity disorder · Frontostriatal system · Virtual reality · Virtual classroom · Virtual human · Neuropsychology · Social skills training

7.1 Introduction

Autism and attention deficit hyperactivity disorder (ADHD) are neurodevelopmental disorders that impact the brain's frontostriatal system and hinder adaptive responses to environmental situations. Since children affected by high functioning autism (HFA) and ADHD often have overlapping symptoms one pressing need is to better understand the syndrome specific pattern of attention problems, and related treatment needs, that differentiate HFA children from those affected by ADHD. Standard measures of executive functions indicate that HFA but not necessarily ADHD children have difficulty with planning and set-shifting or

T. D. Parsons (✉)
Clinical Neuropsychology and Simulation Laboratory,
University of North Texas, Denton, TX, USA
e-mail: thomas.parsons@unt.edu

A. L. Brooks et al. (eds.), *Technologies of Inclusive Well-Being*,
Studies in Computational Intelligence 536, DOI: 10.1007/978-3-642-45432-5_7,
© Springer-Verlag Berlin Heidelberg 2014

cognitive flexibility [1, 2]. Inhibitory control, however, seems to be an executive function that is relatively spared in children with autism [3]. Alternatively, although the executive deficits exhibited by those with ADHD seem to be heterogeneous in nature, their primary executive deficit may be in inhibitory control, specifically suppressing automatic processes or prepotent responses and/or maintain task instructions or representations in working memory [4]. Thus, inhibition is one area of executive functioning that may distinguish children with HFA (no deficit) from children with ADHD (deficit). However, support for this hypothesis has been equivocal.

Previous research has most often relied on paper and pencil-based psychometric tests of executive functions. Although these approaches provide highly systematic control and delivery of performance challenges, they have also been criticized as limited in the area of ecological validity [5]. By ecological validity, neuropsychologists mean the degree of relevance or similarity that a test or training system has relative to the real world, and in its value for predicting or improving daily functioning [6, 7]. Adherents of this view challenge the usefulness of constrained paper-and-pencil tests and analog tasks for addressing the complex integrated functioning that is required for successful performance in the real world. This may be especially true for testing executive function deficits in HFA children because of the social orienting hypothesis of autism. One hallmark of autism is a syndrome specific difficulty with the tendency to attend to and process social stimuli, such as faces or the direction of eye gaze [8, 9]. HFA children may display confusing commonalities with children affected by other frontostriatal developmental disorders such as ADHD. To resolve this issue there is a need to develop ecologically valid measures of social-orienting executive dysfunction in HFA children.

A possible answer to the problems of ecological validity in assessment of executive functioning in HFA children is to immerse the child in a virtual classroom environment where s/he interacts with a virtual human teacher. Research on VR has begun to support its potential for assessment and training of social skills in individuals with ASD. On the assessment side, work has been done to develop a virtual classroom that assesses executive functioning [10, 11]. These virtual environments have been found to offer significant advantages to more traditional methods of diagnosis and observation [12]. The use of virtual environments for training of social skills in individuals with ASD is increasing [13–15]. VR paradigms allow people with ASD to practice their social skills in a safe environment. As they explore the social situation the consequences for their actions (positive or negative) can be carefully controlled by the therapist [16]. Thus, the reaction of the avatars in the VE can be realistic but the therapist can determine the pace and complexity of exposure to social contexts and allow for optimal individualized practice of interaction as many times as is necessary [16]. Thus, VR intervention for children with HFA may also provide a safer environment than "real life" for realistic role playing exercises that maximize learning from mistakes while minimizing their real life consequences [17]. This type of safe role-playing in a virtual social context may be vital to the mastery and generalizability of social skills training with HFA children [18–20].

Recent work by Parsons at the University of North Texas has combined the attentional assessment found in a virtual classroom environment with virtual human technology. The idea is to place a virtual human teacher with verbal and nonverbal receptive and expressive language abilities into the virtual environment to aid in assessment of joint attention. The virtual human teacher acts as a social orienting system that comports well with the social orienting hypothesis of autism. As a result, researchers may differentiate attention deficits that exist regardless of social facilitation from those executive functions that may be alleviated by a virtual teacher. The use of a virtual teacher may be especially helpful in differentiating ADHD and HFA children because children with autism may be especially well motivated in computer-based paradigms. Children with autism have been observed to prefer the computer work relative to regular toy situations [21], to work more diligently on computer tasks [22] and to benefit more from computer enhanced than typical behavioral learning interventions [23]. Computer-based neuropsychological assessments offer a number of advantages over traditional paper-and-pencil testing: increased standardization of administration; increased accuracy of timing presentation and response latencies; ease of administration and data collection; and reliable and randomized presentation of stimuli for repeat administrations [24, 25].

The plan of this chapter will be as follows: In Sect. 1, current approaches to the differentiation of cognitive sequelae in neurodevelopmental disorders will be discussed. Section 2 will review past work using virtual environments for these populations. Next, in Sect. 3, virtual human research will be introduced. Finally, in Sect. 4, there will be a discussion of the promise and potential limitations of a virtual teacher/classroom environment for assessment and treatment of attentional deficits.

7.2 Neurodevelopmental Disorders: Differentiation of Their Cognitive Sequelae

Autism is a neurodevelopmental disorder involving impairments in social and communication skills, as well as repetitive behaviors or thought process [26]. Prevalence estimates have increased dramatically in the last twenty years such that this disorder is now recognized to afflict 1 in 88 children nationwide [27]. An aspect that may help to differentiate attentional processing between persons with ADHD and persons with HFA is the social orienting disturbance wherein children with autism display a syndrome specific difficulty with the tendency to attend to and process social stimuli, such as faces or the direction of eye gaze [8, 9]. The social orienting deficits of persons with autism may limit their capacity for social learning at home and in school and also play a role in their problematic development of social competence and social cognition [28, 29]. Recent research suggests that the social orienting impairments of autism reflect a disturbance of "social executive" functioning that involves frontal motivation, self-monitoring, volitional attention regulation. Further, deficits appear to be found in temporal/parietal systems that involve orienting and processing information about the behavior of other persons [9, 30, 31].

While there are children with autism that are higher functioning, even those that are considered HFA exhibit ongoing and significant deficits in social and communication skills, as well as problematic repetitive behaviors, and excessive focus on isolated areas of interest [32, 33]. Variability in the social and emotional status of these children is often complicated by comorbid ADHD symptoms [34, 35] and other emotional or behavioral disorders including anxiety [36, 37]. An unfortunate limitation of the current diagnostic system is that it does not allow for the combined diagnosis of ADHD with Autism [26]. As a result, there is often potential for HFA children to be misclassified as ADHD and vice versa [35, 38].

Given the overlapping symptoms commonly associated with differential diagnosis of HFA and ADHD, there is need for enhanced understanding of the syndrome specific pattern of attention problems, and related treatment needs that differentiate HFA children from children with ADHD. Both clinical groups are affected by frontal-striatal impairments of attention regulation [39, 40], which likely accounts for part of the symptom overlap, especially with respect to atypical patterns of attention regulation. Nevertheless, social orienting theory and research suggest that HFA children may be expected to display more attention problems in tasks involving social cueing than ADHD children. Standard methods for identifying and assessing social attention problems have been developed for preschool children with autism, such as joint measurement and intervention [41]. However, comparable social attention assessment and intervention methods for school age, higher functioning children have yet to be developed. The development of these methods is more difficult for older higher functioning children because social attention assessment and intervention is best conducted in controlled but ecologically valid social stimulus situations. In older higher functioning children the social stimulus complexity required for ecological validity often impedes the development of sufficiently controlled and transferable methods for social assessment and intervention methods.

Executive attention dysfunction is central to autism [42] especially the social executive demands of coordinating attention with others [9]. The latter refers to joint attention disturbance which can easily be measured in younger children in structured social interaction paradigms. Comparable standard measures are not yet available for older higher functioning children. Instead researchers have most often tried to assess attention problems in non-social executive function tasks or with non-interactive face processing measures [43–45]. The syndrome specific utility of these types of measures is debatable because other disorders such as ADHD involve impairments of attention but little is yet know about the types of attention measures which best discriminate children with HFA or ADHD. Work with non-social executive attention measures suggests that a combination of Stroop and Go-No-Go paradigms may be useful in this regard. ADHD children have difficulty in inhibition prepotent responses on a Stroop Paradigm but children with HFA have significantly less difficulty on these types of task [42]. ADHD children also make more errors of omission and commission on Go-No-Go tasks (e.g. inhibiting a response to "3" in repeated ordered sequences of 1–9) than HFA children indicating that they have more difficulty with attention maintenance and inhibition processes [2]. However, if the task is changed to inhibit responding to a number (3) in a repeated but random sequence of 1–9 numbers, HFA children display a pattern of inhibition errors comparable to ADHD children [46]. Thus ADHD and HFA children may appear similar or different with respect to attention inhibition disturbance based on stimulus presentation parameters.

In addition to Go-No-Go and Stroop methods, recent research suggests that computer-generated flanker tasks lend themselves to discriminant social attention assessment in HFA and ADHD children. In a flanker task children must respond to the direction of a central stimulus surrounded by distracting flankers and the stimuli may be non-social (e.g. arrows) or social (e.g. the direction of gaze in faces). ADHD children display poor performance and abnormal neurocognitive processing on non-social flanker tasks [47, 48]. Alternatively, HFA children display less evidence of behavioral impairment on a non-social flanker task [49, 50] but more evidence of difficult on a social-flanker task than control comparison children [50]. The observations of Dichter and Belger [50] are consistent with the social orienting hypothesis of autism and suggest that there may be a syndrome specific deleterious effect of social attention on cognitive control and executive functions in HFA children.

It is important to note that having a computer (instead of a human) provide feedback on neurocognitive tasks has yielded better inductive reasoning and cognitive flexibility performance in children with autism relative to controls [51, 52]. For example, on neuropsychological tests that assess the ability of a person to display flexibility in the face of changing schedules of reinforcement (i.e., "set-shifting") individuals with autism have been reported to be highly perseverative compared to neurotypical controls and controls with other neuro-developmental disorders: ADHD, language disorder, Tourette's syndrome and dyslexia [3, 53–55]. One example of this is seen on the Wisconsin Card Sorting Task (WCST), in which subjects can be tested for mental flexibility via requirements that the participants sort cards according to one of three possible rules (color, shape, or number). There are a number of component cognitive processes that are required for successful performance on the WCST: generation of a sorting rule, holding the sorting rule in working memory, inhibition of prepotent responding, and the ability to maintain/shift set). An interesting issue is that executive function tasks like the WCST have a high degree of interpersonal interaction between the persons with autism and an experimenter. In a study by [51], individuals with autism were presented with both a standard (examiner gives a paper card to subject and gives instructions) and a computerized version of a card sorting task. Results revealed that group differences found on the standard administration were not present during the computerized administration. Here we see the potential for increased executive functioning performance by children with autism when the social interaction with an experimenter is removed from the testing situation.

7.3 Virtual Environments for Neurocognitive Assessment

One viable approach to the above mentioned problems (e.g., differentiating Autism from other neurodevelopmental disorders) is to capitalize on advances in virtual reality (VR) technology. Virtual environments can provide platforms for child social-attention assessment and intervention that is sufficiently rich in terms of ecologically validity, while also providing scientifically rigorous control, manipulation and bio-behavioral data recording options [10, 56–58]. Virtual Reality is form of human–computer interface that allows the user to "interact" with and become "immersed" in a computer-generated environment [59]. VR offers the potential to deliver systematic social interaction learning opportunities with "virtual people" in precisely controlled, dynamic three-dimensional (3D) stimulus environments [60, 61]. VR paradigms also allow for the sophisticated, objective, real-time measure of participants' behaviors (e.g. visual attention) and training outcomes, such as changes in social attention [10, 61]. Recent cost reductions in VR technologies have led to the development of more accessible, usable and clinically relevant VR applications that can be used to address a wide range of physical and cognitive ailments and conditions [62].

Another reason that virtual environment based assessments may be preferable is that while standard neuropsychological measures have been found to have adequate predictive value, their ecological validity may diminish predictions about real-world functioning. Traditional neurocognitive measures may not replicate the diverse environment in which persons with autism and other neurodevelopmental disorders live. Additionally, standard neurocognitive batteries tend to examine isolated components of neuropsychological ability, which may not accurately reflect the distinct cognitive domains found in neurodevelopmental disorders impacting frontostriatal functioning [63–66]. Although today's neuropsychological assessment procedures are widely used, neuropsychologists have been slow to adjust to the impact of technology on their profession. While there are some computer-based neuropsychological measures (see discussion above) that offer a number of advantages over the traditional paper-and-pencil testing, the ecological validity of these computer-based neuropsychological measures is less emphasized. Only a handful of neuropsychological measures have been developed with the specific intention of tapping into everyday behaviors like interacting with a teacher and peers in a virtual school setting, navigating one's community, grocery shopping, and other activities of daily living. Of those that have been developed, even fewer make use of advances in computer technology.

Virtual environment applications that focus on treatment of cognitive [62, 67] and affective disorders [68, 69], as well as assessment of component cognitive processes are now being developed and tested: attention [10, 70–72], spatial abilities [73, 74], retrospective memory [75], prospective memory [76], spatial memory [77–79] and executive functions [80–82]. The increased ecological validity of neurocognitive batteries that include assessment using virtual scenarios may aid differential diagnosis and treatment planning. Within a virtual world, it is possible to systematically present cognitive tasks targeting neuropsychological performance beyond what are currently available using traditional methods [5, 60]. Reliability of neuropsychological assessment can be enhanced in virtual worlds by better control of the perceptual environment, more consistent stimulus presentation, and more precise and accurate scoring. Virtual environments may also improve the validity of neurocognitive measurements via the increased quantification of discrete behavioral responses, allowing for the identification of more specific cognitive domains [83]. Virtual environments could allow for neurocognition to be tested in situations that are more ecologically valid. Participants can be evaluated in an environment that simulates the real world, not a contrived testing environment [84]. Further, it offers the potential to have ecologically valid computer-based neuropsychological assessments that will move beyond traditional clinic or laboratory borders.

7.4 Assessment of Neurodevelopmental Disorders Using Virtual Environments

Previous research indicates that computer based tasks may be especially appealing to children with autism and encourage future studies [18]. More importantly a small but growing literature indicates that HFA children readily accommodate to virtual environment paradigms and that these paradigms can be effectively used for both social assessment and social intervention. Parsons, Mitchell, and Leonard [18] reported a study in which 12 HFA children displayed comparable competence and enjoyment of a VR environment compared to IQ matched comparison children. The HFA children, though, had more difficulty maintaining appropriate social distances in the VR space. In a subsequent detailed study of phenomenological experience two HFA adolescents were observed to treat VR scenes meaningfully. The adolescents also reported that they enjoyed using the VR platform and provided examples of how they thought the experience could help them in the real world. Also, Trepagnier et al. [20] have described the utility of a VR paradigm for assessing social attention problems (face and eye gaze) in a small sample of HFA children. These studies attest to the feasibility and potential of VR paradigms for assessment and intervention with HFA children.

These prior studies have emulated isolated face presentations or more complex virtual café or bus stop environments [18, 20]. However, one of the most important social contexts for VR emulation with HFA children may be the classroom. HFA children often can regulate their attention well enough to do well in academic classroom requirements. Nevertheless, their inability to deal with the social attention and information demands placed on them by teachers and peers can lead to significant behavior problems that are often mistaken for ADHD, as well as anxious or dysphoric mood that undermine their adaptive skills [32]. Consequently, one optimal ecologically valid approach to diagnosis and treatment of these children may be to use VR methods to simulate classroom social-educational environments under controlled conditions.

A program of basic research by Parsons and Rizzo at the University of Southern California has been the development of a Virtual Classroom over the past several years [10, 11]. Parsons, now at the University of North Texas, has extended the research to include social facilitation of a virtual teacher and other environmental cues (including flanker tasks) to assess social attention. The aim of Parsons's research program has been to develop virtual reality applications for the study, assessment and rehabilitation of attention, cognitive and psychological sequelae of central nervous system dysfunction in children and adults affected by psychopathology or trauma [62]. The original virtual classroom used a head mounted display to present cognitive tasks that appear on a chalkboard and distracters (visual and auditory) that occur both within the virtual classroom and "outside" the classroom window and door. Thus, researchers and clinicians can provide a controlled but rich social stimulus environment where attention and other cognitive challenges can be presented to children along with the precise delivery of and control of distracting auditory and visual stimuli within the naturalistic virtual environment [10, 11]. The validity and utility of the VR classroom has been demonstrated in a study in which response to a Go-No-Go task in the virtual environment differentiated ADHD children from controls on numerous measures of attention and activity [10]. Furthermore, individual differences in virtual classroom attention performance were associated with parent reports of ADHD symptoms [10, 85].

7.5 Virtual Reality for Assessment and Treatment of Social Skills

Numerous studies now indicate that VR methods are applicable with HFA children. Indeed, they may be especially enjoyable and motivating intervention platforms for children with HFA [18]. Children with autism have been observed to prefer the computer work relative to regular toy situations [21], to work more diligently on computer tasks, and to benefit more from computer enhanced or VR intervention than typical behavioral learning interventions [23, 86]. Wallace et al. [87], have observed that many children with HFA report a sense of presence in VR environments that is comparable to that of typical children and other studies have provided preliminary support for VR based social skills training for individuals with ASD [13–15]. As they interact in the virtual social situations the consequences of their actions (positive or negative) can be carefully controlled by the therapist [16]. The realism of VR social interactions can be varied and researchers can control the pace and complexity of exposure to social contexts. This allows for a degree of individualized design with regard to the VR practice of social skills [16]. Thus, VR intervention for children with HFA provide a safer environment than "real life" in which to practice social skills exercises that maximize learning while minimizing risks of failure and negative reinforcement learning [17]. The types of safe role-playing available in virtual social encounters may be especially vital to the mastery and generalizability of learning and social skills training with HFA children [14, 18–20, 88]. Moreover, case study of VR therapy with a child with cerebral palsy suggests that virtual reality treatments may be have sufficient impact on growth and development of children to be associated with functionally adaptive cortical reorganization [89].

A further experimental technology approach to work with children with high functioning autism, includes embodied conversational agents acting as virtual teachers, peers, and tutors. Embodied Conversational Agents (ECAs) are animated virtual agents that interact with users in real-time dialogue through the recogntion and performance of both speech and gesture [90–92]. Tartaro and Cassell [93] as well as Bosseler and Massaro [94] have used virtual animated characters to elicit social skills and language learning. For example, Tartaro and Cassell used a virtual peer used to improve social interaction skills, including turn-taking and gaze behavior. After interaction with the ECA, children improved their scores on the Test of Early Language Development and displayed increased social behaviours, such as improved gaze. They argue that using a virtual human may be preferable to actual human interactions in children with high functioning autism, because virtual tutors have the patience to interact with individuals with these children. Bosseler and Massaro also used ECA tutors for children with autism. For their work, even a month after the intervention with the embodied agent, children were still using their newly acquired vocabulary in everyday situations. The results from these evaluations are very encouraging, and it is hoped that an autonomous social skills tutor aimed at children with autism will likewise lead to improved social outcomes.

7.6 Virtual Teacher/Classroom Environment for Assessment/Treatment of Attention

Thus far this chapter has reviewed some of the issues inherent in differential diagnosis of deficits in children with autism when compared to children with ADHD. The chapter has discussed the ways in which various assessment modalities (paper-and-pencil, computerized, and virtual reality) can be used for differentiating aspects of executive functioning in neurodevelopmental disorders affecting frontostriatal functioning. Of note, the virtual reality environments offer ecologically valid assessment of activities of daily living. Further, through the use of intelligent virtual agents, children with autism can be aided in their development of social skills. In this section, the goal is proffer a potentially exciting new approach to neurocognitive differential diagnosis and social skills training through an integration of ECAs into a virtual schoolroom environment. This approach moves beyond the limitations of past virtual classroom environments that had only limited experience of passive virtual characters to an approach that takes the best of the virtual classroom and merges it with a socially interactive teacher.

The social orienting deficits of persons with autism may limit their capacity for social learning at home and in school and also play a role in their problematic development of social competence and social cognition. Recent research suggests that the social orienting impairments of autism reflect a disturbance of "social executive" functioning that involves frontal motivation, self-monitoring, volitional attention regulation. Further, deficits appear to be found in temporal/parietal systems that involve orienting and processing information about the behavior of other persons. The integration of a virtual teacher into a virtual classroom environment would allow for a more dynamic assessment of both personal and joint attention. While much of the work discussed thus far has focused on an individual's regulation, control, and management of cognitive processes in isolation of others, persons with developmental disorders may have increased deficits in planning, working memory, attention, problem solving, verbal reasoning, inhibition, mental flexibility, task switching, and initiation and monitoring of actions when interacting in a social environment. Of primary interest here is the shared focus (i.e., joint attention) of two or more individuals on an object. Joint attention is achieved when an individual alerts another to an object via eye-gaze, pointing, and/or non-verbal indications. As mentioned earlier, children with autism may have deficits in skills related to joint attention: eye gaze; and identifying intention.

The current iteration of the virtual teacher and classroom includes a battery of neuropsychological measures that can be administered with or without social cues from the virtual teacher: continuous performance test (CPT); picture naming test; and a stroop test. The actual virtual environment includes rows of desks, a teacher's desk at the front, a whiteboard across the front wall, a female virtual teacher between her desk and whiteboard, and peers seated "with" the participant in the room. The virtual teacher instructs the participant to look around the room and to point and name the various objects that they observed. Following this one-minute warm-up period, the virtual teacher tells participants that they are going to "play a game". In the virtual environment, participants view a sequence of stimuli (e.g., CPT; Stroop; or pictures) that appear for brief (a couple seconds) intervals to the left and right of the teacher on the whiteboard. There is a random inter-stimulus interval between the appearance of the stimuli (e.g., CPT; Stroop; or pictures) and the sequence of appearance and disappearance of left and right stimuli is asynchronous. The virtual teacher asks participants to depress a "left" or "right" hand button when any of four target stimuli appears behind her. The virtual teacher also says: "When I look this way (virtual teacher turns left) the target pictures will appear on this side of the board;" "When I look this way (virtual teacher turns right) the target pictures will appear on this side of the board;" and "When I look this way (virtual teacher looks straight ahead) pictures can appear on either side of the board". Two blocks of pictures are presented in fifteen sets of ten pictures. Five sets of ten pictures in each block randomly occur with the teacher looking left, right or forward. Two target pictures are designated in six sets of ten pictures and three target pictures are designated in nine sets of ten pictures.

Distracters are presented across the entire presentation series. Distracters are presented for the entire period of presentation of stimuli. Nine social distracters (e.g., people moving by outside the classroom) and nine nonsocial distracters (e.g., cars moving by outside the window) are presented across the sets of stimuli. Social and non-social distracters occur with the teacher looking to the left, right, or forward.

This virtual teacher and classroom paradigm yields quantitative measures of: (1) Attention to Task: number of targets correctly noted and average reaction time for correct targets; (2) Teacher-Directed Attention to Task: based on virtual teacher's visual regard, assessment of number correct and average reaction times relative to virtual teacher orientation (e.g., teacher looking forward, teacher looking left, and teacher looking right) conditions; and (3) Attention to Tasks during Social and Non-Social Distracter: the number of targets correctly noted in Social, Non-Social and No Distracter conditions and related average reaction times. It is expected that this research paradigm will provide information related to performance with and without social cuing from the teacher. Research with a previous version of the Virtual Classroom indicates performance measures on this task revealed test–retest and construct validity relative to performance on the Conners CPT II task, rs = 0.51–0.79, ps < 0.025 [10]. However, the VR classroom measures were more sensitive to differences in attention among children with ADHD or typical development, ds range from 1.59–1.96, than the CPT attention measures [10, 95].

7.7 Conclusions

Differential diagnosis and treatment of neurodevelopmental disorders that impact the brain's frontostriatal system require assessments that can differentiate the overlapping symptoms. This chapter reviewed the ways in which previous research has most often relied on paper-and-pencil and computerized psychometric tests of executive functions. Again, although these approaches provide highly systematic control and delivery of performance challenges, they have also been criticized as limited in the area of ecological validity. A possible answer to the problems of ecological validity in assessment of executive functioning in HFA children is to immerse the child in a virtual classroom environment where s/he interacts with a virtual human teacher. Recent work by Parsons at the University of North Texas has combined the attentional assessment found in a virtual classroom environment with virtual human technology. The idea is to place a virtual human teacher with verbal and nonverbal receptive and expressive language abilities into the virtual environment to aid in assessment of joint attention. The virtual human teacher acts as a social orienting system that comports well with the social orienting hypothesis of autism. As a result, researchers may differentiate attention deficits that exist regardless of social facilitation from those executive functions that may be alleviated by a virtual teacher.

It is important to note that not all children respond to any given treatment in the same way. Therefore, in beginning a program of research on VR applications to intervention with HFA children it is judicious to anticipate that individual differences in treatment responsiveness will be observed and need to understood. Specific to VR applications it is not yet clear whether HFA all children respond equally to moderately or maximally realistic VR social environments, or VR environments with greater or lesser stimulus complexity. However, research does indicate that, although HFA children self-report a typical level of presence in VR environments, they also report significant differences in sense of presence ranging from mild to strong [87]. "Presence" refers to a sense of "being there" inside a virtual environment and this may moderate the learning effectiveness of VR experiences across individuals [61]. Differences in presence may be related to the type so previously noted stimulus presentation parameters. However, differences in presence have also been related to differences in aspects of cognitive style such visual field independence [96], which is a strength for many but not all HFA children.

It is important to recognize, though, that VR interventions may be best used in conjunction with other in vivo social intervention methods to enhance the probability of improved treatment with pervasive developmental disorders. Used in isolation some children may simply learn how to use or respond to the VR "program" rather than see its relations to the real world. Parsons et al. [18] recommends that VR interactions be practiced in conjunction with intervention provided by a therapist-mentor, rather than in isolated context that entirely replaces real world interactions and therapy. Phenomenological reports suggest this "VR/mentor method" serves to help HFA children maintain the connection between the VR practice and real life scenarios.

References

1. Booth, R., Charlton, R., Hughes, C., Happe, F.: Disentangling weak coherence and executive dysfunction: planning drawing in autism and attention-deficit/hyperactivity disorder. Philos. Trans. R. Soc. B **358**, 387–392 (2003)
2. Happe, F., Booth, R., Charlton, R., Hughes, C.: Executive function deficits in autism spectrum disorders and attention-deficit/hyperactivity disorder: examining profiles across domains and ages. Brain Cogn. **61**, 25–39 (2006)
3. Geurts, H.M., Verte, S., Oosterlaan, J., Roeyers, H., Sergeant, J.A.: How specific are executive functioning deficits in attention deficit hyperactivity disorder and autism? J. Child Psychol. Psychiatry **45**, 836–854 (2004)
4. Castellanos, F.X., Sonuga-Barke, E.J., Milham, M.P., Tannock, R.: Characterizing cognition in ADHD: beyond executive dysfunction. Trends Cogn. Sci. **10**, 117–123 (2006)
5. Parsons, T.D.: Neuropsychological assessment using virtual environments: enhanced assessment technology for improved ecological validity. In: Brahnam, S. (ed.) Advanced

Computational Intelligence Paradigms in Healthcare: Virtual Reality in Psychotherapy, Rehabilitation, and Assessment, pp. 271–289. Springer, Germany (2011)

6. Wilson, B.A.: Cognitive rehabilitation: how it is and how it should be. J. Int. Neuropsychol. Soc. **3**, 487–496 (1998)

7. Chaytor, N., Schmitter-Edgecombe, M.: The ecological validity of neuropsychological tests: a review of the literature on everyday cognitive skills. Neuropsychol. Rev. **13**, 181–197 (2003)

8. Dawson, G., Meltzoff, A.N., Osterling, J., Rinaldi, J., Brown, E.: Children with autism fail to orient to naturally occurring social stimuli. J. Autism Dev. Disord. **28**, 479–485 (1998)

9. Mundy, P., Burnette, C.: Joint attention and neurodevelopment. In: Volkmar, F., Klin, A., Paul, R. (eds.) Handbook of Autism and Pervasive Developmental Disorders, vol. 3, pp. 650–681. Wiley, Hoboken (2005)

10. Parsons, T.D., Bowerly, T., Buckwalter, J.G., Rizzo, A.A.: A controlled clinical comparison of attention performance in children with ADHD in a virtual reality classroom compared to standard neuropsychological methods. Child Neuropsychol. **13**, 363–381 (2007)

11. Rizzo, A.A., Bowerly, T., Buckwalter, J.G., Klimchuk, D., Mitura, R., Parsons, T.D.: A virtual reality scenario for all seasons: the virtual classroom. CNS Spectr. **11**(1), 35–44 (2006)

12. Parsons, T.D., Rizzo, A.A., Bamattre, J., Brennan, J.: Virtual reality cognitive performance assessment test. Ann. Rev. CyberTherapy Telemedicine **5**, 163–171 (2007)

13. Andersson, U., Josefsson, P., Pareto, L.: Challenges in designing virtual environments training social skills for children with autism. In: Proceedings of 6th International Conference Disability, Virtual Reality and Associated Technology. Esbjerg, Denmark (2006)

14. Trepagnier, C., Sebrechts, M., Finkelmeyer, A., Stewart, W., Woodward, J., Coleman, M.: Stimulating social interaction to address deficits of autistic spectrum disorder in children. Cyber Psychol. Behav. **9**, 213–222 (2006)

15. Parsons, S., Mitchell, P.: The potential of virtual reality in social skills training for people with autistic spectrum disorders. J. Intellect. Disabil. Res. **46**, 430–443 (2002)

16. Parsons, S., Beardon, L., Neale, H., Reynard, G., Eastgate, R., Wilson, J., Cobb, S., Benford, S., Mitchell, P., Hopkins, E.: Development of social skills amongst adults with Asperger's syndrome using virtual environments. In: Proceedings of the 3rd International Conference on Disability, Virtual Reality and Associated Technologies, pp. 163–170. Sardinia, Italy (2000)

17. McGeorge, P., Phillips, L.H., Crawford, J.R.: Using virtual environments in the assessment of executive dysfunction. Presence: Teleoperators Virtual Environ. **10**, 375–383 (2001)

18. Parsons, S., Leonard, A., Mitchell, P.: Virtual environments for social skills training: comments form two adolescents with autism spectrum disorder. Comput. Educ. **47**, 186–206 (2006)

19. Trepagnier, C.G.: Virtual environments for the investigation and rehabilitation of cognitive and perceptual impairments. Neurorehabilitation **12**, 63–72 (1998)

20. Trepagnier, C., Sebrechts, M., Peterson, R.: Atypical face gaze in autism. Cyber Psychol. Behav. **5**, 213–218 (2002)

21. Bernard-Opitz, V., Roos, K., Blesch, G.: Computer-assisted instruction in autistic children. J. Child Adolesc. Psychiatry (Zeitschrift für Kinder- und Jugendpsychiatrie). **17**, 125–130 (1989)

22. Hetzroni, O.E., Tannous, J.: Effects of a computer-based intervention program on the communicative functions of children with autism. J. Autism Dev. Disord. **34**, 95–113 (2004)

23. Moore, M., Calvert, S.: Brief report: vocabulary acquisition for children with Autism: teacher or computer instruction. J. Autism Dev. Disord. **30**, 359–362 (2000)

24. Parsons, T.D., Notebaert, A., Shields, E., Guskewitz, K.: Application of reliable change indices to computerized neuropsychological measures of concussion. Int. J. Neurosci. **119**, 492–507 (2009)

25. Schatz, P., Browndyke, J.: Applications of computer-based neuropsychological assessment. J. Head Trauma Rehabil. **17**, 395–410 (2002)

26. American Psychiatric Association: Diagnostic and Statistical Manual of Mental Disorders (4th ed.—Text Revision). Author, Washington, DC (2000)
27. Baio, J.: Prevalence of Autism Spectrum Disorders—Autism and Developmental Disabilities Monitoring Network. Morbidity and Mortality Weekly Report, Surveillance Summaries, **61**(SS03), 1–19 (2012)
28. Dawson, G., Webb, S., Schellenberg, G., Dager, S., Friedman, S., Ayland, E., Richards, T.: Defining the broader phenotype of autism: genetic, brain, and behavioral perspectives. Dev. Psychopathol. **14**, 581–612 (2002)
29. Mundy, P.: Joint attention and social emotional approach behavior in children with autism. Dev. Psychopathol. **7**, 63–82 (1995)
30. Mundy, P.: The neural basis of social impairments in autism: the role of the dorsal medial-frontal cortex and anterior cingulate system. J. Child Psychol. Psychiatry **44**, 793–809 (2003)
31. Mundy, P., Newell, L.: Attention, joint attention and social cognition. Curr. Dir. Psychol. Sci. **16**, 269–274 (2007)
32. Klin, A., Volkmar, F.R.: Asperger syndrome. In: Cohen, D.J., Volkmar, F.R. (eds.) Handbook of Autism and Pervasive Developmental Disorders. Wiley, New York (1997)
33. Gillberg, C.: Asperger syndrome and high functioning autism. Br. J. Psychiatry **172**, 200–209 (1998)
34. Gadow, K., DeVincent, C., Pomeroy, J.: ADHD symptom subtypes in children with PDD. J. Autism Dev. Disord. **36**, 271–283 (2006)
35. Holtman, M., Bolte, S., Poutska, F.: ADHD symptoms in PDD: association with autistic behavior domains and coexisting psychopathology. Psychopathology **40**, 172–177 (2007)
36. Kim, J., Szatmari, P., Bryson, S., Streiner, D., Wilson, F.: The prevalence of anxiety and mood problems among children with autism and Asperger syndrome. Autism **4**, 117–132 (2000)
37. Meyer, J.A., Mundy, P.C., Vaughan, A.E., Durocher, J.S.: Social-attribution processes and comorbid psychiatric symptoms in Asperger syndrome. Autism **10**, 383–402 (2006)
38. Reiersen, A., Constantino, J., Volk, H., Todd, R.: Autistic traits in a population-based ADHD twin sample. J. Child Psychol. Psychiatry **48**, 464–472 (2007)
39. Krain, A.L., Castellanos, F.X.: Brain development and ADHD. Clin. Psychol. Rev. **26**, 433–444 (2006)
40. Rinehart, N.J., Bradshaw, J.L., Brereton, A.V., Tonge, B.J.: Lateralization in individuals with high-functioning autism and Asperger's disorder: a frontostriatal model. J. Autism Dev. Disord. **32**, 321–332 (2002)
41. Kasari, C., Freeman, S., Paparella, T.: Joint attention and symbolic play in young children with autism: a randomized controlled intervention study. J. Child Psychol. Psychiatry **47**, 611–620 (2006)
42. Hill, E.L.: Evaluating the theory of executive dysfunction in autism. Dev. Rev. **24**, 189–233 (2004)
43. Pelphrey, K.A., Morris, J.P.: Brain mechanisms for interpreting the actions of others from biologicalmotion cues. Curr. Dir. Psychol. Sci. **15**, 136–140 (2006)
44. Pelphrey, K.A., Morris, J.P., McCarthy, G.: Grasping the intentions of others: the perceived intentionality of an action influences activity in the superior temporal sulcus during social perception. J. Cogn. Neurosci. **16**, 1706–1716 (2004)
45. Pelphrey, K.A., Morris, J.P., McCarthy, G.: Neural basis of eye gaze processing deficits in autism. Brain **128**, 1038–1048 (2005)
46. Johnson, K.A., Robertson, I.H., Kelly, S.P.: Dissociation in performance of children with ADHD and high-functioning autism on a task of sustained attention. Neuropsychologia **45**, 2234–2245 (2007)
47. Jonkman, L.M., Kemner, C., Verbaten, M.N., Van Engeland, H., Kenemans, J.L., Camfferman, G., Koelega, H.S.: Perceptual and response interference in children with attentional-deficit hyperactivity disorder, and the effects of methylphenidate. Psychophysiology **36**, 419–429 (1999)

48. Vaidya, C.J., Bunge, S.A., Dudukovic, N.M., Zalecki, C.A., Elliott, G.R., Gabrieli, J.D.E.: Altered neural substrates of cognitive control in childhood ADHD: evidence from functional magnetic resonance imaging. Am. J. Psychiatry **162**, 1605–1613 (2005)
49. Henderson, H., Schwartz, C., Mundy, P., Burnette, C., Sutton, S., Zahka, N.: Response monitoring, the error-related negativity, and differences in social behavior in autism. Brain Cogn. **61**, 96–109 (2006)
50. Dichter, G.S., Belger, A.: Social stimuli interfere with cognitive control in autism. Neuroimage **35**(3), 1219–1230 (2007)
51. Ozonoff, S.: Reliability and validity of the Wisconsin card sorting test in studies of autism. Neuropsychology **9**, 491–500 (1995)
52. Pascualvaca, D.M., Fantie, B.D., Papageorgiou, M., Mirsky, A.F.: Attentional capacities in children with autism: is there a general deficit in shifting focus? J. Autism Dev. Disord. **28**, 467–478 (1998)
53. Liss, M., Fein, D., Allen, D., Dunn, M., Feinstein, C., Morris, R., Waterhouse, L., Rapin, I.: Executive functioning in high-functioning children with autism. J. Child Psychol. Psychiatry **42**, 261–270 (2001)
54. Ozonoff, S., Pennington, B.F., Rogers, S.J.: Executive function deficits in high-functioning autistic individuals: relationship to theory of mind. J. Child Psychol. Psychiatry **32**, 1081–1105 (1991)
55. Ozonoff, S., Jensen, J.: Brief report: specific executive function profiles in three neurodevelopmental disorders. J. Autism Dev. Disord. **29**, 171–177 (1999)
56. Parsons, S., Mitchell, P., Leonard, A.: The use and understanding of virtual environments by adolescents with autism spectrum disorders. J. Autism Dev. Disord. **34**, 449–466 (2004)
57. Parsons, T.D., Courtney, C.: Neurocognitive and psychophysiological interfaces for adaptive virtual environments. In: Röcker, C., Ziefle, T.M. (eds.) Human Centered Design of E-Health Technologies, pp. 208–233. IGI Global, Hershey (2011)
58. Parsons, T.D., Reinebold, J.: Adaptive virtual environments for neuropsychological assessment in serious games. IEEE Trans. Consum. Electron. **58**, 197–204 (2012)
59. Parsons, T.D.: Affect-sensitive virtual standardized patient interface system. In: Surry, D., Stefurak, T., Gray, R. (eds.) Technology Integration in Higher Education: Social and Organizational Aspects, pp. 201–221. IGI Global, Hershey (2011)
60. Parsons, T.D.: Virtual simulations and the second life metaverse: paradigm shift in neuropsychological assessment. In: Zagalo, V., Morgado, T., Boa-Ventura, A. (eds.) Virtual Worlds, Second Life and Metaverse Platforms: New Communication and Identity Paradigms, pp. 234–250. IGI Global, Hershey (2012)
61. Bailenson, J.N., Yee, N., Blascovich, J., Guadagno, R.E.: Transformed social interaction in mediated interpersonal communication. In: Konijn, E.A., Tanis, M., Utz, S., Linden, A. (eds.) Mediated Interpersonal Communication, pp. 77–99. Lawrence Erlbaum Associates, United Kingdom (2008)
62. Parsons, T.D., Rizzo, A.A., Rogers, S.A., York, P.: Virtual reality in pediatric rehabilitation: a review. Dev. Neurorehabil. **12**, 224–238 (2009)
63. Dodrill, C.B.: Myths of neuropsychology: further considerations. Clin. Neuropsychol. **13**, 562–572 (1999)
64. Parsons, T.D., Rizzo, A.A., Buckwalter, J.G.: Backpropagation and regression: comparative utility for neuropsychologists. J. Clin. Exp. Neuropsychol. **26**, 95–104 (2004)
65. Parsons, T.D., Rizzo, A.A., van der Zaag, C., McGee, J.S., Buckwalter, J.G.: Gender and cognitive performance: a test of the common cause hypothesis. Aging, Neuropsychol., Cogn. **12**, 78–88 (2005)
66. Wilson, B.A.: Ecological validity of neuropsychological assessment: do neuropsychological indexes predict performance in everyday activities? Appl. Prev. Psychol. **2**, 209–215 (1993)
67. Rose, F.D., Brooks, B.M., Rizzo, A.A.: Virtual reality in brain damage rehabilitation: review. Cyberpsychol. Behav. **8**, 241–262 (2005)
68. Parsons, T.D., Rizzo, A.A.: Affective outcomes of virtual reality exposure therapy for anxiety and specific phobias: a meta-analysis. J. Behav. Ther. Exp. Psychiatry **39**, 250–261 (2008)

69. Powers, M.B., Emmelkamp, P.M.: Virtual reality exposure therapy for anxiety disorders: a meta-analysis. J. Anxiety Disord. **22**, 561–569 (2008)

70. Law, A.S., Logie, R.H., Pearson, D.G.: The impact of secondary tasks on multitasking in a virtual environment. Acta Psychol. **122**, 27–44 (2006)

71. Parsons, T.D., Rizzo, A.A.: Neuropsychological assessment of attentional processing using virtual reality. Ann. Rev. CyberTherapy Telemed. **6**, 23–28 (2008)

72. Parsons, T.D., Cosand, L., Courtney, C., Iyer, A., Rizzo, A.A.: Neurocognitive workload assessment using the virtual reality cognitive performance assessment test. Lect. Notes Artif. Intell. **5639**, 243–252 (2009)

73. Beck, L., Wolter, M., Mungard, N.F., Vohn, R., Staedtgen, M., Kuhlen, T., Sturm, W.: Evaluation of spatial processing in virtual reality using functional magnetic resonance imaging (FMRI). Cyberpsychol., Behav., Soc. Networking **13**, 211–215 (2010)

74. Parsons, T.D., Larson, P., Kratz, K., Thiebaux, M., Bluestein, B., Buckwalter, J.G.: Sex differences in mental rotation and spatial rotation in a virtual environment. Neuropsychologia **42**, 555–562 (2004)

75. Parsons, T.D., Rizzo, A.A.: Initial validation of a virtual environment for assessment of memory functioning: virtual reality cognitive performance assessment test. Cyberpsychol. Behav. **11**, 17–25 (2008)

76. Knight, R.G., Titov, N.: Use of virtual reality tasks to assess prospective memory: applicability and evidence. Brain Impairment **10**, 3–13 (2009)

77. Astur, R.S., Tropp, J., Sava, S., Constable, R.T., Markus, E.J.: Sex differences and correlations in a virtual Morris water task, a virtual radial arm maze, and mental rotation. Behav. Brain Res. **151**, 103–115 (2004)

78. Goodrich-Hunsaker, N.J., Hopkins, R.O.: Spatial memory deficits in a virtual radial arm maze in amnesic participants with hippocampal damage. Behav. Neurosci. **124**, 405–413 (2010)

79. Parsons, T.D., Courtney, C., Dawson, M., Rizzo, A., Arizmendi, B.: Visuospatial processing and learning effects in virtual reality based mental rotation and navigational tasks. Lect. Notes Artif. Intell. **8019**, 75–83 (2013)

80. Armstrong, C., Reger, G., Edwards, J., Rizzo, A., Courtney, C., Parsons, T.D.: Validity of the virtual reality stroop task (VRST) in active duty military. J. Clin. Exp. Neuropsychol. **35**, 113–123 (2013)

81. Parsons, T.D., Courtney, C., Arizmendi, B., Dawson, M.: Virtual reality stroop task for neurocognitive assessment. Stud. Health Technol. Inform. **143**, 433–439 (2011)

82. Parsons, T.D., Courtney, C., Rizzo, A.A., Edwards, J., Reger, G.: Virtual reality paced serial assessment tests for neuropsychological assessment of a military cohort. Stud. Health Technol. Inform. **173**, 331–337 (2012)

83. Gaggioli, A., Keshner, E.A., Weiss, P.L., Riva, G.: Advanced technologies in rehabilitation: empowering cognitive, physical, social and communicative skills through virtual reality, robots, wearable systems and brain-computer interfaces. IOS Press, Amsterdam (2009)

84. Gorini, A., Gaggioli, A., Vigna, C., Riva, G.: A second life for eHealth: prospects for the use of 3-D virtual worlds in clinical psychology. J. Med. Internet Res. **10**, 1–21 (2008)

85. Adams, R., Finn, P., Moes, E., Flannery, K., Rizzo, A.A.: Distractibility in attention deficit/hyperactivity disorder (ADHD): the virtual reality classroom. Child Neuropsychol. **15**, 120–135 (2009)

86. Self, T., Scudder, R.R., Weheba, G., Crumrine, D.: A virtual approach to teaching safety skills to children with autism spectrum disorder. Top. Lang. Disord. **27**, 242–253 (2007)

87. Wallace, S., Parsons, S., Westbury, A., White, K., Bailey, A.: Sense of presence and atypical social judgments in immersive virtual environments. Autism **14**, 199–213 (2010)

88. Strickland, D., Marcus, L.M., Mesibov, G.B., Hogan, K.: Brief report: two case studies using virtual reality as a learning tool for autistic children. J. Autism Dev. Disord. **26**, 651–659 (1996)

89. You, S.H., Jang, S., Kim, Y.-H., Kwon, Y.H., Barrow, I., Hallett, M.: Cortical reorganization induced by virtual reality therapy in a child with hemiparetic cerebral palsy. Dev. Med. Child Neurol. **47**, 628–635 (2005)

90. Kenny, P., Parsons, T.D.: Embodied conversational virtual human patients. In: Perez-Marin, C., Pascual-Nieto, I. (eds.) Conversational Agents and Natural Language Interaction: Techniques and Effective Practices, pp. 254–281. IGI Global, Hershey (2011)

91. Parsons, T.D., Kenny, P., Ntuen, C., Pataki, C.S., Pato, M., Rizzo, A.A., St-George, C., Sugar, J.: Objective structured clinical interview training using a virtual human patient. Stud. Health Technol. Inform. **132**, 357–362 (2008)

92. Parsons, T.D., Kenny, P., Cosand, L., Iyer, A., Courtney, C., Rizzo, A.A.: A virtual human agent for assessing bias in novice therapists. Stud. Health Technol. Inform. **142**, 253–258 (2009)

93. Tartaro, A., Cassell, J.: Playing with virtual peers: bootstrapping contingent discourse in children with Autism. In: Proceedings of International Conference of the Learning Sciences. Utrecht, Netherlands, 24–28 June 2008

94. Bosseler, A., Massaro, D.: Development and evaluation of a computer-animated tutor for vocabulary and language learning in children with autism. J. Autism Dev. Disord. **33**, 653–672 (2003)

95. Pollak, Y., Weiss, P.L., Rizzo, A.A., Weizer, M., Shriki, L., Shalev, R.S., Gross-Tsur, V.: The utility of a continuous performance test embedded in virtual reality in measuring ADHD-related deficits. J. Dev. Behav. Pediatr. **30**, 2–6 (2009)

96. Hecht, D., Reiner, M.: Field dependency and the sense of object-presence in haptic virtual environments. Cyberpsychol. Behav. **10**, 243–251 (2007)

Chapter 8
Engaging Children in Play Therapy: The Coupling of Virtual Reality Games with Social Robotics

Sergio García-Vergara, LaVonda Brown, Hae Won Park and Ayanna M. Howard

Abstract Individuals who have impairments in their motor skills typically engage in rehabilitation protocols to improve the recovery of their motor functions. In general, engaging in physical therapy can be tedious and difficult, which can result in demotivating the individual. This is especially true for children who are more susceptible to frustration. Thus, different virtual reality environments and play therapy systems have been developed with the goal of increasing the motivation of individuals engaged in physical therapy. However, although previously developed systems have proven to be effective for the general population, the majority of these systems are not focused on engaging children. Given this motivation, we discuss two technologies that have been shown to positively engage children who are undergoing physical therapy. The first is called the *Super Pop VR*™ game; a virtual reality environment that not only increases the child's motivation to continue with his/her therapy exercises, but also provides feedback and tracking of patient performance during game play. The second technology integrates robotics into the virtual gaming scenario through social engagement in order to further maintain the child's attention when engaged with the system. Results from preliminary studies with typically-developing children have shown their effectiveness. In this chapter, we discuss the functions and advantages of these technologies, and their potential for being integrated into the child's intervention protocol.

S. García-Vergara (✉) · L. Brown · H. W. Park · A. M. Howard
School of Electrical and Computer Engineering, Georgia Institute of Technology,
85 5th Street NW, Atlanta, GA 30308, USA
e-mail: sergio.garcia@gatech.edu

L. Brown
e-mail: lavonda.brown@gatech.edu

H. W. Park
e-mail: haewon.park@gatech.edu

A. M. Howard
e-mail: ayanna.howard@ece.gatech.edu

A. L. Brooks et al. (eds.), *Technologies of Inclusive Well-Being*,
Studies in Computational Intelligence 536, DOI: 10.1007/978-3-642-45432-5_8,
© Springer-Verlag Berlin Heidelberg 2014

Keywords Serious games · Physical therapy and rehabilitation · Play therapy · Social robotics · Darwin-OP · *Super Pop VR*^{TM}

8.1 Introduction

Upper-arm motor impairments affect a number of population demographics, from children living with cerebral palsy [1] to adults recovering from stroke. In the clinical setting, the most effective means to improve recovery of motor function is through rehabilitation which involves intense, repetitive engagement of the respective limb. Unfortunately, due to a number of factors, including increases in medical costs, reduction in paid benefits, and limits on time available for therapists to provide quality one-on-one sessions, there is a growing need to introduce low-cost rehabilitation systems into the home environment. These systems must not only be designed to engage patients into the rehabilitation protocol established by the therapist, but also provide accurate and appropriate feedback on and tracking of patient performance. This desire for engaging home-based rehabilitation systems is especially prevalent when addressing the therapeutic needs of children with physical and/or cognitive impairments [2].

Pediatric physical therapy differs from adult therapy in that children typically cannot (or may not be willing to) follow direct instructions required of a therapy routine. As such, clinicians typically incorporate therapy in play to provide an engaging and motivational intervention that may enhance the child's participation in the therapy session. No one will argue about how important play is during childhood. The role of play in the development of children has been extensively studied, and a large body of work exists to discuss the importance and nature of play in children [3]. As such, these alternative technologies for engaging children with disabilities in rehabilitation should have a key requirement of incorporating concepts of play within their design.

One of the key factors in play, which is also shown to be a determinant for effecting compliance in rehabilitation is engagement. To effect engagement and/or motivation, one such promising technology that has been gaining momentum in recent years is the coupling of virtual reality games with robot-assisted rehabilitation. Virtual reality (VR) refers to a computer technology that creates a three-dimensional (3D) virtual context and virtual objects that allow for interactions by the user [4]. These gaming scenarios enable robust changes in motor task difficulty level, as well as effect the quantity/quality of the feedback on performance, which have been shown as key factors that influence engagement and thus adherence. In [5], Colombo et al. showed that game like scenarios in conjunction with a robot-assisted rehabilitation device helped motivate users through score keeping, in which the scoring mechanism was coupled with the individual's achieved range of motion (ROM). In [6], a case study provided preliminary evidence that using custom-made VR rehabilitation games with a robotic ankle orthosis can be clinically more beneficial than the same rehabilitation in the absence of the VR games. Similarly, in [7] a ten patient study showed that a VR robot-assisted therapy approach induced a motor output effect that was considered comparable to those obtained with conventional approaches in the presence of a human therapist for patients different neurological gait disorders. Finally, in [8], the Gentle/s system was shown as an appealing device that, when coupled with VR technology, can provide robot mediated motor tasks in a three dimensional space. Although this is just a sampling of the current efforts in this domain, the common theme has been to increase motivation for robot-assisted rehabilitation through the use of interactive gaming. Although preliminary evidence shows that most of these VR-robotics coupled systems are effective, what is missing is their focus on engaging children. As such, this chapter discusses the use of VR and robotics to assist in the rehabilitation process of children who are undergoing physical therapy.

We segment this chapter into two primary technologies that, once integrated, provide an integrated system for this domain. Section 8.2 provides an overview of the state-of-the-art in gaming and robotics. Section 8.3 discusses a VR system that can provide feedback and tracking of patient performance during game play. Section 8.4 details a pilot study that integrates robotics into the VR gaming scenario through social engagement, whereas Sect. 8.5 provides concluding remarks and a summary of next steps.

8.2 Related Work

Although there are very few research efforts focused on using integrated virtual reality-robotic systems for children, there has been growing interest in research involving therapeutic play between robots and children [2]. KASPAR [9], a child-sized robot for engaging children with autism, utilizes expressions and gestures to communicate with its human partner. Another robot designed to teach social interaction skills is CosmoBot [10], a commercially-available telerehabilitation robot that enables a therapist to record robot movements to enable the performance of repetitive and predictable motions, which adheres to a specified behavioral skill. In [11], researchers utilize a humanoid robotic doll, named Robota, to engage children with autism in imitation-based games. In a related domain, teleoperated robots have been shown to enable achievement of play-related tasks that go beyond the child's own manipulation capabilities. In [12], a teleoperated robot called PlayROB was developed to enable children with physical disabilities to play with LEGO bricks. The "Handy" robot in [13] was used to assist children with cerebral palsy in performing a variety of tasks such as eating and brushing teeth. In a pilot study, the authors showed how the robot could be used to enable drawing. Cook et al. [14] also showed the use of robot arms for assisting children in play related tasks.

With respect to virtual reality (VR) systems, alone, there have been a number of pilot studies that have focused on children in recent years. Reid [15] conducted a study to show the benefits of a VR system for children with cerebral palsy.

Her studies suggest that a virtual environment allows for increased play engagement and the opportunity for children to practice control over their movements. Bryanton et al. [16] showed that using VR systems to guide exercises may enhance exercise effectiveness. This work focused on the rehabilitation of lower-body motor skills (i.e. ankle dorsiflexion movements in chair-sitting and long-sitting). The results of these studies reported that children have better control of ankle dorsiflexion and show greater interest in doing the same exercise when presented to them through a VR system than as a stand-alone exercise. Golomb et al. [17] investigated whether an in-home remotely monitored VR videogame can help improve hand function and forearm bone health in an adolescent with hemiplegic cerebral palsy. In [18], researchers used a Wii console to augment the rehabilitation of an adolescent with cerebral palsy, whereas in [19], a motion-capturing product called the EyeToy was used to provide a relatively low-cost in-home virtual environment.

Although the feasibility of VR systems has shown to have positive outcomes in the children-rehabilitation domain, there is still a lack of automating feedback on patient performance through these systems. On the other hand, robotic devices, which has been shown to provide a concrete method for objectively recording and assessing the performance of a patient through repeatable and quantifiable metrics (position, trajectory, interaction force/impedance) [20], has not been well-integrated in the child-rehabilitation domain. As such, in this chapter we discuss technologies that enable the design of an integrated VR-robotic system for child-rehabilitation.

8.3 A Virtual Reality Game for Upper-Arm Rehabilitation

8.3.1 Introduction

Virtual reality (VR) environments play an important role in the rehabilitation field. Therapists and researchers have studied its importance in physical therapy interventions for people with different conditions such as stroke, Parkinson's disease, and cerebral palsy. Unfortunately, most of these VR systems do not integrate clinical assessment of outcome measures as an automated objective of the system. In addition, most of these systems do not allow real-time adjustment of the system characteristics that is necessary to individualize the intervention. Previous research has shown that VR environments present many benefits in the rehabilitation of individuals with motor skill disorders. Not only do they improve compliance for individuals working with their exercises [21], but they also enhance exercise effectiveness [16]. In this section, we discuss a VR system that integrates clinical assessment of outcome measures, as well as allows individualization of the rehabilitation protocol through real-time adjustment of game parameters. In the subsequent section, we then segway into discussion of the robotics platform, and show methods for robot integration into the gaming scenario.

8.3.2 Objective

While there have been a number of VR systems developed for use as part of physical therapy interventions for children with motor skill disorders, most do not incorporate a formal method of evaluating the subjects' upper-body motor skills in real-time and in the comfort of their own home. Taking into consideration the limitations of previously developed systems, the goal of our low-cost VR gaming system is to function as an in-home rehabilitation tool for individuals with any motor skills disorder that is user-friendly for both the user and the healthcare professional. The two key features of the system are: (1) the ability to individualize the rehabilitation protocol through adaptation of game settings, and (2) the ability to autonomously record and assess rehabilitation outcome measures for providing feedback to the therapist in real-time. Individualization is achieved through an adaptable user interface that allows the therapist to select desired game settings to match the rehabilitation objectives customized to the user's capabilities. In addition, unlike common entertainment systems, while users are engaged in repetitive movements during game play, the system is capable of analyzing the user's upper-body movements in real-time. Interaction with the game yields an assessment of outcome measures by quantifying the kinematic parameters that describe human movement. Some of these parameters include range of motion (ROM), movement smoothness, and deviation from path.

8.3.3 Description of Overall System

The VR system consists of a virtual game developed to work on any general-purpose computer system running a Windows 64-bit operating system, and a three-dimensional (3D) depth camera. The camera is used to track the user's upper-body movements and map them into the presented virtual environment. For our application, we utilize the Microsoft Kinect 3D camera [22] along with an open source SDK that provides the necessary functions for tracking upper-body human movement. Beyond the basic requirements of a computer system to run the game and a 3D depth camera, there is no need for additional equipment like gloves or helmets. In addition, the users are free to move their entire bodies without being restricted to traditional computer inputs (e.g. keyboard and mouse).

The developed VR application is called *Super Pop VR*[TM] [23]. When playing, the user is immersed in a virtual world where virtual bubbles (represented by colored circles, squares, and/or triangles) appear on the screen surrounding the user. The goal of the game is to pop as many bubbles as possible in a certain amount of time by moving a hand over the center of the bubble. The 3D depth camera is used to track the skeleton of the user in order to determine the coordinates of the hand joints. Two blue markers follow the user's hands in order to provide feedback to the user on the exact position of their hand as recognized by

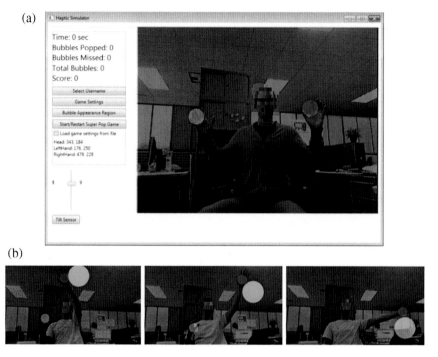

Fig. 8.1 **a** Main graphical user interface of the *Super Pop VR*™ game. **b** Example of a 90° trajectory created by the position of the three super bubbles

the system (Fig. 8.1a). The user is instructed to pop the yellow (good) bubbles and to avoid the red (bad) bubbles. Moreover, there is a set of green bubbles called Super Bubbles (SBs) that are worth double the points as the yellow bubbles. Based on the user's intervention protocol established by the therapist, a certain amount of time is specified in which all yellow and red bubbles on screen get erased and a set of two or three SBs appear on their own one by one. Each set of SBs highlights the trajectory that the therapist will use to evaluate the users rehabilitation outcome metrics. For example, three SBs may be placed such that a 90° motion is created forcing the user to follow this 90° trajectory (example shown in Fig. 8.1b). These sets of SBs are used to determine the point in time where the system captures and stores the user's upper-body joint coordinates. This information is used to evaluate the user's movements by calculating the relevant kinematic parameters. After playing the game for a given period of time, the therapist can analyze the results of the assessment in order to track the user's progress and to evaluate areas that may need improvement.

Figure 8.1a also shows the main graphical user interface (GUI) that the user sees once a game starts. Besides showing the virtual environment, the main interface also depicts the user's progress during game play. In addition, four main buttons are located at the left side of the GUI: 'Select Username', 'Game Settings',

'Bubble Appearance Region', and 'Start/Restart Super Pop Game'. When pressed, the first three buttons access secondary GUIs that provide the therapist options for customizing the intervention protocol of the game (Fig. 8.2).

The 'Select Username' GUI lets the therapist assign individual usernames or IDs in order to enable the system to be used by multiple users (Fig. 8.2a). The 'Game Settings' GUI offers the option to choose from three different game difficulties with hardcoded parameters (Easy, Normal, and Hard) as well as a Custom option (Fig. 8.2b). The Custom option enables the therapist to provide their own combination of game settings depending on the needs of the user. Finally, the 'Bubble Appearance Region' GUI shows a snapshot of the subject taken by the camera when the corresponding button is pressed. In this interface, the therapist can select the workable region in which regular bubbles will appear and the position of the SBs based on the placement of the subject from the shown snapshot. Figure 8.2c shows the workable region as a red rectangle and the SBs as green circles. Given that all users are of different heights and all have different arm reach, this interface allows for personalized sessions accommodating the different body structures of the users. The therapist can also select the SB display/appearance interval duration, the number of SBs used for the protocol, and identify the arm to be assessed.

The combination of options and features provided by the different interfaces provide the therapist the freedom to match the level of difficulty of the game to the user's capacity. For example, if the experimental protocol is designed to improve the user's maximum range of motion (ROM), the therapist would position three SBs such that they are spaced with a slightly greater angle than the user's effective ROM. This way, through practicing the specified repetitive motion that will appear throughout the game, the user will progressively increase his/her ROM given that he/she will need to reach the next SB.

Game sessions can also be individualized to the capabilities of the user by customizing the difficulty level. The difficulty of each game can be set by selecting different combinations of the following parameters: game duration in seconds, total number of levels, game speed in bubbles per second (rate at which the bubbles appear on screen), bad bubble ratio, bubble size, good bubble score, and bad bubble score. These parameters serve different purposes in the rehabilitation protocols. For example, the size of the bubbles is linked to the accuracy of the user. Intervention protocols designed for users with poor accuracy will include larger bubbles. Similarly, the speed of the bubbles is linked to how developed the symptoms of the users are. Users with more developed symptoms usually have slower movements, thus their intervention protocols will include games with a slower pace.

All the game levels have equally distributed durations determined by dividing the total game duration by the total amount of levels. At each passing level, the game increases its difficulty by: increasing the game speed by the selected value, increasing the bad bubble ratio by the selected value, and/or decreasing the bubble radius by the selected size value. The shape and the scores for the bubbles remain constant throughout the game. It's important to mention that all selected settings

(a)

(b)

(c)

Fig. 8.2 Secondary graphical user interfaces accessed by pressing the corresponding buttons on the main GUI: **a** 'Select Username', **b** 'Game Settings', and **c** 'Bubble Appearance Region'

Table 8.1 Kinematic parameteres used for assessinf user's upper body movements

Kinematic parameters	Definition	Method
Range of motion	The angle created by the corresponding joint	Dot product
Movement time	The total amount of time needed to move between the initial and final positions	Fitt's law [37]
Movement smoothness	Measures how jittery the user's movements are	Movement units [38]
Deviation from path	Measures how close/far the user's movements are from the defined path between the initial and final positions	Robot kinematics
Angular velocity	Measures how fast/slow the user moves the corresponding joint	Jacobian matrix
Movement speed	Measure's how fast/slow the user's movements are. The system measures the speed of the wrist	Jacobian matrix

are saved for future games and associated with the username/ID such that the therapist doesn't have to change the settings for each game and can later correlate the results and progress to the corresponding individual.

8.3.4 Description of Real-Time Kinematic Assessment

In addition to the customization feature, the game has the ability to assess the user's upper-body movements by analyzing the trajectory of the upper-body joints in time. This information is not only used to track the user's progress, but also to identify the parameters that the user may need improvement on.

The user's upper-body movements are mathematically described by certain kinematic parameters related to limb movements. The parameters of interest are: shoulder and elbow ROM, movement time, movement smoothness, deviation from path, shoulder and elbow angular velocity, and movement speed. All parameters are calculated using the user's joint coordinates that are captured and stored at each frame while he/she pops the SBs. The system starts capturing the relevant data when the user pops the first SB in a given sequence. Similarly, the system stops capturing the data after the user pops the last SB in the same sequence. These two points in time define the initial and final positions of the user's joints. Each SB sequence containing the user's relevant kinematic data is assessed, and the algorithm returns the result for each one.

The methods that are used for calculating the different parameters depend on the definition and their purpose. Table 8.1 shows a brief description of the parameters and the corresponding general method for making the calculations.

Through individualization and feedback, the resulting VR game is not only user friendly and provides motivation for users to practice their recommended exercises in their homes, but it also provides a kinematic algorithm that assesses the user's

Table 8.2 Selected *Super Pop VR*™ game settings for the preliminary experiments

Game duration	60 s
Total levels	3
Game speed	0.6 bubbles/second
Bad bubble ratio	10 %
Bubble size	3
Good bubble points	5 points
Bad bubble points	−5 points
Super bubble time interval	20 s
Super bubble duration	5 s

movements without interrupting the progress of the game. An example of its use in a pilot study with children is now discussed.

8.3.5 Pilot Study with Children

Preliminary experiments were conducted to show that the *Super Pop VR*™ game is enjoyable, encouraging, and user-friendly. Given that this work is primarily focused in the rehabilitation for children who have cerebral palsy, the selected demographic for these experiments were children. Seven children (mean age $m = 7.71$, standard deviation $\sigma = 1.48$, Male: 2, Female: 5) played the game and answered some questions regarding their experience. The participants were instructed to play the game for 60 s each. Keeping in mind that the purpose of these experiments was to show that the game is motivating and user-friendly, the game settings were selected such that the game was not too hard yet not too easy. Table 8.2 shows the overall selected game settings. The instructions given by the experimenter was strictly scripted to avoid any influence it might cause to the participant's experience. The script was as follows:

> Hello, today we're going to play a game with the Kinect camera. The purpose of the game is to pop as many bubbles as you can in one minute. To pop a bubble just hover one of your hands over it using the blue markers that are following your hands. You want to pop the yellow and green bubbles which are worth five and ten points respectively, but avoid the red bubbles because these will take away five points from your score. After you complete the game, I will ask you some questions about your experience with the game.

On a 5-point Likert scale, from disagree (1) to agree (5), post-experiment surveys report that children participants, in general, enjoyed playing the game. Table 8.3 shows the statements made in the survey. Moreover, Fig. 8.3 shows the averages and standard deviations of the participants' responses to these statements. It is important to recognize that only seven children participated in this preliminary study and the results were used primarily as feedback. The survey includes positive and negative statements about the game, which have the goal of identifying potential areas that could be improved for a better experience when playing the game.

Table 8.3 Survey stataments presented to the participants after the played the game

#	Statements
1	I could see all my movements from the screen very well
2	I found the objects in the game very interesting
3	The objects I saw in the game were very attractive
4	I could hear all music in the game very well
5	The music I heard out of the game was very attractive
6	I could not hear where all of the sounds out of the game came from
7	The movements to play the game were too hard
8	The movements used to touch objects in the game were fast, they were not too easy, but also were not too hard
9	I must still learn a lot before I can play the game well
10	I could predict what was going to happen after I had made a movement
11	I had the feeling I could accomplish the game
12	I would find it nice if I could play the game together with more friends at the same time
13	The game was so attractive that I lost all count of time
14	I would like to play the game more often
15	The game training is less fin than regular computer/video games
16	The request from the game was easy to understand
17	The request from the game was easy to follow
18	It was very logical playing the game by popping the objects
19	I found it hard to follow the game by moving my hands
20	I became more tired from playing with the game than from the regular computer/video games
21	I like playing the game

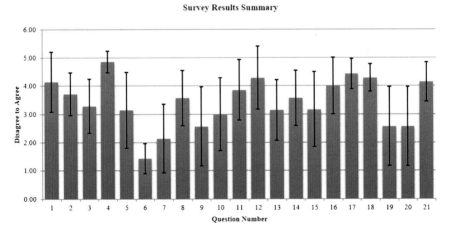

Fig. 8.3 Averages and standard deviation results from the participant's responses to the survey's statements

The most noted positive feedback was obtained from questions 1, 4, and 21. The participants felt that they could see their movements very well in the screen (mean response $m = 4.1$, standard deviation $\sigma = 1.1$), hear all the music in the game very well ($m = 4.9$, $\sigma = 0.4$), and liked playing the game overall ($m = 4.1$, $\sigma = 0.7$) respectively. Given that the mean values are relatively high and standard deviation values are relatively low, these results suggest that the participants not only enjoy the game, but also recognize that the game is functioning properly—at least in terms of tracking the user's movements and playing the sounds when the user pops the bubbles.

On the other hand, questions 5 and 12 pointed out some areas where we can improve the functionality and likability factor of the game. These questions revealed that the participants didn't find the music very attractive (mean response $m = 3.1$, standard deviation $\sigma = 1.3$), and would like the capability to play the game together with more friends at the same time ($m = 4.3$, standard deviation $\sigma = 1.1$) respectively. There were scattered responses for the statement concerning the attractiveness of the music played when popping the different bubbles. To deal with this variation in response, our current version of the game now provides the option of selecting any desired sound file from the user's hard drive, in addition to the already provided sound options from different known songs such as: 'Twinkle, Twinkle, Litle Star', 'Row Row your Boat', and 'Für Elise'. Regarding the multiplayer option, we're convinced that adding the capability for two or more people to play at the same time will increase the game's motivation factor. Moreover, the user will see better results as opposed to playing the game alone. Hidding et al. [24] reported that group therapy yields better results than individual therapy in improving thoracolumbar mobility and fitness. Based on these results, we hypothesize that playing the game with other people at the same will also yield better therapy results in terms of increasing movement speed and accuracy, and decreasing movement jitteriness.

In addition to being motivating and user-friendly, the *Super Pop VR*^{TM} game is also capable of outputting accurate outcome measures. A separate study showed that the system is able to correctly measure the user's shoulder ROM with an error of less than 5 % [23]. The system's ability to output accurate results, the system's ability to individualize the intervention protocols of the users, and the fact that the system is user-friendly and enjoyable, results in a system that can serve the therapy needs of individuals with upper-body motor impairments such as children who have cerebral palsy.

Given these positive outcomes, we now discuss approaches based on prior efforts in the social robotics domain to incorporate robots in the gaming scenario as a method to increase the engagement factor for long-term adherence.

8.4 Integration of Social Robotics in Gaming Scenarios

Socially assistive robotics, defined as robots that provide assistance to human users primarily through social interaction [25], continues to grow as a viable method for a multitude of assistive tasks ranging from robot-assisted therapy to eldercare. Through the use of social cues, socially assistive robotics can enable long-term relationships between the robot and the child that drastically increases the child's motivation to complete a task [26]. In addition, ample studies have shown that the effect of being perceived as a social interaction partner can be enhanced by a physical robotic embodiment [27]. These characteristics are ideally suited for providing motivation to a child interacting with a robot in a therapeutic gaming environment. Generally speaking, children are more attracted to a robot when the robot exhibits positive feedback [28, 29] and are more motivated when the robot uses appropriate behavioral techniques to reengage [30]. As such, we follow the theme of socially assistive robotics by utilizing a robotic system to engage the child during a gaming scenario through social interaction. In order to accomplish this goal, we examine two techniques: engagement through behavioral interaction and learning from gaming demonstration.

8.4.1 Engagement Through Behavioral Interaction

In most clinical settings, therapists are able to observe a child's engagement in real-time and employ strategies to reengage the student, which, in effect, improves attention, involvement and motivation in the rehabilitation protocol. In general, clinicians are able to engage by implementing behavioral cues such as direction of attention, facial expressions, proximity, and responsiveness to the child's activity. This behavioral engagement is deemed as a crucial component in home-based rehabilitation, especially given absence of the clinician in the child's home environment.

For the socially assistive robotic agent, we utilize the DARwIn-OP platform (Darwin) (Fig. 8.4) [31]. To enable interaction with the human, Darwin is programmed with a range of verbal and nonverbal behaviors. The nonverbal behaviors, or gestures, for the robotic agent included eye gaze, head nods/shakes, and body movements. Table 8.4 shows a sample of the nonverbal behaviors used in this investigation, and Fig. 8.5 shows three snapshots of the *head scratch* gesture. A total of eight gestural behaviors were programmed onto the humanoid platform. The verbal behaviors enable Darwin to provide socially supportive phrases for reengagement as the child interacts during a virtual scenario. During the utterance of verbal phrases, Darwin turns his gaze towards the child when speaking; otherwise, he remains looking at the virtual screen. The goal of the verbal phrases is to encourage the individual based on their current performance within the virtual scenario. It is very important that the phrases are socially supportive and convey the message that the child and Darwin are interacting together as a team. There is a dialogue established between the individual and Darwin, and not a unidirectional knowledge flow (i.e. Darwin is not giving instructions or issuing commands to the child). This open dialogue integrating socially supportive phrases between teacher and individual is ideal for optimal learning and engagement [28]. A sample of these socially supportive phrases is shown in Table 8.5.

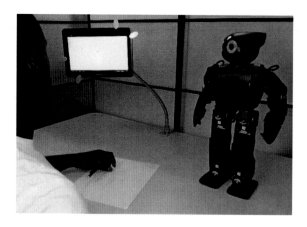

Fig. 8.4 The robot agent Darwin

Table 8.4 Sample of nonverbal behaviors from the robotic agent

Gesture	Behavioral meaning	Description of motion
Conversation	Body movements used to engage children while talking	Head nods and both arms move outward while maintaining eye contact
Head nod	Back-channel signal meaning continue; okay; yes	Head moves in an up and down motion
Head shake	Negative connotation; sad; no	Head moves from side to side while facing the ground
Tri-gaze	Eye contact distributed between three things	Eye contact the screen, child, then workstation for 3 s each
Head scratch	Confusion; lost	Arm/hand moves back and forth next to head
Fast arm	Positive connotation; approval; excitement	Arm is bent and raised next to head; arm then quickly moves downward
Hand wave	Hello; goodbye	Arm is bent and raised next to head; forearm moves back and forth
Eye contact	Attention is directed towards an object	Head (eyes) is aligned with a specified target

8.4.2 Pilot Study with Children

To evaluate the ability of the robotic agent to engage children during interaction with a virtual environment, we employed a between-groups design for this study [32]. To guarantee that the skills are evenly distributed between the robot engagement groups (None, Verbal, Nonverbal, Mixture), the children were selected at random. A total of 20 children between the ages of 15 and 16 years old (mean age $m = 15.5$, standard deviation $\sigma = 0.51$, Male: 12, Female: 8) took part in this experiment. Our experiment involved one factor type of behavioral interaction, with four levels. Each level is defined as follows:

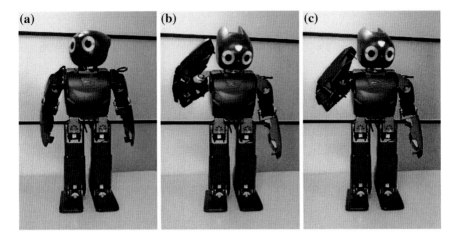

Fig. 8.5 The head scratch gesture broken down into three parts. **a** Initial position—Darwin is standing and has eye contact with the tablet-based test. **b** Darwin's *right* arm scratches his head. His head is down and eye contact is with the pencil and paper. **c** Darwin's arm stops moving, and his head moves up to make eye contact with the subject. He then returns to the initial position

Table 8.5 Sample of verbal responses from the robotic agent

Interaction	Speed	Phrase
Correct	Fast	"Fantastic!"
		"Awesome!"
		"You're really good at this"
	Slow	"This is hard, but we're doing great"
		"Thanks for all your hard work"
		"You're doing great! I had trouble with that one too"
Incorrect	Fast	"Hang in there. We're almost done"
		"Can you slow down a little so we can do it together?"
		"Please wait for me. You're leaving me behind"
	Slow	"This part is very challenging"
		"Don't sweat it. We'll get the next one"
		"Don't worry. I had trouble with that one too"
None	Inactive	"Are you still there?"
		"Don't forget about me over here"
		"Don't give up on me. Come on, let's keep going"

- **None**: Represents the control group. No agent is present.
- **Verbal**: The agent will say socially supportive phrases for reengagement as the child navigates through the virtual environment. He will gaze towards the child when speaking to him/her.
- **Nonverbal**: The agent will use only gestures for reengagement as the child navigates through the virtual environment.
- **Mixture of Both**: The agent will use both gestures and phrases for reengagement as the child navigates through the virtual environment.

Table 8.6 Statistical analysis of survey responses

Question	Verbal	Nonverbal	Both	No agent	p value
I was frequently bored	1.8	3.4	1.8	4.6	0.002*
I enjoyed the virtual environment	4.0	3.2	4.4	2.2	0.07
Darwin reacted appropriately	4.2	1.8	4.4	n/a	0.002*
Darwin distracted me	2.4	2.8	1.8	n/a	0.53

*Statistically significant

At the start of the virtual scenario, Darwin gives a verbal introduction along with gestures to introduce himself and the activity that the children are about to perform. As each child advances through the scenario, his or her progress is communicated to Darwin. Essentially, every action completed is sent to Darwin, as well as the time intervals taken to navigate through the scenario. In the cases where the child may take a long time to complete a task, it is necessary to interrupt this inactivity (eliminate idle time) and effectively increase engagement. Once the child's progress and speed are communicated to Darwin, he will respond appropriately based on the behavioral interaction type (verbal, nonverbal, or both).

Depending on the child's state, Darwin provides the children cues that are either verbal, nonverbal, or a combination of the two (depending on the experimental group). For both verbal and nonverbal behaviors, the behavior was selected at random based on the message sent to Darwin. For the engagement type that incorporates both verbal and nonverbal cues, the gestures and phrases were scripted and paired prior to Darwin's random selection. As such, we were able to expand Darwin's library of verbal and nonverbal cues by pairing the same phrase with multiple gestures. Although a phrase when it stands alone may mean one thing, by adding a gesture, the underline meaning of the message can be altered. Upon execution of a pair, both the gesture and the phrase are performed simultaneously. For example, if the message sent to Darwin states that the virtual-child interaction behavior was completed too slowly, he may say, "You're doing great! I had trouble with that one too," while nodding his head.

We look to validate the hypothesis that the use of a robot agent can increase the quality of interaction in a virtual environment by adaptively engaging with the child. Adaptive engagement is based on the concept that the engagement model is driven by identification of the child's behavioral state. To prove or disprove this hypothesis, we looked at the responses from an exit survey. After task completion, we asked them to rate their agreement with a series of statements on a 5-level Likert scale that ranged from 1 (Disagree) to 5 (Agree). Table 8.6 shows the average response to each question and the p-values from the ANOVA tests, which are separated by test groups.

By monitoring the child, Darwin was able to effectively maintain the child's attention, although there was a statistically significant variance in how appropriate the children deemed Darwin's reactions to be during the interaction. The nonverbal group thought Darwin's actions were not appropriate with a score of 1.8 (Slightly Disagree = 2; $\sigma = 0.84$), while the remaining groups had an average score of

4.3 (Slightly Agree = 4; $\sigma = 0.99$). The lack of understanding of Darwin's actions was interpreted as him not giving any feedback at all, which resulted in a more unpleasant virtual reality (VR) experience.

Because boredom is often associated with poorer engagement [33], it is important to note that there was a statistically significant variance in how bored the subject deemed him- or herself to be throughout the scenario. For both the verbal group and the group with a mixture of verbal and nonverbal cues, the average response to the question on boredom during the test was 1.8 (Slightly Disagree = 2; $\sigma = 1.07$). The nonverbal group followed with a score of 3.4 (Neutral = 3; $\sigma = 1.52$), while the group with no agent was the most bored with a score of 4.6 (Agree = 5; $\sigma = 0.55$). This shows that the verbal group and the group with both verbal and nonverbal cues were able to minimize boredom the best when compared to the other groups.

Interestingly enough, although two of the children stated that Darwin was a distraction, the survey question that asked if Darwin was a distraction showed otherwise across these groups. The average score across all groups with Darwin present was 2.3 (Slightly Disagree = 2; $\sigma = 1.35$). Overall, the children enjoyed interacting with the system when Darwin was present. The children in the group with both verbal and nonverbal cues enjoyed interacting with the virtual environment the most with a score of 4.4 (Slightly Agree = 4; $\sigma = 0.89$). The verbal group followed with a score of 4.0 (Slightly Agree = 4; $\sigma = 1.41$). Next, the nonverbal group followed with a score of 3.2 (Neutral = 3; $\sigma = 1.48$). However, when Darwin was not present, the children did not seem to enjoy the virtual environment as much with a score of 2.2 (Slightly Disagree = 2; $\sigma = 1.30$).

In conclusion, across all behavioral interaction types—verbal, nonverbal, and both—the children enjoyed Darwin's presence. A mixture of both cues and verbal cues only tend to have the least amount of boredom associated with it, which is ideal for a richer virtual environment and higher levels of engagement. On the contrary, the group having no robot agent present enjoyed the scenario the least and experienced the most boredom. Overall, the use of only nonverbal cues such as gestures shows no significant trends when compared to verbal cues; therefore, this works suggests that verbal behavioral cues is ideal for enhancing performance and increasing engagement in a virtual environment.

8.4.3 Learning from Gaming Demonstration

The role of robot learning for child-based engagement in a therapy scenario is to increase the duration of the child's interaction by incorporating the concept of turn-taking. Studies have shown that when children are required to teach others, they themselves become more engaged in the task [34]. In this work, we utilize a case-based learning approach in which a robotic platform observes the child's motions during game play, generates an appropriate behavior, and then engages with its child partner as a learner. This learning response is accomplished by utilizing a

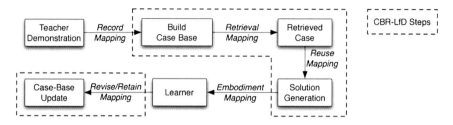

Fig. 8.6 Steps of case-based reasoning (CBR) incorporated within the overall structure of recording and encoding demonstrations, retrieving and reusing cases, and mapping a generated behavior to the robot's embodiment. CBR steps are depicted inside *dashed boxes*

mimicking process in which the child and robot take turns in accomplishing a goal, thereby motivating and stimulating the social behavior of the participant.

Our robot learns from the user by first observing the user, storing information about their situation-action responses (further defined as a case), and then retrieving these cases to execute a corresponding behavior. The child engages to teach the gaming task to the robot in a shared workspace and intuitively monitors the robot's behavior and progress in real time. In this setting, the teacher (child) is able to interrupt and correct the robot's behavior at the moment the learning is taking place, thus providing a means to continuously engage the child in the protocol of the game. We utilize a method called case-based reasoning to enable this collaborative teaching/learning process.

Case-based reasoning (CBR) is a human memory and cognition methodology that solves new problems based on the solutions of similar past problems [35]. By comparing the current task to some past task cases stored in memory, the best solution is retrieved and adapted to the current task. The first phase of CBR is acquiring knowledge, i.e., training the case base. During this phase, the system observes the game performed by the child and generates a case (problem–solution pair) for each demonstration, which is then saved in the case base. The problems are given as game states, such as game-object information and game score. The solution is extracted from the person's behavior towards a given problem. In the second phase when a new problem is introduced, the most similar past problem and its solution are retrieved from the case base. When measuring a similarity between two cases, the distance is computed as a sum of weighted distances between each problem feature. Our system provides the tool to autonomously train this similarity metric through pattern recognition. Next is the reuse step in which the retrieved solution is adapted to the current task. We use a method of averaging the solution over multiple retrieved solutions with a Gaussian distance kernel. As the case base expands and demonstration improves, deviation of the retrieved solutions decreases. During the last phase, the new problem–solution pair is revised and retained in the case base. The full algorithm is as depicted in Fig. 8.6 [36].

For task interaction, the retrieved case and its solution are used to reproduce the task behavior on a robotic platform through a mapping from the adapted solution to the robot's state and action space. This includes generating synthesized gestures

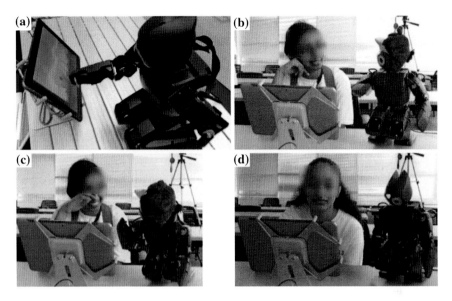

Fig. 8.7 Darwin's behavior is reproduced from the retrieved experiences. **a** Darwin interacting with a virtual environment through wireless communication, **b** making eye contact and providing feedback after the participant's demonstration, **c** encouraging the participant, and **d** expressing sadness after an unsuccessful attempt

that triggers actions within the virtual environment. Darwin also generates a combination of speech and gesture primitives that, as discussed in Sect. 8.4.1, enable engagement through behavioral interaction (Fig. 8.7).

8.4.4 Pilot Study with Children

For validation, we engaged 33 participants (mean age $m = 18.27$, standard deviation $\sigma = 8.56$) including 19 children ($m = 12.26$, $\sigma = 4.24$), some with special needs, to teach a virtual game shown in Fig. 8.8 to our robot, Darwin. We analyzed data collected during various trials during a two-month period.

For the experiment, participants were asked to teach the robot how to interact with a virtual game. The instruction given by the experimenter was strictly scripted to avoid any influence it might cause to the participant's experience. The script was as follows:

> Now, I'd like you to teach Darwin to play the same game. Just teach him in the same manner you would teach your younger sibling. Provide Darwin with demonstrations how to solve each level. Whenever you reach out to provide demonstration to Darwin, he will wait for his turn. Continue teaching each level until you are satisfied that Darwin had learned the level well enough, or think Darwin had stopped learning. Later, I want you to show me what you have taught Darwin, and collaboratively solve each level with him.

Fig. 8.8 **a** The game used in the experiment, and **b** shots of experiment conducted in an open-house styled setting with a group of local school children

> Darwin may or may not try to communicate with you, and he may not use human language. Afterwards, I will ask you some questions about your experience teaching a task to Darwin.

The growth progress of the case base and any interaction with the tablet was logged, and two video cameras were placed to record the whole evaluation session. Later, the log was used to evaluate the system, and the videos were analyzed for interaction studies.

First, the learning performance of the robot is determined. In Table 8.7, the performance of generated solutions is compared with varying k (number of retrieved cases). Distances are computed between a newly introduced problem and problems in the case base using the robot's retrieval method. Then the performance of each retrieved and adapted solution is evaluated using a logarithm of the earned game score. As computed, the average number of demonstrations given to the robot is: $m = 29.17$, $\sigma = 10.25$. It was also observed that the participants utilized other forms of natural interactions though the robot only could learn from

Table 8.7 Performance of case-retrieval method using k-nearest neighbors measured by the resulting game score when the generated solution was applied

k	log(score)
1	4.14 ± 2.23
2	4.02 ± 2.02
3	4.13 ± 1.72
4	3.96 ± 1.46
5	3.11 ± 1.87
6	2.79 ± 0.92

Mean performance [log(score)] of case-retrieval methods

Table 8.8 (a) Average time and ratio of social-interaction occurrences with and without the robot. (b) Detailed social cues exhibited towards the robot

	(a)		(b)	
	Without robot	With robot	Social interaction	Percentage
Average total time of interaction	342 s	1,445 s	Eye contact/gaze	22.72
Average time of social interaction	11 s	503 s	Gestural interaction	14.20
Percentage of social interaction	3.22 %	34.81 %	Vocal interaction	28.66
			Instructive	36.50
			Non-instructive	63.50

actual demonstrations of game play. These natural forms of interaction were measured as the length of time when an eye contact was made or when vocal/gestural-interaction behaviors were observed. These interactions were then categorized into instructive and non-instructive interactions in Table 8.8a. On average, participants, spent 5 min and 42 s without the robot and 24 min and 5 s with the robot playing the game. The more significant measurement is the ratio of how much social interactions were initiated during these sessions. When the robot wasn't present, these interactions were observed as forms of utterances or calling out to other people. Compared to 3.22 % social-behavior occurrence without the robot, participants dedicated 34.81 % of their time exhibiting social cues when the robot was present. Detailed break down of the social interactions toward the robot is depicted in Table 8.8b. Note that these cues are often observed simultaneously with one another, and the measurement ratio is calculated against the total time of the interaction. On a 5-point Likert scale, from strongly disagree (1) to strongly agree (5), post-experiment survey reports that the participants felt their robot was socially interacting with them ($m = 4.7$); was socially communicating with them ($m = 3.72$); thought Darwin was learning from them ($m = 4.33$) similar to their friends ($m = 4.01$); and thought the robot enhanced their overall experience with the virtual game ($m = 4.8$) (Fig. 8.9).

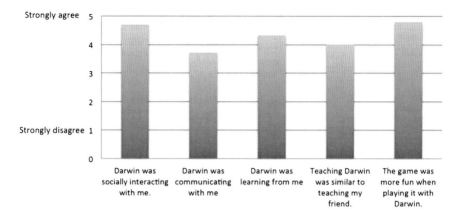

Fig. 8.9 On a five-point Likert scale, from strongly disagree (*1*) to strongly agree (*5*), post-experiment survey reports that the participants felt their robot was socially interacting with them and enhanced their overall experience with the task

8.5 Discussion and Future Work

Although there are very few research efforts focused on using integrated VR-robotic systems for children, the work presented in this chapter discusses various approaches and preliminary results showing the use of these systems in therapeutic play. There are many compelling reasons for utilizing robots in virtual reality (VR) gaming scenarios, ranging from augmenting the capabilities of children with motor impairments to increasing engagement of such children. Although much of the presented work is encouraging, we still need more quantitative results to validate the benefits of utilizing VR-robotic systems in pediatric therapy settings. These quantitative results should show the clear benefits achieved from children with disabilities interacting with the coupled virtual-robotic system. Additional substantial quantitative evidence, as well as longitudinal studies that demonstrate the effectiveness of the system, is still necessary.

The overall research presented herein brings up several interesting observations regarding the use of VR-robotic systems in pediatric therapy. Many prior papers in the domain of assistive technologies for children with disabilities discuss the difficulty of performing studies involving children. Common reasons included distraction from outside stimuli and the wide variances found in children's abilities. Another observation is the emphasis that many other researchers placed on robustness and iteration in design. For example, in many prior studies, perhaps due to the novelty of the robot, children would interact with the robots in unexpected ways. Although we emphasize individualization with respect to our system, these identified issues still need to be considered in improving the design of the system discussed in this chapter.

The pilot studies with children, as discussed in this chapter, have provided us sufficient baseline evidence to understand both the limitations of the system, as well as those attributes that are essential for establishing long-term adherence to a rehabilitation protocol. Future efforts will focus on enhancing the autonomy of the system such that the virtual system adapts in direct correlation to adaptation of the robot's social behaviors. This will ensure that both components correlate and grow with the capabilities of the child, as well as ensure the system is continuously engaging. Also, since our focused demographic is children with disabilities, our next set of trials will focus on engaging children with cerebral palsy in the experimental protocol.

Acknowledgments This research was supported by the National Science Foundation Graduate Research Fellowship under Grant No. DGE-1148903 and under Grant No. 1208287. Any opinions, findings, and conclusions or recommendations expressed in this material are those of the authors and do not necessarily reflect the views of the National Science Foundation.

References

1. Bax, M., Goldstein, M., Rosenbaum, P., Leviton, A., Paneth, N.: Proposed definition and classification of cerebral palsy. Dev. Med. Child Neurol. **47**(8), 571–576 (2005)
2. Howard, A.: Robots learn to play: Robots emerging role in pediatric therapy. In: 26th International Florida Artificial Intelligence Research Society Conference, May 2013
3. Piaget, J.: Play, Dreams and Imitation in Childhood. Routledge and Kegan Paul Ltd., London (1951)
4. Chen, Y., Kang, L., Chuang, T., Doong, J., Lee, S., Sai, M.T., Jeng, S., Sung, W.: Use of virtual reality to improve upper-extremity control in children with cerebral palsy: a single-subject design. Phys. Ther.: J. Am. Phys. Ther. Assoc. **87**(11), 1440–1454 (2007)
5. Colombo, R., Pisano, F., Mazzone, A., Delconte, C., Micera, S., Carrozza, M., Dario, P., Minuco, G.: Design strategies to improve patient motivation during robot-aided rehabilitation. J. NeuroEng. Rehabil. **4**(3), 1–12 (2006)
6. Cioi, D., Kale, A., Burdea, G., Engsberg, J., Janes, W., Ross, S.: Ankle control and strength training for children with cerebral palsy using the Rutgers Ankle CP: A case study. In: IEEE International Conference Rehabilitation Robot (2011)
7. Brütsch, K., Schuler, T., Koenig, A., Zimmerli, L., Koeneke, S.M., Lünenburger, L., Riener, R., Jäncke, L., Meyer-Heim, A.: Influence of virtual reality soccer game on walking performance in robotic assisted gait training for children. J. Neuroeng. Rehabil. **7**, 15 (2010)
8. Loureiro, R., Amirabdollahian, F., Topping, M., Driessen, B., Harwin, W.: Upper limb robot mediated stroke therapy gentle/s approach. Auton. Robots **15**, 35–51 (2003). doi:10.1023/A:1024436732030
9. Dautenhahn, K., et al.: KASPAR—a minimally expressive humanoid robot for human-robot interaction research. Appl. Bionics Biomech. **6**(3), 369–397 (2009)
10. Brisben, A., Safos, C., Lockerd, A., Vice, J., Lathan, C.: The cosmobot system: Evaluating its usability in therapy sessions with children diagnosed with cerebral palsy. Retrieved on 3, No. 5 (13) (2005). http://web.mit.edu/zoz/Public/AnthroTronix-ROMAN2005.pdf
11. Dautenhahn, K., Billard, A.: Games children with autism can play with robota, a humanoid robotic doll. In: Cambridge Workshop on Universal Access and Assistive Technology, pp. 179–190 (2002)
12. Kronreif, G., Prazak, B., Mina, S., Kornfeld, M., Meindl, M., Furst, F.: Playrob—robot-assisted playing for children with severe physical disabilities. In: IEEE 9th International Conference on Rehabilitation Robotics (2005)

13. Topping, M.: An overview of the development of handy 1, a rehabilitation robot to assist the severely disabled. J. Intell. Rob. Syst. **34**(3), 253–263 (2002)
14. Cook, A.M., Meng, M.Q., Gu, J.J., Howery, K.: Development of a robotic device for facilitating learning by children who have severe disabilities. Neural Syst. Rehabil. Eng. **10**(3), 178–187 (2002)
15. Reid, D.: Benefits of a virtual play rehabilitation environment for children with cerebral palsy on perceptions of self-efficacy: a pilot study. Pediatr. Rehabil. **5**(3), 141–148 (2002)
16. Bryanton, C., Bossé, J., Brien, M., McLean, J., Mccormick, A., Sveistrup, H.: Feasibility, motivation, and selective motor control: virtual reality compared to conventional home exercise in children with cerebral palsy. CyberPsycology Behav. **9**(2), 123–128 (2006)
17. Golomb, M., McDonald, B., Warden, S., Yonkman, J., Saykin, A., Shirley, B., Huber, M., Rabin, B., AbdelBaky, M., Nwosu, M., Barkat-Masth, M., Burdea, G.: In-Home virtual reality videogame telerehabilitation in adolescents with hemiplegic cerebral palsy. Arch. Phys. Med. Rehabil. **91**(1), 1–8 (2010)
18. Deutsch, J., Borbely, M., Filler, J., Huhn, K., Guarrera-Bowlby, P.: Use of a low-cost, commercially available gaming console (Wii) for rehabilitation of an adolescent with cerebral palsy. Phys. Ther.: J. Am. Phys. Ther. Assoc. **88**(10), 1196–1207 (2008)
19. Jannink, M., Van Der Wilden, G., Navis, D., Visser, G., Gussinklo, J., Ijzerman, M.: A low-cost video game applied for training of upper extremity function in children with cerebral palsy: a pilot study. CyberPsychology Behav. **11**(1), 27–32 (2008)
20. Brooks, D., Howard, A.: Quantifying upper-arm rehabilitation metrics for children through interaction with a humanoid robot. Appl. Bionics Biomech. **9**(2), 157–172 (2012)
21. Reid, D.: The influence of virtual reality on playfulness in children with cerebral palsy: a pilot study. Occup. Ther. Int. **11**(3), 131–144 (2004)
22. Zhang, Z.: Microsoft kinect sensor and its effect. IEEE Multimedia **19**(2), 4–10 (2012)
23. García-Vergara, S., Chen, Y.P., Howard, A.: Super pop VR^{TM}: An adaptable virtual reality game for upper-body rehabilitation. In: 5th International Conference on Virtual, Augmented, and Mixed Reality held as part of HCI International, vol. 8022, pp. 44–49 (2013)
24. Hidding, A., van der Linden, S., Boers, M., Gielen, X., de Witte, L., Kester, A., Dijkmans, B., Boolenburgh, D.: Is group physical therapy superior to individualized therapy in ankylosing spondylitis? A randomized controlled trial. Arthritis Rheum. Am. Coll. Rheumatol. **6**(3), 117–125 (1993)
25. Feil-Seifer, D., Matarić, M.J.: Defining socially assistive robotics. In: IEEE International Conference on Rehabilitation Robotics (ICORR-05), Chicago, IL, pp. 465–468 (2005)
26. Kidd, C., Breazeal, C.: Robots at home: Understanding long-term human-robot interaction. In: IROS, pp. 3230–3235 (2008)
27. Powers, A., Kiesler, S., Fussell, S., Torre, C.: Comparing a computer agent with a humanoid robot. In: HRI, pp. 145–152, March 2007
28. Saerbeck, M., Schut, T., Bartneck, C., Janse, M.: Expressive robots in education: Varying the degree of social supportive behavior of a robotic tutor. In: CHI, pp. 1613–1622 (2010)
29. Park, E., Kim, K., Pobil, A.: The effects of a robot instructor's positive vs. negative feedbacks on attraction and acceptance towards the robot in classroom. In: ICSR, pp. 135–141 (2011)
30. Szafir, D., Mutlu, B.: Pay attention! Designing adaptive agents that monitor and improve user engagement. In: CHI, May 2012
31. Ha, I., Tamura, Y., Asama, H., Han, J., Hong, D.: Development of open humanoid platform darwin-op. In: SICE, pp. 2178–2181 (2011)
32. Brown, L., Howard, A.: Engaging children in math education using a socially interactive humanoid robot. In: IEEE-RAS International Conference on Humanoid Robots, Atlanta, GA, 2013
33. Baker, R., D'Mello, S., Mercedes, M., Rodrigo, T., Graesser, A.: Better to be frustrated than bored: the incidence, persistence, and impact of learners' cognitive–affective states during interactions with three different computer-based learning environments. Int. J. Hum Comput. Stud. **68**, 223–241 (2010)

34. Gartner, A.: Children Teach Children: Learning by Teaching. Harper & Row, New York (1971)
35. Kolodner, J.L.: An introduction to case-based reasoning. Artif. Intell. Rev. **6**(1), 3–34 (1992)
36. Park, H.W., Howard, A.: Case-based reasoning for planning turn-taking strategy with a therapeutic robot playmate. In: IEEE International Conference on Biomedical Robotics and Biomechatronics, Japan, pp. 40–45, Sept 2010
37. Mackenzie, I.S.: Movement time prediction in human-computer interfaces: a brief tour on Fitt's law. Proc. Graph. Interface **92**, 140–150 (1992)
38. Hofsten, C.: Structuring of early reaching movements: a longitudinal study. J. Mot. Behav. **23**(4), 280–292 (1991)

Part II
Technologies for Music Therapy and Expression

Chapter 9
Instruments for Everyone: Designing New Means of Musical Expression for Disabled Creators

Rolf Gehlhaar, Paulo Maria Rodrigues, Luis Miguel Girão and Rui Penha

Abstract The purpose of this project was the development of tools that facilitate the musical expression of a well-defined group of physically and mentally challenged people. Knowing those people and becoming familiar with their capabilities, tastes and ambitions was the departing point for the design of solutions. Once developed, they remained in their institutions, to become a part of their lives, contributing to the range of occupational activities they can be involved with. The composition of a piece of music that uses those resources and its performance by these people in the main concert hall of Casa da Música in April 2010 was more than a proof of concept, it was a truly engaging, motivating experience and, indeed, a proper artistic challenge.

Keywords Disability · Bespoke musical instruments · Assistive technology · Computer-aided performance · Modern performance practice

R. Gehlhaar (✉)
School of Art and Design, Coventry University, Coventry, UK
e-mail: rolf.gehlhaar@gmail.com

P. M. Rodrigues · R. Penha
Department of Communication and Art, University of Aveiro, Aveiro, Portugal
e-mail: pmrodrigues@ua.pt

R. Penha
e-mail: rui@ruipenha.pt

L. M. Girão
Artshare, Aveiro, Portugal
e-mail: luis.miguel.girao@artshare.pt

L. M. Girão
Planetary Collegium, Plymouth, UK

L. M. Girão
CESEM-FCSH-UNL, Lisbon, Portugal

R. Penha
INET-MD, Aveiro, Portugal

9.1 Introduction

Active/participatory musical activities, such as music-making or creating/composing music, are important aspects in the development of an individual's fertile relationship with music and of self-awareness in the self's relationship to others. These activities contribute to a general understanding and appreciation of music but, above all, they allow self-expression and communication within "languages" whose rules can be rewritten at any given moment and with tools and means that can be tailored to the needs and capacities of individuals and groups.

However, the disabled encounter many obstacles in their quest for self-expression through music. Most musical instruments are difficult to use: they are the result of hundreds of years of an evolutionary process that has favored able-bodied skilled performers. Consequently, the development of musical activities with people that have restrictions in their physical or mental capacities can be severely hampered by the lack of instruments that can produce proper musical results without needing to be mastered by complex bodily and mentally processes. The same applies to the processes that underlie the expression and structure of "musical thoughts", either through improvisation or composition—either dependent on notation of music or not. Not many composers or musicians have developed "musical languages" that are accessible to non-musicians to perform and yet make musical sense and offer a proper challenging, aesthetically rewarding experience for the listener and the player. The purpose of the I4E project was the development of tools that allow musical expression by a well defined group of physically and mentally challenged people. Knowing those people,—their personal characteristics, capabilities, tastes, ambitions—was the departing point for the design of solutions that, once developed, remained in their institutions to become a part of their lives to contribute to the range of occupational activities they could be involved in. After the development and prototype testing stage, one of the authors composed an 'orchestral' piece that uses those resources. Its performance by the participants in the main concert hall of Casa da Música, Porto, in April 2010 was more than a proof of concept, it was a truly engaging, motivating experience and, indeed, a proper artistic challenge.

There are two sets of reasons why we believe this project is important: one has to do with the people for whom the instruments were created, the general cause of empowering people with disabilities, giving them opportunities and access to activities that we believe contribute to their happiness, dignity, well-being and human development. Making and creating music is a source of immense joy. Disabled people have the need and right of access to this experience. Although this project might have reached just a few people in a universe of many, we believe that we have established a model and are attributing the project significance and projection that will raise consciousness and yield further results.

The second set of reasons is related to the musical technology community: the history of music has been a permanent search for new tools of expression, both at the level of instruments and ways of structuring the musical discourse.

We believe that working under constraints and requirements such as the ones that are imposed by disability will lead to new paradigms. The history of inventions has shown several times that ideas developed for disabled people became essential tools in the everyday life of the able bodied. In the 21st Century we are witnessing significant progress in the development of new interfaces for musical expression. Perhaps working with people with disabilities can catalyze the emergence of new ideas for musicians and artists in general.

9.2 The Evolution of Robotic and Technology Assisted Musical Instruments

The earliest recorded mechanized musical instrument is the 'hydraulis', a set of organ pipes invented by Ctesibius of Alexandria in the 3rd Century BC. The pipes are semi-submerged in a vat of flowing water, the moving water induces a drop in pressure inside the pipe, and, when a pipe valve is opened via a keyboard, air flows through the pipe. Usually, the water was kept flowing by one or two persons operating a pump [1, 2].

However, in later instruments such as barrel organs (ca. 1502) [3] and the player piano (the earliest, Claude Seytre's French patent of 1842) a player was no longer required; only someone who would provide the power, usually air from a bellows system. Later developments also allowed the 'operator' of these instruments some expressive control, such as over loudness and small fluctuations of tempo [4].

As the technology of electronics developed at the beginning of the 20th Century, electronic instruments began to appear. One of the earliest was the theremin, originally known as the etherphone, thereminophone or termenvox/thereminvox. It is an electronic musical instrument controlled without physical contact from the player, named after its Russian inventor Prof. Leon Theremin, who patented the device in 1928. Originally the product of Russian government-sponsored research into proximity sensors, the theremin fascinated audiences and came to be used in numerous composition and movie soundtracks. However, it was not a commercial success [5]. It is still in use today; the most current version of it being offered in kit form by Moog Music [6].

With the advent of the computer—and the microprocessor—robotic musical instruments rapidly became an art form. Generally, however, they do not allow for expressive control by player. They usually operate automatically and independently, often within an art installation context. One of the earliest exponents of these kinds of 'musical robots' was Trimpin, whose early installations included a six-story high microtonal xylophone running up a spiral staircase in a house, with computer-driven melodies running up and down it [7]. Eric Singer founded in 2000 the LEMUR (League of Electronic Musical Urban Robots) [8], responsible for a number of notorious robotic instruments, including the Orchestrion (for Pat

Metheny), a robotic version of George Antheil's 'Ballet Mécanique', and the Gamelatron, a robotic Gamelan. Another group of artists—EMMI—is presently gaining considerable notoriety building and composing music for robotic instruments [9]. These instruments require no or little previously acquired musical skill of the 'player' and are thus generally considered to be autonomous instruments. Instruments which allow significant expressive input from a user are often entirely software instruments and have become quite commonplace. There are, of course, also software music systems designed for educational purposes, such as Expresseur, which promises to allow the user to *experience the pleasure of playing music, without having to worry about the complex problem of notes* [10]. Although most instruments used in music therapy are traditional idiophones or plucked strings of many varieties, technologically assisted toys, instruments and interfaces specifically designed to be used in musical or educational therapy are also commercially available. The "quest" for new interfaces for musical expression is a very actual trend, as seen by the vitality of a conference such as NIME (New Interfaces for Musical Expression) [11]. Only a few examples can be referenced here: Soundbeam, an ultrasonic sensor that can trigger sounds [12], the soothing and stimulating environments by Snoezelen [13], numerous developments by Anthony Brooks, a pioneer in the development of alternative sensor-based non-invasive perceptual controllers for creative expression [14], numerous significant projects by Tod Machover who has worked in many specifically disabled-people oriented applications of his instruments [15] and the recently developed SKOOG, a soft, squeezable object that plugs into a computer's USB port [16].

Wanderley [17] is a key figure in the field of gestural control of sound synthesis and new musical instrument design, and one of the instruments used in our final performance was the T-Stick [18, 19], developed by Joseph Malloch under Wanderley's supervision.

The authors' interests lie primarily in developing musical assistive systems for persons with a disability. The aim is not therapeutic; it is to make the pleasure of playing music, particularly within a social context, accessible to such persons. The authors defined five basic requirements for the designs:

1. to produce high quality sound;
2. to allow for realistic modifications of the performance parameters;
3. to be learnable—at least the basics—in a relatively short time, even for non-musicians;
4. to promote communal musical interaction;
5. to be more or less portable.

In order for something to be considered as an 'instrument' of musical expression it naturally must require some physical and musical training and acquired skill. Instruments with assistive technology may allow the disabled user to overcome many common limitations. They may be mechanically assisted, such as the SuperString and the mechanized Gamelan instruments developed by Miguel Ferraz, José Luís Azevedo and one of the authors (Penha) at Casa da Música [20], or consisting entirely of software with appropriate human–computer interfaces, such as HeadSpace [21], or the computer-vision based Sound = Space, or Instrument A, as developed for this project.

9.3 A Description of the Process

In this section we describe the context in which the project was developed and provide detailed information about the nature of the specific instruments that were developed, as well as the workshops in which the techniques of their playing was communicated to the 'clients'. We thank Joana Almeida, Teresa Coelho and Anabela Leite of Casa da Música, the administrators and producers of this project, and the members of Factor E—Ana Paula Almeida, Joaquim Branco, Filipe Lopes, Paulo Neto, Nuno Peixoto, Jorge Prendas, Jorge Queijo, without whose dedication and effort it would not have been possible.

9.3.1 Context

The project Instruments for Everyone (I4E) was commissioned by the Education Service of Casa da Música, in 2008. The Casa da Música (CdM) is a leading cultural Portuguese institution with an inventive and wide cultural remit that promotes the national and international musical scenario. Education, inscribed into the philosophical roots of the project, is regarded as an essential part of its mission. The Education Service offers a broad range of activities for a broad range of participants from the public. From 2006 to 2010 one of the authors (Rodrigues) was coordinator of the Education Service. A major emphasis was placed on the development of new ideas that promote accessibility to music making and musical creation. As a part of this effort, a core team of ten young musicians/educators was assembled to develop, test and implement educational activities specifically designed to meet the requirements of the Education Service at Casa da Música and to guarantee a substantial part of the regular educational activities. The work of this group, Factor E, had a laboratory character that arose from a series of creative, exploratory residencies, in which the technological aspect was frequently a strong element. The group would also work under the leadership of international artists and educators, innovating and reflecting upon new ideas that could inspire and be used by other educational agents. This fertile environment allowed the development of several pioneering projects. The communication of these experiences to the artistic, scientific and academic community was considered of major importance [22, 23].

A significant effort was devoted to promoting musical activities for people with special needs. Regular activities (such as workshops, concerts or creative projects) as well as an annual festival, *Ao Alcance de Todos* (Within Everyone's Reach), offered specially prepared events for and with groups of people with special needs.

Some of these activities were musical or cross-arts performances that resulted from participatory processes developed over periods of several months [24]. Other activities within the festival *Ao Alcance de Todos* included seminars, conferences and workshop-like experiences, commissioning artists to develop ideas aimed at musical expression and participation of people with physical and/or mental disabilities [22]. The I4E project was conceived as a 2-year project for very specific people from institutions that responded positively to a challenge issued by the Education Service, in relation to the general *Ao Alcance de Todos* in order to strengthen the implications of the project. The first year of the I4E project would be devoted to the design and development of new instruments for specific disabled people. Some of these designs were to be shown and discussed in the 2009 edition of the festival. The second year would be devoted to the implementation of musical practice with the instruments in the institutions. The composition of a piece of music and its performance by the people for which the instruments were developed was regarded as an important aspect of the project and programmed for the 2010 edition of *Ao Alcance de Todos*. This coincided with the hosting of a RESEO [25] conference devoted to opera, dance and disability; workshops using the newly developed instruments were also part of the program.

The participants in this project were recruited from eight different institutions for the disabled located in the Porto and Coimbra area. Some of these institutions deal with people with a specific type of disability (for example, cerebral palsy or Down's syndrome) whereas others support people with broad-spectrum disabilities. The potential clients were identified by two methods: teacher recommendation and a questionnaire designed by the team. The purpose of this questionnaire was to identify the specific skills, disabilities and musical interest of the clients. It contained questions such as "What kind of an instrument would you like to be able to play? What kind of music would you like to play? Would you like to be part of a group of musicians playing together? Would you like to perform in a concert? Are you willing to commit yourself for a longer period? Are you willing to rehearse to develop greater musical skills?". Our musical tradition implies this close, long-term relationship with one's instrument, so these last two questions were of utmost importance to identify the clients that would regard this as an opportunity to develop skills that would last after a first tentative approach to deal with the instrument.

After evaluating the returned questionnaires the authors were able to estimate the scope of the project and identify and meet the individuals that wanted to participate. The following list shows an overview of the results and allows a quick insight into the main groupings of skills and disabilities identified in the clients as well as an identification of the mode of use of instruments and interfaces that were to be designed:

1. Special Needs Typologies

    (a) Physical difficulties

        (1) Head movement as sole input for computer interface
        (2) Semi-controlled movements of arm, excluding hands and fingers

(3) Semi-controlled movements of fingers
(4) Impaired/lack of vision

(b) Mental difficulties

(1) Mild learning/attention difficulties
(2) Severe learning/attention difficulties

2. Instrument Typologies

(a) Physical instruments
(b) Physical instruments with mechanical assistance via sensors
(c) Physical instruments with a programmable robotic element, played via sensors
(d) Digital interfaces requiring only simple physical manipulation
(e) Digital hands-off interfaces

3. Application Typologies

(a) For individuals but also applicable also in a communal context
(b) For several players simultaneously (communal).

9.3.2 Designing the Instruments

On the basis of the above information and classification, the authors developed the specific design considerations for the project. Initially the authors proposed 8 different designs. During the prototype production phase of the project, due to temporal and financial constraints, this number was reduced to four; however, several instances of each design were produced, so that some institutions would receive them all. These instruments were as follows:

9.3.2.1 Instrument A

Instrument A is a computer loop-based sequencer for composing and performing developed by one of the authors (Penha). By the incorporation of automatic 'rotating' menus, all actions that are normally available to the user of a conventional sequencer may be accessed by a single action such as a simple push-button switch, a head switch, blow switch, etc.—see Fig. 9.1 (large soft push-switch) and Fig. 9.2 (foot switch—blue pad on table leg). Specifically, the program is divided into two parts: the graphic interface (constructed in Processing [26]) and the audio driver (constructed in Max/MSP [27]).

The GUI (graphic user interface) of Instrument A evolved from previous experiences with circular-based sequencers, as the open-source software Políssonos by one of authors (Penha) [28, 29], shown in Fig. 9.3.

Fig. 9.1 Instrument A with
push switch

Fig. 9.2 Instrument A with
foot switch

In this sequencer, which uses circle subdivisions as a visualization of the inner rhythms of layered loops, the user can choose sounds, compose loop-based musical phrases with them and layer them in sequences to form a song.

When Instrument A is launched, a sequence of circular rotating menus appears in the GUI, which enables the user to start a new composition—see Figs. 9.4 and 9.5.

This sequence of menus allows the user to choose a (sampled) rhythmic loop, which will then serve both as the basis for the new composition as well as define the tempo that will synchronize all the other loops.

As the pointer of the 'rotating' menu arrives at the user's choice, he/she can make a selection by clicking (with the large button switch, the head- or blow-switch, etc.) The speed of rotation is adjustable to the reaction time of the user. The first click chooses the action, the blue border of the menu changes color to yellow and a second click confirms the choice. While it is possible to switch off, this 'double-click' feature, our initial experiences with the users showed that the

Fig. 9.3 GUI of políssonos

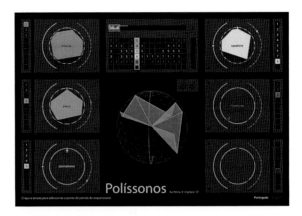

Fig. 9.4 An initial aspect of instrument A

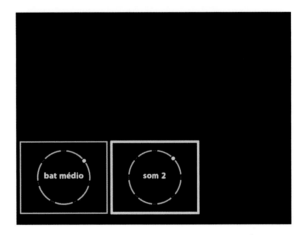

number of involuntary selections justified such a strategy to avoid unwanted actions. Figure 9.6 shows two users of Instrument A, playing as a duet, performing their compositions in an improvised process of construction and deconstruction. The player on the right is able to click a mouse pad, requiring no special adaptor; the player on the right is using a long-levered head switch.

Further rotating submenus allow the user to choose a sound—its pitch, its position within the loop and its loudness—that will be synchronized to this quantized loop. Figure 9.5 shows three completed horizontal systems, 'voices' of the composition. At the left is the chosen sampled loop, to its right are five sound samples (indicated by the polygons) that accompany it, which can be chosen from several samples organized in the rotating sub-menus by timbral characteristics. Each vertex of a polygon indicates a temporal point of the loop where the sample will be triggered. All of these polygons have been constructed by making choices in the sub-menus mentioned above. When performing the composition, the player can, in real-time, alternate at will between the 'polygons of sounds' of the voices

Fig. 9.5 A further aspect of instrument A

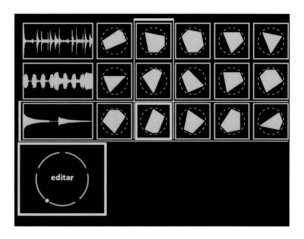

and between the three voices themselves. This important characteristic allows the composers using Instrument A to become performers, constructing or deconstructing their song in real time.

Two client groups—four users in total—were provided with a complete system consisting of a laptop computer, a small PA system and the specific switch suited to their need.

9.3.2.2 Sound = Space

Sound = Space consists on a web-cam based surveillance system that captures the presence and movement of persons in the view of a camera suspended from the ceiling of a relatively high (>3 m) space and triggers sounds. Based upon a similar system—using ultrasonic echolocation—developed by the first author, Gehlhaar in 1985 [30] and modified for this implementation by the team, the recognition and tracking information is conveyed to a topology of sounds which are triggered by persons entering the space. A musical topology results from the analysis and processing of information gathered about the movement of a body, or several bodies, in a sensorized space. This information, fed as control variables to a computer executing compositional algorithms and, via some digital synthesis routines, produces sounds. Thus, the audience becomes the performers. The images conveyed by a web-cam that captures the position and movement of persons in the space are processed by proprietary software (in Processing) written by one of the authors (Girão). The tracking information is then handed to the simultaneously running Ableton Live program, which is used to create the musical topology (*which* sounds located *where* in space) and to trigger the sounds. Figures 9.7 and 9.8 show graphic instances of two classic Sound = Space (S = S) topologies; the outer diamond indicates the active space; the inner, its inner zone; the arrow, the orientation of the control functions.

Fig. 9.6 Two users of instrument A playing as a duet

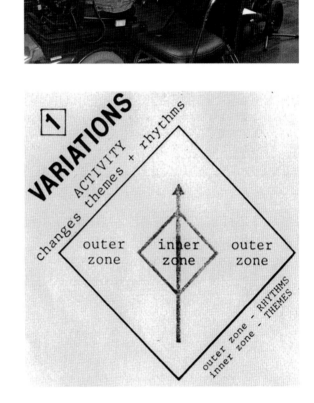

Fig. 9.7 Aspect of a classic S = S topology

The GUI and structure of this program is such that even modestly computer literate persons can produce their own topologies, on the fly even, using pre-existing samples or samples they have recorded or created themselves. Performing and composing with Sound = Space is a communal activity, well suited for group musical expression. It encourages thinking topologically with sound, communicating these thoughts to fellow creators and then performing by designing 'choreographies' which bring about the expression of communal musical intentions.

All three of the client institutions that had access to a suitable space (with a ceiling height >3 m) were supplied with a complete Sound = Space system consisting of a web-cam, a computer, software and a PA system. Figure 9.9 shows

Fig. 9.8 Aspect of another
classic S = S topology

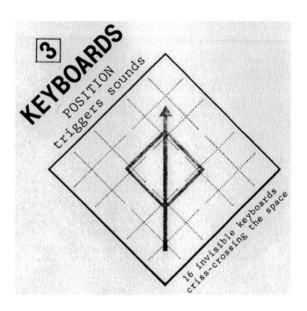

Fig. 9.9 Aspect of the
Sound = Space

an installation of Sound = Space being 'played' by several clients; it is permanently installed in the music room of two of the participating institutions (the
camera on the ceiling and dedicated computer are out of shot.).

9.3.2.3 Matrixx

The Matrix is an interface consisting of a physical box of 'nests' and a number of
each of five types—and colors—of 'eggs', objects which may be inserted into nests
in the box; each specific egg, when inserted into a 'socket', triggers one or two
pitches and an additive rhythmic pattern—see Fig. 9.10. Rhythmic phrases are

Fig. 9.10 Aspect of the matrixx

constructed by inserting the eggs into the sockets. There are 16 'empty' sockets; each representing a time slot—with the duration of one tempo unit—in a cycle of 16. The characteristics of the different egg types are:

- type 1, triggers a short sound;
- type 2, triggers a long sound 2× the tempo unit;
- type 3, triggers an indeterminate rising interval, the duration of each note equaling one tempo unit and the pitch of the first note repeats that of the last note of the closest key in an ascending sequence, independently of the silence between them;
- type 4, triggers an indeterminate falling interval; each note equalling one tempo unit—first pitch repeating the last pitch of the closest key in ascendant sequence, independently of silence between them;
- type 5, triggers an indeterminate sequence of 3 notes; each note one tempo unit—first note repeats the last note of the closest key in ascendant sequence, independently of silence between them.

The program reads the input from a tempo knob and uses that value to subdivide the Second. That then becomes the unit of tempo. All the rest is produced with adding and multiplying that unit. Furthermore, if an egg is inserted on the 3rd bar there will be 2 pauses and then the corresponding sound of the key will sound. If another egg is inserted on bar 5 a pause between the 2 sounds will happen. The loop ends at the last egg inserted.

In this manner, complex rhythmical and melodic sequences can be easily constructed, deconstructed and altered in real time, according to the demands of the player and the musical context. Every time an egg is inserted it selects a different note or sequence of notes. Once the egg is inserted, the sequence of notes remains the same. In order to choose a different sequence the user only has to unplug and re-plug the same egg in the same slot. This feature is what makes of The Matrixx a real compositional tool. The Matrixx has a set of adjustable controls:

- TEMPO, a knob, continuously variable, that controls the tempo of the rhythmic sequence, from very slow to very fast;
- TAP, a push button which overrides the built-in trigger, immediately triggering a sound, making synchronised play possible;
- VOLUME, a knob, continuously variable, controlling the output volume, from silent to loud;
- ON/OFF, a single-throw switch.

It is both a stand-alone device, powered by a standard PSU, with a small in-built loudspeaker, and a networkable device (for group music making), with MIDI in/outs, USB in out, output for headphones and amplifier.

The Matrixx is intended as a general easy-to-use interface, with a special focus upon persons with broad-spectrum special needs—such as impaired hearing, impaired vision impaired motor skills, autism—who enjoy music and would appreciate a means of musical expression. Although Girão designed it with the various special needs of the prospective users in mind, it is easy to use and learn but it is not just a sound toy. It is a genuine musical instrument on which proficiency can be developed and which also lends itself to synchronised group music making, either with several Matrixxes at the same time, or in the context of the other instrument designed. The two Matrixxes built for one client group are usually played together.

9.3.2.4 SuperString

The SuperStrings are a pair of long string instruments employing harpsichord strings of medium thickness amplified by a magnetic pick-up. It is an idea inspired in earlier work by the Gehlhaar [31]. The strings are stretched across a 1.6 m long narrow plank set on small 'feet', which makes it completely accessible from both sides and allows for 'bending' the plucked or struck notes. One of each pair was fitted with servomotor driven frets and a 'plucker', so that even severely disabled users could control it via ultrasonic sensors monitored by a micro controller. Figure 9.11 shows a player using a prototype of the mechanized instrument.

There are many different ways that the unmechanized SuperString can be played: plucked, struck with metal or wood, drummed with finger tips, bowed, etc., each mode of playing generating sounds of different qualities. All sounds played can be easily transposed downwards by pressing on the handle with the knob (which causes a downward bending of the plank and thereby a relaxation of the string tension). Due to the high degree of amplification of the long, quite soft steel strings, the instrument is extremely sensitive and responsive, making it very easy to play. Its length makes it ideally suited to be played by two persons at the same time, thereby encouraging musical interaction. Furthermore, due to its simplicity, it is also extremely robust, able to withstand the enthusiastic 'bashing' that its sometimes 'physical' nature evokes from the players. It is, however, also a very

Fig. 9.11 Prototype testing
the mechanised SuperString

fine instrument, able to produce a myriad of gentle sounds, harmonics, etc., especially when tapped with the fingertips or stroked with a bow or a thin steel rod.

The mechanized version, employing 6 servo-motors to change the pitch and one to pluck the strings, is basically played without touching it, via two ultrasonic sensors, that measure the distance to any object they are facing (the hand or arm, for example) and thereby trigger the plucking servo-motor and the six 'rocker arms' that change the pitch by shortening the strings, giving the instrument a range of 12 semitones, as in the tail piece illustration below, Fig. 9.12. The plucking servo of the instrument need not necessarily be activated as the rocker arms strike the strings and already cause them to sound, albeit quite delicately.

Three different institutions received a pair of these instruments. The participants in each institution formed a quartet of players of the pair of instruments, as illustrated in Fig. 9.13.

9.3.2.5 Ernst

Ernst was developed not as an instrument, but as a reactive experience that aimed to stimulate the user's interaction. It is an automated improviser that re-uses the sounds produced by the user to make music and it was developed by one of the

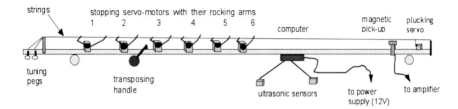

Fig. 9.12 An illustration of the mechanized version of SuperString

Fig. 9.13 A quartet playing
a pair of SuperStrings

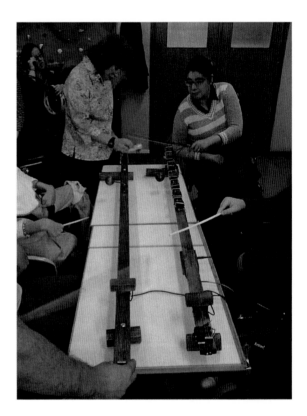

authors (Penha) using Max/MSP. The input from the microphone is recorded and segmented into several snippets of audio. These snippets are analysed for their pitch, volume envelope, rhythm characteristics and metrical stage within a larger musical phrase. This information, along with its relation to the surrounding snippets, is fed into 4th order Markov chains [32] that are responsible for outputting responses to the audio input, simulating a cognitive stage inside the computer. As soon as the characteristics aimed for each response sound are calculated, the system searches for the recorded snippet that more closely resembles these features and triggers it after all the necessary processing (e.g. pitch and timeshifting, envelope morphing), in a process known as concatenative synthesis [33].

Fig. 9.14 The GUI of ernst

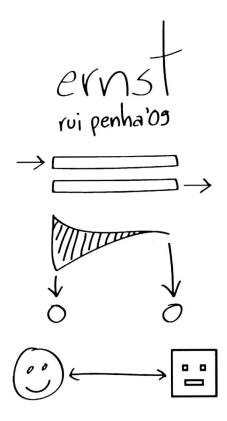

As the user sings or talks to the microphone, the software starts giving back phrases built on the musical characteristics of the user input. The sounds used to build these phrases are exclusively manipulations of the input sounds, so the user easily recognizes his or her own voice. Whilst the results were sometimes too unpredictable, due to limitations on the recognition of musical characteristics on the input audio, Ernst was capable of providing rewarding interaction experiences to some of the clients. However, as it is not a fully predictable musical instrument, it was not used in the final concert (Fig. 9.14).

9.4 Developing Musical Activities with the Newly Created Instruments

The development of the instruments described above took place mostly during the first year of the project. There was a process of interaction between the authors and the clients, with a set of programmed visits by the team and also the regular accompaniment by the project manager of the Education Service designated for the project. The final versions of the instruments resulted from adaptations and improvements made after direct observations and/or direct reports by clients and music teachers who were working in the project within the institutions. The role of music teachers in the institutions was of paramount importance. They were in many cases, due to impairments of the final clients, the true interface with the authors. In some other cases, however, like the Instrument A, a long-term collaboration between the person developing the application and one of the clients was established. At *Ao Alcance de Todos* 2009 the instruments/interfaces were presented to the institutions in an individual workshop held at Casa da Música, where members of Factor E also participated [34]. This started the second phase of the project, which had the objective of incorporating the newly created resources within the regular musical activities held at the home institutions. In the case of institutions where music was not part of the regular activities offered to the disabled people, but also gradually as the project entered the second year, several elements of Factor E started to work within the institutions, developing musical activities that would make full use of the instruments that had been received, in particular developing new techniques of making music together. This phase of integrating the new instruments in the regular activities of the institutions overlapped with the composition of the piece to be performed at *Ao Alcance de Todos* 2010. The development of musical activities with the newly created instruments was therefore nurtured by both the interest in incorporating them in the regular activities as well as the prospect of participating in a performance for which particular skills had to be developed through exercises proposed by the first author. Clients and teachers alike received the instruments with great enthusiasm and every institution adopted them into their regular (daily-weekly) musical activities. The institutions that received several instruments developed 'ensembles' in which the teachers often participated as players.

9.4.1 Composing for and Performing with I4E

As the project entered its second year, Gehlhaar was addressed by a new challenge: the composition of an 'orchestral' piece that would use the instruments that had been developed, as well as other resources, namely a Choir, the Robotic Gamelan and the T-Sticks. The idea of involving a choir came from the observation of an unrelated project going on during the first year of I4E. Choir singing was implemented with great success in a psychiatric hospital, involving patients and members of staff, and a new idea for 2010 was needed. The above-mentioned Robotic Gamelan [20] had been tested but a proper artistic challenge involving disabled people was yet to be developed. The Robotic Gamelan is a set of gamelan instruments that can be played by mechanical arms controlled by a computer that responds to measurements made by ultrasonic sensors. It had been used in early experiments in 2009 with a group of people suffering from cerebral palsy that would use their arms and heads to control the instruments and produced promising results.

The T-Stick [18, 19], an instrument that had been shown by Joseph Malloch at one of Casa da Música's Handmade Music events, was thought to be an interesting tool to experiment with blind people, due to its characteristics of haptic feed-back. A new software synthesizer for the T-Stick was developed by one of the authors (Penha) using Max/MSP. It could be controlled using two different gestures: a thrusting of the T-Stick would trigger a Gong sample, with its pitch-bend controlled by waving the T-Stick afterwards; caressing the embossed stripes of capacitive sensors would generate a slowly-evolving texture, with its volume controlled by the speed of the motion and its spectral characteristics controlled by the position of the hand. These sounds where both tuned to the same note in each instrument, with each T-Stick being tuned to a different note of Casa da Música's Gamelan slendro scale.

The novel *A viagem do elefante* (The Elephant's Journey) by the Portuguese Nobel-Prize winner José Saramago was chosen as starting material to structure a narrative. The idea was not the composition of an opera—although it became evident that there was a very strong theatrical sense in the final performance due the physicality of some of the instruments and players. A loose sense of narrative was conveyed by text elements, sung or spoken by the choir, as well as by the projection of very simple graphical elements and text. The duration of the piece is about 55 min. It is a very complex soundscape of electronic sounds (T-Sticks; Instrument A, Matrixx), pre-recorded acoustic samples, including environmental sounds (Sound = Space), amplified strings (SuperString), amplified gamelan (Robotic Gamelan) and voices (Choir). The score contains a great deal of graphic notation and involves both specific instructions for certain sound textures to be produced as well as directions for processes or improvisations. The piece was performed at Sala Suggia, the main venue of Casa da Música. Around 80 people participated, each group having a leader, either a member of Factor E or a teacher from the institution. It was the final concert of 2010 edition of *Ao Alcance de Todos* and truly a memorable artistic achievement as well as, of course, a remarkable moment in the life of the project and the participants.

9.4.2 The Final Composition: Viagem

The composition for the final phase of this project was entitled: **Viagem** *for many different instruments, pre-existing and invented, physical and electronic, manual and robotic, imaginary and real, and choir*, with subtitles by Virginia Capoto Araújo and simultaneous visuals created by Vahakn Gehlhaar-Matossian. As already mentioned, the piece referenced Sarramago's *A Viagem do Elefante*. The authors were of the opinion that a background story, accompanied by real time visuals and subtitles of significant fragments of the text (being sung or spoken by the choir) would make the quite abstract music more accessible to the audience.

The original idea for the piece—and the disposition of the choir and instrumental forces—was that of an *encampment*, reflecting Jose Saramago's 2008 story based on historical fact. In the 16th Century King João III of Portugal gifted an elephant to Archduke Maximilian of Austria. On its ensuing journey from Lisbon to Vienna a company of soldiers and priests accompanied it. The sequence of musical events of which the piece is composed, inspired in part by this story, is essentially based upon the extensive musical exercises that Gehlhaar developed for the Factor E team to take to their groups of musicians during the 6 months that he was composing. The content of these exercises was structured such that the players would not only learn to play their instrument but also learn structured improvisation and make music together. For example, two SuperStrings were always played by four people; the players learned to play together, to engage with one another according to the exercises. These focused on the invention of a musical 'event' and the subsequent guided improvisational cycle of transformation of that event. As a result, the musicians already knew the score—more or less—after they had learnt their exercises during the 6 month's preparation for the concert. Gehlhaar had decided quite early on that the reading of parts by the musicians would be impractical (in some cases impossible as the participants were blind). The whole composition was to be performed from memory under guidance of group leaders, the very same persons of the Factor E team that tutored the individual groups.

In the performance of *Viagem* there were eight groups:

1. twelve persons moving in the Sound = Space;
2. four SuperStrings played by eigth musicians;
3. two Matrixxes played by two musicians;
4. four composers playing Instrument A;
5. seven musicians playing the mechanized gamelan instruments and four gongs;
6. four musicians playing the T-sticks;
7. two choirs of about 25 singers each.

Every group had its own director who could read from score and take co-coordinative directions from the master conductor, in this case, the composer. Generally the groups were autonomous. However, often a players or singers of one group were required to listen to events that were played outside their own group and import them into their own cycle of imitations and transformations.

The main organizing principle of the composition—reflecting the idea of the progression of the elephant and a large accompanying force across the face of Europe—was a broad arch composed of sections of high activity (moving) alternating with sections of little activity (camping). Musically speaking, there are many short local, foreground events embedded into an overall, almost cumbersome background stream. Locally, instrumental colors and densities change rapidly or slowly within a broad framework of transitory interrelationships, weaving a changeable tapestry of sounds and words.

9.5 Feedback from Those Involved

Feedback from educators working in the institutions and their 'clients' was a constant part of the project from the very start. The idea of the project was to develop instruments tailored to the specific needs and motivations of clients and institutions and a first type of methodology was a questionnaire, as described above. The evolution of the instruments was determined by the experiments with the early versions and this type of feedback was gathered mostly by informal interviews at meetings and music sessions held with the team of developers. An independent visit by other educators working for the Education Service at an intermediate stage of the project gave rise to a written report that was also used as a feedback source for the developers. Finally an overall feedback on the project was requested at two different stages: shortly after it was concluded (as an open question via email and also as informal interviews in the context of the making of a documentary) and about 4 years after the project had been concluded (again as an open question via email). This section analyses the overall feedback and tries to identify different types of reactions and points of view.

The initial feedback received from participants—disabled performers and teachers—showed that the main goals were achieved. We received an enthusiastic reception to the project by the staff and participating institutions: they were able to provide access to meaningful musical (and physical) activity for students who are generally excluded from it, they displayed willingness to come to terms with new technologies and to undergo some training to expand their musical horizons. As stated by one of the teachers that participated in the project (Anna Petronella):

> Casa da Música has become 'our House', and fills us with pride and appreciation. We achieved much in terms of motivation, new materials and perspectives for intervention. These gains are not restricted to the group directly involved, but also extend through time to a much broader group of users and professionals.

The effect in self esteem and motivation of the participants was reported also by members of staff from institutions that attended to the final performance (Cláudia Gonçalves):

> … I was moved to watch people with disability working towards a common goal with a professional attitude and highly motivated,

a feeling shared by other members of the institution even if not participating directly in the project (Pedro Fernandes):

> we all felt our ego 'on high' for having the privilege of being part of such great initiative.

The success of the project seems to be related to the instruments and what they allow, but also to the human implications of the process, as it can be inferred from the feedback of one the educators working in an institution (Paulo Jacob):

> the project was very valuable due to the visits to the institution, to the professionalism, sensitivity and knowledge of the technical team involved. The products developed were also important, and in our case, Instrument A, we observe an actual response to the needs

of our costumers (yes, we are talking about improving their Quality of Life), something that will lead to the emergence of new groups of users [...]. The continuity of Instrument A usage is guaranteed!

It is quite clear that the success, or lack of success, of projects like this depend not only on the tools that are developed but also, or above all, in the contexts that surround their creation and use. The final performance, and in particular the implications of performing a specially composed piece that would integrate the diversity of instruments developed, gave a sense of direction and an extra motivation for all participants. In our opinion—and this is supported by evidence in many other projects—this sense of goal and the intrinsic artistic implications of sharing the results of a laborious process with an audience are at the root of the meaning that people give to these experiences. In the particular case of this project, it is fascinating to look at the statements of the clients for whom the instruments were developed. The "happy faces" described as the "expression of success" by one of the educators from APPC (Paulo Jacob) has particular compelling foundations in the interviews that are part of the documentary published by Casa da Música—*Ao Alcance de Todos 2010*, directed by Tiago Restivo and Tiago Caiano. A set of carers/technicians and disabled users were interviewed and, even if sometimes disability (mental or physical) can hamper the communication skills of some of the people interviewed, it is obvious that the project had a profound impact in their lives. The analysis of this material reveals three major groups of ideas: reflections on what can be broadly defined as "empowerment", reflections about the learning process, reflections about aesthetic aspects of the experience. The first group has to do with the impact the project had on the sense of worth, belonging, connectedness or self-esteem. It expresses the effects of "having a voice", "being able to", "achieving", the emotional/psycho-sociological impact. The second group is a set of colorful descriptions of how the learning took place or considerations about the importance of learning as an outcome of the experience. The third type of ideas expresses the sheer enjoyment of the sounds, music and other artistic facets of the project and reveals the emotional/aesthetic effects. The following is a selection of statements that illustrate the variety of impressions expressed.

"These people have no musical education and this experience gave them the opportunity to experiment with instruments, to create their own music, and the most exciting part for me, as a technician, was to see their emotion and their involvement in the experience, and the enthusiasm with which they played the various instruments. They worked as a group which meant that they interacted with each other through this musical experience and through the sounds they created." [...] "I think we all try to develop our full potential during our lives, to enjoy things and to look for those interests which will give us a sense of achievement and fulfillment, a meaning to life." [...] "These are new ways for them to express themselves, to communicate, to feel a sense of achievement and to generally be more active as people." Anabela Aguiar, technician

I never thought it would be like this, I thought that other people played the music and the instruments, but it was the other way round. We played the instruments and the others taught us how. Pedro Alves, client

"As far as playing new instruments, or innovative instruments, I feel I can identify myself with the spirit of those instruments. I went there and I played them, and I believe I learnt a lot from these instruments." "[...] I would like to play one of those instruments so that I could belong with other people, be with other musicians to make quality music." "These instruments create a bond between people. They bond them in a different way, in a new way, and so I believe that more of these experiences should take place for people to learn more about this type of music, these instruments [...]." Aristides Andrade, client

He played first and I immediately played the same thing. Everything he played I played. [...] He played, I played. He played first with his hands and then he slid a little iron bar back and forward, left and then right. He did tum-tum-tum and then tum-tum-tum and then tuuuuummm [...] I am there to play and Sérgio is there to help. What is a low, high and medium? Sérgio asked me: where is low?—Low is on the other side. And high?—On the other side. And medium?—In the middle where that star is. You have to play where the star is. Toooomm and that's high! People feel what we are playing. Really. Jorge Ribeiro, client

"It was really a stick, a round circular stick, more or less this long with little bits sticking out, and when you rubbed them they made a beautiful sound. [...]. If we held it one way it made a certain sound, if we made certain movements with our hands it made a totally different sound, so it was a fantastic way to make us understand and learn how to use our sense of touch." [...] "And during our rehearsal with the pianist we truly felt we were making music. It was at that moment that I thought to myself: we were actually doing something, this was music." [...] "I felt an enormous sense of achievement; I was part of a production, of something important." Maria Velho, client

"Initially I was not very interested in this instrument. But when I heard the end result, when all the instruments came together, all in the right place, coming in at the right time, it was fabulous" [...] "Instruments are beautiful when they are played properly, but when some instruments are combined with other instruments they aren't beautiful, they are spectacular and this is exactly what happened with us. Together with the choir, the dance and the Gamelan, we were creating something beautiful; the final result was not beautiful, it was spectacular." "[...] it was really good to feel that, at least for that moment, all these people can be useful to society." Nuno Rodrigues, client

I was rather nervous at first, but then as the show went on the nerves disappeared. I have been to shows before but I had never been on stage. It was a new experience for me. I looked at the room and I thought: my God what is this? But then as the show went on I was fine and it went really well. I could have stayed there another two or three hours if necessary; it was an unforgettable experience for me. [...] In the end this was a job done by all, not just one, by all." [...] "I felt like a star, although I know I am not a star, but at that moment I truly felt like one. Márcio Moreira, client

This was undoubtedly a groundbreaking experience for us technicians and for the young people who work with us, because all of these instruments were unknown to us, so not only did we learn something new, we were also able to create as we went along. New stuff was always coming up, new sounds—it was a very enriching experience for us because it gave us the freedom to experiment. A number of unexpected things happened during the show, and we had to improvise—but it all turned out really well in the end, which was fantastic. In that sense it was really amazing. Sérgio Silva, technician

The most important moment for me and the one I enjoyed the most, was undoubtedly the show because I was working with a number of people and not on my own. Alexandra Dias, client.

"When the conductor indicated, I played the hammers, and the music I played combined with the music from other instruments, made beautiful music."[...] "I felt much more useful, more important. I felt I was an artist. I enjoyed learning more music and my love of music grew." Maria Manada, client

To receive the appreciation they received for having participated in a musical performance, to have been a part of the scenery together with everyone on the stage was ecstasy, they went wild; it was like opening the borders. New horizons were opened up to them, and I believe that from now on they know that music is not only for musicians." Pedro Fernandes, institution member

When we talk about recovery, we mean promoting healthy attitudes towards life in terms of self-esteem, self-concept, understanding that one is capable of doing things. It's obvious that when people participate in a project, they feel equal to others. José João, technician

"Really hard but very gratifying once we saw the final result, once we had mixed in all the other instruments [...] The interesting and also gratifying part of this is that we met the requests made to us. It gave us an enormous amount of work and we spent many months working on this together [...] we always tried to work as a team and I think we succeeded in the end [...] I feel a great sense of fulfillment. To have played at the Casa da Música is wonderful and very gratifying, and I think we can be proud of ourselves because it's a different world." [...] "We outdid our expectations and we impressed the public. We showed them that, within our limitations and with the help of those who make us our instruments and our software, we can do what everyone else does. We put on a fantastic show just like any other musician or any other actor." Adelaide Almeida, client

[...] to be able to show others that we also know how to play, how to present a show. Liliana Neto, client

[...] our relation to music should not be a passive one, music should be for everyone. I believe it can only be good for you. Paulo Jacob, educator

Approximately 4 years after the project had been concluded an email was sent to the participating institutions and educators inviting them to write about the "reverberations" of the project. Although several institutions replied that after the performance at Casa da Música the instruments gradually stopped being used, due to changes in personnel or facilities, a few answers pointed out some long-term effects.

The educator Anna Petronella, for example, reported that some of the exercises and instruments (Sound = Space, Matrixx and SuperString) were regularly used in Drama activities and even in a touring performance of an original music theatre project developed at APPACDM in Coimbra. She also reported that her usage of the Soundbeam had been enriched by the learning experience with the Sound = Space as well as by the integration of Ernst and Matrix. She reports:

"I used the technology that was available and adapted it to the specific interests of the clients (42) attending the area". After July 2012 the institution had to adapt to new financial restrains that forced the most part of these activities to cease. She concludes, however, by saying that "the experience of the project was so deeply absorbed that, in one way or another, it still inspires, nourishes and strengthens relations"

The educator Paulo Jacob (APPC, Coimbra) also reported continuing activities and analyzed the response by the clients he worked with, concluding that the project was important because it provided an opportunity to demonstrate capacities; an opportunity to have a first significant stage experience; an opportunity to work with instruments that facilitate the access to musical composition and performance (Instrument A being referred to as very friendly); personal reward due to the integration in a group made of disabled and non-disabled people; recognition for the work (expressed as "applause" by the general public); valuable experiences in the exploratory processes and in the creation of the performance; recognition by family and friends.

One of the clients from this group, Adelaide Almeida, reported above, reflected on the project, confirming her first impressions, namely emphasizing the hard work she went through, and she concludes by saying: "I still use the software [Instrument A] regularly, including for the development of music/sound for theatre plays, Christmas parties, etc., so the project only brought me good, positive things."

Paulo Jacob also reported a boost in interest by the institution in technologically assisted musical activities, with the acquisition of computers and the development of a specialized response for a group of severely disabled clients (with cerebral palsy and degenerative diseases). Every client with severe motor limitations is now entitled to have a personalized response, individual and in groups, within the framework of a program entitled Music and Technology. He has developed a plan of work, specifically for Instrument A, that includes five phases:

(1) promotion of autonomy (a process of personalization of switches, initiation and adaptation to the program); (2) content development (selection of sounds and composition); (3) content personalization (the clients are encouraged to create a personal list of sounds and musical ideas for further development). After this list has been conceived the client participates in the process of collecting the sounds he/she has chosen by: (a) field recording, (b) accessing the internet, (c) introducing the new sound files in the program; (4) experimentation and development of personalized Instrument A; (5) integration of the client in a group situation (with other personalized Instrument A).

He also reports the development of a new project based on a 'Makey Makey' [35] as a way to "develop new forms of musical interaction and find new paths for a more fulfilling life."

Some of the educators of Factor E also sent their "reverberations" at this stage. Ana Paula Almeida highlights the "development of an intuitive communication between the different musicians, the joy to be seen and heard, a final result of an unexplainable quality defying any preconception that could still exist." Joaquim Branco, philosophically reflects about this experience (and others with disabled people) and identifies three repercussions: "(1) love is our right (2) communication is made of intention in silence (3) it is possible to reach where life simply changes." He concludes: "I was one before Ao Alcance de Todos, today I am another! Better." Filipe Lopes emphasizes the fact that the process of composing Viagem allowed time and room for development of ideas within the group he was working with (Robotic Gamelan) "there was time to work on details, to move backwards and forwards, to refine the patches and compose together"[…] Rolf attended some of the rehearsals and incorporated some of these ideas in the score, and that conferred a bigger sense of belonging to the piece."

For the authors this was a particularly complex and demanding project and, yet, one of the most fulfilling. The entire project was an opportunity to work on a fundamentally worthwhile undertaking: to assist in developing access to musical activity; to encourage participation in group musical activity to persons who generally are excluded from it; to exercise and develop our own creative skills in an inspiring atmosphere, both within the team and with enthusiastic clients. We organized ourselves with an embedded flexibility of approach, providing us with opportunities for our own personal growth, allowing us to respond in very individualistic ways, always with possibilities for revision. Despite the relatively short duration of the project we believe that we were still able to respond with original projects—involving both the development of bespoke hardware and software— that met the demands (and in many cases, went beyond the dreams) of the 'clients', such as increasing the level of student participation in already existing musical activity and expanding the pedagogical techniques and instrumental resources of the teachers while broadening the conception of what music making can be.

9.6 Advice to Practitioners

The following is not to be considered as an exhaustive list; it simply contains aspects that are sometimes overlooked when working on such a comprehensive project.

- ensure that the estimate of costs is accurate and flexible and foresees contingencies;
- assess the clients carefully and precisely;
- get to know both the staff and clients as individuals;
- design the instruments having the staff's skills and sensibility in mind, as they will be the primary catalysts of the client's success;
- design over a broad range of abilities/disabilities—something for everyone;
- design so that at least a moderate competence on the instrument(s) may be achieved within the time frame of the project;
- design so that instruments could be programmed, constructed, debugged and repaired by the core team and afterwards, maintained by the clients themselves;
- develop written and illustrated user and maintenance instructions for each instrument;
- instruct the staff of the institutions to which the instruments will be handed over at the end of the project; assist them to take ownership of what can be done with them;
- design musical exercises—both individual and group—which will encourage and assist the players in learning to play their instrument or how to control the interactive/reactive environment;
- create a fertile ground for musically fulfilling experiences to arise from early stages of getting acquainted with the instruments;

- do sectional rehearsals before tutti ones; this gives the musicians a bit of time to get used to being in large groups, hearing the others, working with strangers, on a professional stage;
- compose the ensemble piece in such a way that the players may take ownership of it.

9.6.1 Problems Encountered and Methods for Arriving at Solutions

The problems we encountered may be divided into the following categories: methodological, practical/logistical, financial. We will discuss them in that order.

9.6.1.1 Methodological

All members of the core team—the authors—were working primarily from the perspective of composers. We had had some experience in musical projects such as interactive multimedia installations, music theatre works and workshops related to them, many specifically aimed at persons with a disability. There are many sources and paradigms [36–38] for some of these projects, especially those of therapeutic value of musical activities education but we could find none specifically for such a project as ours. We decided to organize the project as if it were one of educational research, to find and make contact with as many relevant institutions—schools, day centers and clinics—as possible. Then we reduced that number to those where both staff and clients seemed to be most excited by the idea.

The scope of our project, requiring 2 years for its completion, posed a fundamental difficulty that had to be overcome: finding clients who would seriously commit to such a long term project. We knew that it was imperative that the staff of these institutions be brought on side, as their committal would provide the project the long term home it required. Not only is their committal crucial but also their willingness to be educated in the use and maintenance of the new instruments—some primarily digital—we were going to develop. Even though we are well into the computer age, there still exists resistance and general phobia towards technology among educators within special needs institutions. Fortunately we were able to overcome any resistance and also to develop instruments that were reliable and easy to maintain and a pleasure to operate. One of the contributing factors to our success was the excitement with which our project was received by the potential participants—they couldn't wait to get their hands on the instruments. In every instance this helped overcome any hesitation on the part of staff.

9.6.1.2 Practical-Logistical

Essentially, right from the beginning we were under pressure: the scope of the project was large and there was a pretty certain cutoff point approximately 20 months down the road. Efficiency was of the essence, keeping everyone involved in the project informed. This meant that a functional communications systems had to be established; this was done via a document sharing protocol. Our first task was quickly to organize the client base, develop a projects 'manifesto' and decide upon the number of clients we could serve within the limits of the project. The we could recruit the participants, do the research as to their needs and desires, implement the designs, 'manufacture' the prototypes, test them, manufacture the final instruments and present them to the clients. All of this needed to be accomplished in approximately 14 months.

Due to the geographic spread of the institutions involved—some 60 km—it proved at first difficult to organize meetings with the clients at their institutions; we always needed to find a slot in their busy schedule. In the end we had to meet with them on four consecutive days approximately every 6 weeks. This allowed us to go into depth, both with the staff and the clients, ensuring that no one was overlooked and that everyone was catered for and that the end result would be able to be combined into some sort of 'ensemble' for a performance.

The remaining 8 months of the project were dedicated to musical instruction on the instruments, carried out by Factor E, the educational force of Casa da Música. and of course, the composition of an ensemble 'piece' and the final 4 days of rehearsals and the performance.

9.6.1.3 Financial

Because this was such a novel project, it was difficult to budget it. In the end the administration decided that we had a fixed amount and would have to make do with that. We knew there were many unforeseeable such as breakdowns, software glitches, hardware problems and replacements, all requiring extra time and travel.

One of the great advantages of realizing the project with Casa da Música was the fact that the cost of the one of the authors (Rodrigues), the educators (Factor E) and the administrators were met outside the project budget. This allowed us to be much more ambitious and to give the project greater scope; otherwise the budget would not have been sufficient.

9.7 Concluding Remarks

Among the myriad design challenges of this project was the following: striking a balance between ease of access (to 'draw the user in') and yet achieving enough depth to encourage longer-term use and more involved creative engagement.

In some cases, discovery is the key: an instrument in which the questions 'what can it do?' and 'what can I do with it?' and 'what can it teach me?' and 'what can I teach it?' are significant. The main points for this project were to make something to be played rather than something to be played with and using technological means in order to effect a transfer of skills, replacing some of the generally required traditional skills for playing musical instruments.

The main challenge in the design of new musical instruments is how to integrate and transform an 'apparatus' into a coherently designed channel for the—communication and expression—of meaningful musical experiences with emotional depth.

And, furthermore, how to educate the users so that musically they may take the greatest advantage of their new instruments? The accessibility of such an approach can come at the expense of limiting the musical range and possible gestures associated with creating sound. But, does accessibility need to come at the expense of range of expression? We thought not, but we knew that the education of the user, both individually and communally, was going to be crucial. We were not looking to develop musical instruments with which any standard repertoire, be it classical or popular, could be recreated. Rather, we were looking for ways in which the participants could musically express themselves, both privately and in a group.

We believe this project was important for everyone that participated in it. As far as we are concerned, it made us (re)discover an essential aspect of our activity as artists: the possibility to create effective channels for expression and human development.

References

1. Wikipedia: water organ: http://en.wikipedia.org/wiki/Water_organ
2. Hydraulis: http://www.hydraulis.de
3. Ord-Hume, A.: Barrel Organ: The Story of the Mechanical Organ and its Repair. George Allen & Unwin, London (1978)
4. Player Piano History: http://www.playerpianos.com/2009/04/player-piano-history.html
5. Glinsky, A.: Theremin: Ether Music and Espionage. University of Illinois, Urbana (2000)
6. Moog Music: Etherwave theremin, http://www.moogmusic.com/products/etherwave-theremins/etherwave-theremin-standard
7. Trimpin: Computers and Music, http://www.aes.org/sections/pnw/pnwrecaps/2002/trimpin/
8. Lemur: League of electronic music urban robots, http://lemurbots.org/
9. EMMI: Expressive machines musical intrsuments—music robots, http://www.expressive machines.com/
10. Expresseur: http://www.expresseur.com/intro/index.php
11. NIME: http://www.nime.org
12. Soundbeam: http://www.soundbeam.co.uk/
13. Snoezelen MSE Therapy: http://www.snoezeleninfo.com/
14. Brooks, T., Camurri, T., Canagarajah, N., Hasselblad, S.: Interaction with shapes and sounds as a therapy for special needs and rehabilitation. In: Fourth International Conference on Disability, Virtual Reality and Associated Technologies, pp. 205–212. Veszprém, Hungary (2002)

15. TED talks: Tod Machover and Dan Ellsey play new music, http://www.ted.com/talks/tod_machover_and_dan_ellsey_play_new_music.html
16. Skoogmusic: http://www.skoogmusic.com/skoog
17. Wanderley, M.M.: http://www.idmil.org/people/marcelo_m._wanderley
18. The T-Stick, http://www.idmil.org/projects/the_t-stick
19. Malloch, J., Wanderley, M.M. The T-Stick: From musical interface to musical instrument. In: International Conference on New Interfaces for Musical Expression 2007, New York City, USA (2007)
20. Gamelão Robótico: http://vimeo.com/8843986
21. Clarence Adoo: Headspace, http://www.clarence.org.uk/headspace.html
22. Petersson, E., Brooks, A.: ArtAbilitation: An interactive installation for the study of action and stillness cycles in responsive environments. In: Computers in Art and Design Education Conference, Perth, Australia (2007)
23. Almeida, A., Girão, L., Gehlhaar, R., Rodrigues, P.M., Rodrigues, H., Neto, P., Mónica, M.: Sound = space opera: choreographing life within an interactive musical environment. Int. J. Disabil. Human Develop. 10(1), 49–53 (2011)
24. Rodrigues, P.M.: Beauty and expression: music and musical theatre productions. In: What's Special? Opera, Dance and Music Education for and With People With Special Needs Across Europe, pp. 38–44. RESEO, Brussels (2010)
25. RESEO: http://reseo.org
26. Processing: http://processing.org
27. What is Max ?: http://cycling74.com/whatismax
28. Penha, R.: Towards a free, open source and cross-platform software suite for approaching music and sound design. In: Méndez-Vilas, A., Solano Martín, A., Mesa González, J., Mesa González, J.A. (eds.) Research, Reflections and Innovations in Integrating ICT in Education, pp. 1204–1208. Formatex, Badajoz (2009)
29. Penha, R.: Políssonos, http://ruipenha.pt/software/polissonos
30. Gehlhaar, R.: SOUND = SPACE. Contemp. Music Rev. 6, 59–72 (1992)
31. Gehlhaar, R: Superstring. Feedback Papers no. 2. Feedback Studio Verlag, Cologne (1971)
32. Markov Chain: http://mathworld.wolfram.com/MarkovChain.html
33. Schwarz, D.: Current research in concatenative sound synthesis. In: International Computer Music Conference 2005, Barcelona, Spain (2005)
34. Ao Alcance de Todos 2009, http://vimeo.com/7285810
35. MaKey MaKey, http://www.makeymakey.com
36. Sze, S.; Yu, S.: Effects of music therapy on children with disabilities. In: 8th International Conference on Music Perception and Cognition, Adelaide, Australia (2004)
37. Gehlhaar, R.: SOUND = SPACE workshops for disabled children http://www.gehlhaar.org/x/doc/resfelr.doc
38. Wigram, T.: A method of music therapy assessment for the diagnosis of autism and communication disorders in children. Music Ther. Perspect. 18(1), 13–22 (2000)

Chapter 10
Designing for Musical Play

Ben Challis

Abstract Though sensory spaces are a common feature within many special needs schools, the way in which they are designed and resourced varies greatly between provisions as do the types of activities that take place within them. A short series of case studies has been carried out across a cross section of UK special needs schools to demonstrate this contrast whilst also attempting to better understand the reasoning and motivation behind their design and usage. In 2012, eight schools were visited in England and Wales with the aim of documenting the types of sensory space that were available, the resources that were featured within each space and the types of sensory activities that were being used. The key themes that emerged during the case studies are discussed alongside an historical overview of the conception and evolution of the multisensory environment.

Keywords Sensory · Play · Sound · Music

10.1 Introduction

This chapter aims to explore the use of music and sound through assistive improvised-play in sensory spaces. The term 'assistive' is used here by way of reference to any technology (specialist, commercial, novel) that is in some way enabling within a context of creating and manipulating sonic-landscapes or musical-ideas. Whilst acknowledging that there are clear benefits to encouraging children and adults with perhaps quite profound individual-needs to engage with sensory focused activities, it is also suggested that there is a need for a more coherent approach to designing such spaces. A recent series of visits to special

B. Challis (✉)
Department of Contemporary Arts, Manchester Metropolitan University, Cheshire Campus,
Crewe Green Road, Crewe, CH CW1 5DU, UK
e-mail: B.Challis@mmu.ac.uk

A. L. Brooks et al. (eds.), *Technologies of Inclusive Well-Being*,   197
Studies in Computational Intelligence 536, DOI: 10.1007/978-3-642-45432-5_10,
© Springer-Verlag Berlin Heidelberg 2014

needs education (SNE) schools has shown there to be a wide spectrum of perceptions on offer as to the kinds of activities that might take place in sensory spaces and that this, in turn, is reflected in the considerable variety of layouts and technologies that can be observed in use. This spectrum ranges from the ad-hoc (or perhaps organic) approach to design through to commercially commissioned spaces with feature-rich content. A broad range of sensory stimuli and technologies exist within this spectrum, offering contrasting levels of emphasis across the types and nature of interactions being explored (tactile, haptic, auditory, visual). In addition, there are certain technologies (e.g. the bubble-tube) that can be regarded as common features and others (e.g. the resonance board) that are not and there are good examples of 'found' technologies that perhaps surpass the potential offered by many specialist technologies.

Sensory spaces can perhaps now be regarded as ubiquitous within SNE provision with designs for new buildings and extensions identifying areas for use as a sensory room or sensory garden and with existing provision identifying available space to be offered over for such use. However, there is little literature available on either the design of sensory spaces or how to get the most out of their potential when they are available. So this raises the fundamental question as to what different combinations of layout, technologies and activities can be regarded as 'good'? There is unlikely to be an absolute answer to this question and the very nature of individual needs dictates that different provisions are likely to have different solutions based on the specific therapeutic requirements of the individuals they are working with. However, there is much to share here and this chapter attempts to detail some of this within a discussion that is non-judgemental; the aim being to elaborate on contrasting approaches with a view to helping practitioners make informed design choices that suit their own needs. Within this, there will be significant emphasis placed on the notion of interaction through improvised-play. The suggestion being that, although cause and effect style interactions are of obvious benefit (the bubble tube being a good example of this), there could be more opportunities for free-play within sensory spaces; where outcomes are perhaps less deterministic whilst still being mapped in a meaningful way to an individual's gestures.

Alongside this key theme of design for 'play', the chapter will also propose that interaction with music and sound in sensory spaces is perhaps under-explored. A key observation within the survey is that music and sound are often employed in a passive sense within sensory spaces; providing an ambient backdrop to other sensory activities for example. Even when the interaction is more active, the use of sound and music can often be restricted to a cause-and-effect style approach. For example, a sensory garden may contain an ornamental flower that when hit or touched produces an appropriate sound (e.g. the buzz of a bee); the single action is rewarded with a single sound. In contrast, it is a rarity to find examples where the subtleties of more prolonged movement (moving a hand or body within a field or across a surface) are captured and mapped to meaningful sonic events. These two key themes (design for play and interactivity for music and sound) will be maintained throughout the following observations and discussions.

10.2 Snoezelen and the Evolution of the Sensory Space

Schools for special needs education will often have one or more areas that can be identified as being a sensory space,[1] yet there is little consensus on how such a space might be designed, resourced and, ultimately, used. There are commercial companies that will design and equip such spaces, offering a range of specialist technologies and furnishings along the way yet there is little research available to suggest why one combination of resources might be more effective than another. So, where does the concept of the sensory space come from and what transformations have occurred throughout its evolution?

The original concept behind the sensory space can be attributed to the work of Hulsegge and Verheul [1] who suggested that fundamental sensory stimulation could be a more direct and meaningful way of reaching out to individuals with profound and multiple learning difficulties (PMLD) than that of focusing on intellectual ability alone. In their book 'Snoezelen: Another World', Hulsegge and Verheul describe the creation and use of sensory rooms for this purpose at the De Hartenberg Centre in the Netherlands. The rooms were essentially controlled sensory environments where a care-assistant could work alongside an individual with PMLD as he or she interacted with various sensory stimuli. Importantly, the emphasis was not on learning in any traditional sense, more on recreation and play with contrasting sensory experiences readily on hand for the individual to explore. Indeed the term 'Snoezelen' is derived from two Dutch words: 'snuffelen' to sniff and 'doezelen' to doze and though learning might be achieved within these sessions it was not regarded as a primary aim.

Pagliano [2] describes how these multisensory environments were, in part, born from the coming together of specific technological and social changes. The late 1970s saw the emergence of the discotheque, a high-tech environment dedicated to the creation of powerful audio-visual experiences for the purpose of entertainment alone. New technologies were being developed to enhance these sensory experiences (mirror-balls, sound-to-light, projector-wheels), all against a backdrop of amplified beat-based music. At this same time, the manufacturing industry was exploring the potential for various new plastic technologies. Soft-play furnishings could now be produced easily using PVC; not only cheap to manufacture but waterproof and easy to wipe-clean. Velcro was being used to produce simple but effective instant-access fasteners and vacuum-formed plastics were enabling the creation of lightweight playground equipment. The arrival of these technological novelties (discotheque and versatile plastics) happened to coincide with a number of sociological changes that were evolving in the way in which individuals with disabilities were being perceived and supported; a progressive movement away from institutionalisation and towards mainstream integration. The Snoezelen effectively embodied all three of these notable happenings, creating safe

---

[1] Sensory spaces are also referred to as multisensory environments or MSEs.

environments where individuals with PMLD could experience rich sensory experiences in a recreational and, ultimately, therapeutic context.

The Snoezelen model was adopted relatively rapidly within special needs education and on quite a global scale leading, in turn, to the rapid commercialisation behind the manufacture of many associated technologies. Ultimately, Snoezelen became the registered trademark of one specific company and though this remains to be the case there are numerous other companies (in design, consultancy and manufacturing) that very much owe their existence to this original concept. There are various names and permutations that are now in use and besides the registered trademark of Snoezelen, any special needs school might describe how it has one or more multisensory environments, though the term used might be sensory room, a sensory garden, maybe a sensory corridor or even a sensory corner. Portable sensory trolleys exist too such that a dedicated room or area is not required and activity rooms (hydrotherapy pools, rebound rooms etc.) might be enhanced to also function as some form of sensory environment. With this in mind, the term sensory space is being used within this chapter as a catch-all term for any such realisation or adaptation of the original Snoezelen concept.

Though the design and installation of sensory spaces continued steadily, the underlying philosophy was often confused, leading to mixed interpretations and experiences along the way. Indeed, the originators of the Snoezelen concept acknowledged this in their own writings identifying a general lack of solid theoretical basis from which to understand and guide Snoezelen use. Pagliano [2, 3] highlights this further by describing specific contradictory examples such as the suggestion by Hulsegge and Verheul that expertise is not absolutely necessary even though they also emphasised the need for careful observation and, more significantly, that learning is not a must even though the individuals should be given the opportunity to gain experience. Ultimately this lack of clarity led to the branching of the original concept into two separate ideologies, one that stayed true to the original philosophy where the emphasis would be placed on recreation only and one that moved more towards an educational model where emphasis would be placed on learning through cause and effect. In a study into the perception and purpose of sensory spaces, Bozic [4] formalises these two contrasting approaches as being part of either a child-led repertoire or a developmental repertoire. He also points out that both repertoires may well coexist within the same school as it is generally individual members of staff who are ultimately responsible for interpreting the function and use of sensory spaces.

Though sensory spaces can be regarded as widely available within SNE provision in the UK, there is very little research available that questions whether there are measurable benefits from using such resources. Where research has been carried out the results have tended to be inconclusive (e.g. [5, 6]) or perhaps not open to generalisation (e.g. [7]). Indeed, in a review of available research in this area Hogg et al. [8] also identify examples of negative outcomes commenting that

> Snoezelen is not a cure-all for everyone—even those with apparently similar characteristics...

suggesting that

> ...efforts must be made to establish what the individual brings to the situation in terms of personality and sensitivities that makes Snoezelen beneficial, a non-event, or positively detrimental.

Whilst a more recent meta-study by Lotan and Gold [9] concluded that

> ... weaknesses in the examined research methodologies, the heterogeneity between research designs, the small number of available research projects, and the small number of participants in each research project, prevent a confirmation of this method as a valid therapeutic intervention at this time.

So, the long-term benefits of working with sensory spaces are yet to be fully assessed and this places considerable reliance on anecdotal evidence for identifying any potential positive or negative impact. There is still substantial value to be attached to personal observations and experiences though, as the special needs educator is typically working at an individual needs level where the opportunity to generalise rarely arises anyway. This is an aspect that Mount and Cavet [10] identify in their review of similar studies into the relative merits of sensory spaces, ultimately arguing that there is likely to be as much significance to be placed on the quality and abilities of the individual member of staff as the equipment and spaces they are operating within.

The Snoezelen philosophy identified music and sound as being an important part of the sensory environment but was not prescriptive as to how this might be achieved. In many of the sensory spaces visited for this study, interaction with music was typically a passive activity yet music is also used in very active ways within other therapeutic contexts. Though more active interaction with sound and music did occur elsewhere within the schools visited it was seldom observed directly within the sensory spaces where one-to-one sessions might be more likely to occur. This observation is discussed in greater depth later on but as the emphasis of this chapter is on musical-play within sensory spaces the following section outlines a few key points in connection with the therapeutic potential of music and improvised play.

10.3 Music and Sound in Therapeutic Contexts

For many people, the idea of engaging in music-making activities may well also suggest a need to adhere to formal structure; harmonic progressions, rhythmic patterns, melodic phrases and so on. However, music can also operate at a much more abstract level where there are less formal rules to follow. Free jazz, for example, explores this notion whilst retaining some of the sounds and textures that perhaps instil the 'feel' of jazz. At a more fundamental level, free improvisation offers a considerable sense of freedom. However, the absence of agreed structure in improvised music does not negate the emergence of shared ideas, or the development and reiteration of phrases and patterns, nor does it exclude the exploration of tonality, meter and harmony. Perhaps more liberating is the notion that such formal musical constructs do not have to be predefined or even visited at all for music to happen. In essence, free improvisation encourages us all to play with sound, to offer new ideas and perhaps reflect those we hear around us, to embrace or resist the musical flow and to develop or simply abandon a train of musical thought.

This same concept of musical play is fundamental to key applications of music in therapeutic settings. Musical play exists in a clinical setting where a Music Therapist will use improvised music as a tool for creating a relationship with a client for therapeutic purposes; it also exists in Community Music where group-based musical activities can have therapeutic outcomes. Both Music Therapy and Community Music can employ game-like strategies for creating, sharing and generally communicating musically.

Clear definitions for both Music Therapy (and Community Music) are understandably difficult to create given their potential breadth and reach but the following is a meaningful working definition.

> Music therapy is a systematic process of intervention wherein the therapist helps the client to promote health, using music experiences and the relationships that develop through them as dynamic forces of change. Bruscia [11, p. 20]

Typically, music therapists are expected to be technically able musicians with good improvisatory skills. They will use these musical skills, perhaps over a prolonged period of time, to develop a relationship that can be described as being either a client-therapist relationship or a client-music relationship. In the client-therapist relationship music will be used in therapy, and the emphasis will be on the therapist as musician whereas in the client-music relationship, the music will be used as therapy with the emphasis being on the musician as therapist [11].

More an ideology, Community Music places significant emphasis on the individual values that can be gained from taking part in collective music making. Though there may not be specific therapeutic aims within Community Music as such there can, and often will, be therapeutic outcomes that can be observed (e.g. benefits from social inclusion, physical health and mental well-being). As with Music Therapy, the Community Music practitioner is likely to be a good improviser but though high technical skill might be an added strength it is not essential. The Community Music practitioner's real skills lie in being able to quickly respond to the needs and diverse abilities of a possibly unfamiliar group of people whilst constructing ways of making music in the moment.

There are philosophies for free improvisation in music (e.g. [12, 13]) and in the arts in general (e.g. [14]) that encourage us to embrace the notion of play in music and the applications and benefits of musical play in both Music Therapy and Community Music are quite apparent. In Community Music in particular, considerable emphasis is placed on understanding how best to encourage such play through introducing groups to game-like exercises using all manner of sound sources including those we can make with just our bodies alone. Community Music embraces the idea that anyone can create sound and therefore everyone will have something to offer as music, at its most fundamental level, can be thought of as organised sound. Moser and McKay [15] provide a comprehensive resource for Community Music practitioners along with various approaches to musical game-play [16, 17]. However, there are clear barriers for individuals with, for example, restricted mobility or perhaps learning difficulties that will impact on the range of instruments and therefore opportunities that might be available. It is here that assistive music-technologies can contribute greatly by enabling such individuals to engage in as expressive a way possible; translating limited movement into larger musical gestures, or mapping complex musical patterns onto less complex and more intuitive actions. In this respect, assistive music-technologies will include adapted and novel acoustic and digital 'instruments' including those that are almost toy-like, where the interaction is as important as the sounds the object can produce. Assistive music-technologies feature within both Music Therapy [18] and Community Music [19] and contrasting technologies were observed in use in the following study though a more comprehensive overview is provided in the section after this.

10.4 Sensory Spaces in Practice

In an attempt to cast some light on the contrasting perceptions of what a 'sensory space' can be, a short series of case studies has been carried out across a number of special needs schools in England and Wales. As a survey, both the choice of schools to visit and the subsequent interviews with practitioners has been deliberately open and as such there is no suggestion that any of the observations being offered here should form the basis for generalisation. There are some recurring themes but this is a small study and care should be taken not to suggest that these represent the norm and, whilst acknowledging that there may be occasional commonalities, greater emphasis has perhaps been placed on those comments and observations that are unexpected or novel. So, the study should perhaps be thought of as an exploratory journey; an opportunity to appreciate the variety of attitudes, techniques and resources that are currently in use. The findings are being shared here simply as examples of contrasting practice and though individual opinions and choices might inspire some discussion this is not meant to be judgmental or critical in any way. However, out of respect to those who have been helpful enough to share both their time and experience, no references are made to either individuals or the schools where they work.

In total, nine schools have been visited with many of these having a significant proportion of children with PMLD children though two of the schools had a higher proportion of autistic children. Each visit involved an open interview and a tour of any resources and spaces where sensory learning regularly takes place. On most occasions it was possible to see staff and pupils engaged with sensory activities though this was very much dependent on whether an activity happened to be timetabled at that same time. The case studies are not discussed separately, instead they are presented here alongside each other in reference to a number of themes that emerged across the study and though an identifying number is attached to each school this is only for reference purposes such that connections can be made between schools and themes. Although there is an underlying focus on the passive and active uses of sound and music within sensory spaces in the collected observations the approach to questioning was quite open, considering all approaches to working with sensory activities in purpose designed spaces.

10.4.1 Perceptions and Attitudes

Of the schools visited, six out of nine were generally positive about the use of sensory rooms within their provision. Within these six, if there were negatives identified, these generally related to the practicalities of maintenance and scheduling of activities rather than any underlying philosophical issues. In contrast to this, the key contact at School 8 was keen not to start a tour at the dedicated sensory spaces, instead preferring to focus on aspects of the general environment that had been designed to encourage a wide variety of 'organic' sensory experiences. As an IT coordinator, the interviewee was happy to promote the use of technology and virtual activities but against a backdrop of understanding where and why this might be appropriate. A key example that she offered was the opportunity for a child to interact with an animal (the school has chickens) with meaningful sensory stimulation from stroking, smelling, seeing the bird move, hearing it 'cluck' and so on. Her concern was that in gravitating towards the technological, the environmental can be overlooked and that this is a fundamental aspect of life. In this sense music is sound and any sound can be music. Much of the sensory work at School 8 was based on the Touch Trust[2] model and music and sound featured heavily within this but not necessarily within dedicated sensory spaces. Interestingly, technology was seen as being an asset within this for the simple reason that many musical instruments are quite simply inaccessible for the children at the school. Here, assistive music-technology was seen as having an appropriate and enabling role to play.

In a similar vein, the member of staff at School 1 was also cautious about a tendency to overly rely on technology and generic sensory equipment expressing a preference for Intensive Interaction. This approach to developing basic communication does not require a sensory space as such and though the school has two sensory trolleys there is no dedicated sensory space. The Intensive Interaction Institute[3] describes Intensive Interaction as:

> First and foremost, Intensive Interaction is highly practical. The only equipment needed is a sensitive person to be the interaction partner. The approach works by progressively

---

[2] http://www.touchtrust.co.uk

[3] http://www.intensiveinteraction.co.uk

developing enjoyable and relaxed interaction sequences between the interaction partner and the person doing the learning.

Nind and Hewett describe this in practice:

> The practitioner attempts to engage the learner in one-to-one interactive games with the emphasis being on pleasure first and foremost. This involves practitioners in modifying their usual body language, voice and face in order to make themselves attractive and interesting to their less sophisticated partners. A central principles is that the content and the flow of the activity follows the lead of the learner through the practitioners responding to her/his behaviour. Nind and Hewett [20, p. 8]

Employing this approach does not exclude additional sensory activities and music and sound are strong features of the daily activities that the children take part in. This same school has a visiting music therapist who also employs Intensive Interaction techniques by mirroring the movements and gestures of the child with complementary melodic and rhythmic phrases.

The other school that did not wholly embrace sensory spaces was School 4 where there was considerable criticism on the generic designs that exist within many SNE schools. The observations by members of staff (including the IT coordinator) at School 4 were that, in general, sensory rooms (gardens, trolleys and so on) are not designed and equipped by educators, more 'specified' by companies that have an interest in selling specialist equipment. This particular school was in the process of moving into a new building and staff had designed the plans for the sensory spaces after much reflection on a number of negative experiences of working with generic designs in a previous building. Two sensory rooms were being designed and, though a commercial company would ultimately install the specialist resources, the school's staff were the driving force behind the equipping of the spaces. This design strategy is elaborated further in the following section and the message was a simple one—the educators and practitioners need to have substantial input into the design of these spaces else the same issues will keep resurfacing due to the generic nature of sensory space design specifications.

One common theme that appeared was that, where a school had a 'traditional' sensory room similar to the Snoezelen model, the sensory space was generally regarded as simply being a good place for one-to-one activities. This was frequently reflected in the timetabling structure where individual members of staff would be offered set times within the space which they could then use in whichever way suited their needs.

10.4.2 The Spaces

At the time of the visits, all the schools had specialist commercial sensory equipment but only School 1 did not have a dedicated sensory space, instead having two sensory trolleys. Schools 7 and 8 both had light and dark sensory rooms though both the activities and resources being used suggested that these were mainly regarded as general-purpose sensory spaces. The remaining schools all had at least one dedicated sensory space and possibly some additional sensory corners set up within one or more classrooms. Schools 2 and 6 stood out as having considerably more investment in sensory spaces and commercial resources than the other schools. Both were relatively new buildings containing one or more sensory rooms, a sensory corridor, sensory garden and a hydrotherapy-pool with additional sensory resources. Though the individual types of resources are discussed in the following section, it is worth mentioning that the sensory spaces in these two schools were technology rich yet, in contrast to this, these were not necessarily the spaces that became the most memorable. Again, this is perhaps partly the result of recognising similar technologies in many different places such that when a space is seen that contains none of this same technology it simply stands out; someone has gone to considerable effort to create and equip a space with all the right objects and materials for a very specific group of users. Such a space was seen at School 7 where the backroom to an arts-classroom had been converted into a DIY sensory space. More will be made of the kinds of resources that were in this space in a later section but the point of interest here is the notion of generic designs versus small-scale ad-hoc design where the latter immediately seems more in touch with its users. One final observation worth discussing is that most staff mentioned taking additional sensory activities into the dedicated spaces rather than relying on the specialist equipment within. Without undertaking a comprehensive survey on specific usage it would be difficult to determine the individual worth of commonly included technologies but this practice does reinforce the earlier observation on one-to-one activities.

10.4.3 Specialist Technologies

As mentioned earlier, the sensory resources in all the schools visited tended to be drawn from roughly the same set: bubble-tubes, fibre-optic lighting, light-wheel, infinity tunnel, interactive switchboard, mirror-ball, audio playback. In talking to staff though, it was not clear why these particular technologies should be included specifically beyond being part of the original fit-out. With Pagliano's observations on the evolution of the Snoezelen in mind and how the emergence of the discotheque was influential within this there is at least some explanation as to why these technologies might be in use but their inclusion also appears both prescriptive and inflexible. The same was true for sensory gardens in that the same technologies tended to be present including interactive switches that trigger sounds, wind-chimes, interactive water-fountains and bubble-machines. The IT coordinator at School 5 gave an account of the original installation of the equipment for the school's sensory garden whilst also pointing out that most of the technology no longer worked. When first installed, there were two artificial plants with switches that when pressed would produce a sound. However these sounds were relatively unconnected (an animal noise for example) but when the system was checked to see how the sounds could be changed (e.g. to make the sound of a bee), it became clear that only one of the company's engineers could really alter the sound set; the sounds have not been changed as a result.

Some of the schools had interactive floor-projectors; perhaps most notably the two schools mentioned earlier that had clearly invested quite substantially in sensory equipment throughout the entire school. The IT specialist at School 7 was using interactive projections but was also not keen on the way in which they tended to be set up. He suggested that there is a tendency to use them in their default or demo settings (disturbing leaves on a surface, revealing parts of pictures etc.) even if this may not be a meaningful interaction for the child taking part. He described how he had set up such a device to track a ball that a child would throw such that a trail could be added visually. In this context, he added, the child is having his or her own actions reinforced in a meaningful way, helping them understand the path of the ball in relation to their own movement and all within quite playful and enjoyable interaction. So he was using technology but tailoring its use to very specific needs and learning outcomes.

Before moving on to talk about repurposed technologies, it is worth reflecting on the observation offered whilst at School 5 that much of the sensory garden equipment no longer worked. This was quite a common observation across the different visits with references to equipment failing easily and ultimately being costly to replace and/or maintain. However, there was another factor as to why some equipment was not being used and this related to complexity. Certain technologies are easy to use and work with whilst others require some level of training to allow the activities to move away from the pre-programmed or default settings. The control software for the hydro-pools is a good example of this with consensus from all schools that this particular resource can be incredibly complex to work with. Anecdotally, this same software has also been trialled by two specialist theatre technicians[4] with similar results. Regardless, this software still regularly forms the basis for control within new installations of this type. On a smaller scale, it is often the case that a school will have one or more Soundbeam systems yet these will not always be in use within sensory activities. There is an element of training required to fully understand how to use these devices and this perhaps leads to them being perceived as highly specialist; it is not uncommon to find one member of staff being identified as either the main or only user of the system. As described earlier, assistive music-technologies can be hugely enabling and expressive devices, allowing children to create and control music and sound in a quite magical way. However, the IT coordinator at School 5 explained that, in her experience, the devices were often allowed to fall into disuse as a result of inexperience.

---

[4] Two members of specialist technical staff working in a university theatre space were asked to trial and appraise the software in an effort to better understand how complex the system is. Even with professional knowledge of controlling theatre lighting and sound, the system was quickly identified as being unintuitive to both program and operate.

10.4.4 Repurposed Technologies

There have been a number of examples of adapted, found and repurposed technologies in use within some of the spaces visited. The art-classroom sensory space in School 7 was laid out like a fantasy grotto, with Christmas fairy lights adorning the side wall and all manner of tactile materials attached to surfaces and objects. Plastic drinks bottles were hanging in a row from the ceiling to make a percussion instrument and the girl that was in the space at the time was clearly enjoying herself with the various sensory opportunities that were being offered to her. In the same school, the IT coordinator had been experimenting with off-the shelf gaming devices like the Microsoft Kinect to capture movements and map them to sound and images. Drawing partly on the work of Keay-Bright [21] whilst also repurposing software intended for the VJ market, this tutor was clear to point out why he was turning to these types of technologies. Some of the children he works with have difficulty working with the kind of switch-technology that might normally be used to trigger sound and music, so rather than sidestep those sensory opportunities he had been looking for non-contact approaches to interaction. As with a number of the other schools, he was familiar with Soundbeam for music making activities of this type but had found a highly cost-effective and versatile alternative platform in Kinect. He was also exploring the use of smartphones and tablets for similar reasons, not so much because of the touch-sensitive surface but more for the touch-free opportunities that are offered from image tracking using the inbuilt camera. There are now a number of apps that use this technique to provide Theremin and Soundbeam style interaction with music and sound. Indeed, this interest in mobile technologies as tools for mapping actions to sound and music was echoed by practitioners at many of the other schools though all of them acknowledged that given the context in which they are being repurposed, they are generally not robust enough.

In School 5, the IT coordinator explained that of all the technology that the school had within its sensory spaces, the most successful was probably an old disco sound-to-light unit; this was housed within the main sensory room and was known to be a popular activity with many of the children. The reasoning behind this seems to be that the child is much more influential in the final outcomes as it is their own actions that appear to create the light patterns. There is a common picture here as many of the activities that are built into the generic sensory rooms are generally quite passive or involve basic cause-and-effect actions (press a switch, hear a sound). The IT coordinator at School 5 was keen to identify this as being an outmoded attitude that perhaps dates back to a period where switch-based access to control/communication was seen as being an essential component of developing basic life-skills. She added that technology had progressed so rapidly that there is now a far richer set of assistive tools available such that approaches to interaction ought to be richer and more flexible yet this often does not seem to be the case.

One of the teachers in School 3 demonstrated a collection of early learning toys and instruments that had been repurposed for use by some of the children with particularly complex needs. Though the school is well equipped with sensory spaces and specialist technologies, the teacher described how some of the toys she is using are perhaps more likely to make a connection as they were simply fun to use. Her emphasis was very much on encouraging interaction through play and she also suggested that the tangible nature of the objects and their natural multisensory nature were all appealing within this. Some of the children in her group have visual impairments in addition to profound learning needs so toys and gadgets that make sound are of great interest, but particularly so if the tactile and haptic stimulus offered is engaging too.

10.4.5 Activities (Music and Sound)

The auditory interactions observed (and heard) can broadly be defined as being either passive or active. Passive activities would use music and sound as a listening experience, perhaps to accompany another activity but possibly as a passive activity in its own right; listening to an audio story for example. As an active interaction, the sound or music was being created or influenced at some level by the individual engaging with the object (instrument, toy, gadget). Of interest here is the level of freedom the individual was afforded whilst playing with the sound objects. For example, were any choices available or was the interaction perhaps limited to a small and discrete set of outcomes?

For many of the dedicated sensory spaces, music and sound appeared to be either passive or very limited cause-effect style interaction. It was observed that music was often used as a background component to create a relaxing atmosphere rather than offering opportunities for the sound to be either created or controlled. This was understandable in an environment like a hydrotherapy-pool where another activity was being facilitated but was perhaps a little surprising in the dedicated sensory spaces. Even in the dedicated spaces, the emphasis on active interaction seemed to lie predominantly with tactile and visual stimuli rather than auditory; where active interaction was observed it appeared to be in the larger classroom areas. A common theme that emerged across most of the schools was the idea that sensory activities could occur quite easily and fluidly outside of a dedicated-space such that these spaces were only really used where one-on-one interaction was an absolute requirement. The following are a few examples of ways in which staff used music and sound based interaction outside of sensory spaces but within structured sensory activities. It should be added that all schools were exploring sound and music at some level and that staff connected with these activities were keen to point out that PMLD children respond quickly and enthusiastically to sound-focused interaction. It was also suggested that an auditory-stimulus is likely to receive a stronger and more positive response than that from visual-stimulus.

In School 1 (the school without a sensory space), a key member of staff described how music and sound featured throughout the day within a range of sensory activities that also included Intensive Interaction; this was a relatively small class of PMLD children some of whom were described as being pre-speech. Each morning, the group would take part in a musical activity where a beat would be created, perhaps as a rhythm to a piece of music, perhaps as part of a story, or perhaps as part of a name-game. The group would use hand-percussion and, where this was too difficult for the child, they could work with a support-teacher to create a sound. The same was true of speaking names in the name game where support-teachers might work with the child to say their name with and/or for them perhaps using technology (e.g. BIGmack® switch) to enable the child to trigger a sound or a recording of their name. The member of staff was keen to point out how meaningful it was for the children to be able to hear and 'say' their name in this way adding that each of the children had a recorded personalised story that could be played to them, often with key sounds that would bring additional meaning to the story. She also reinforced how important rhythm was in the daily activities describing how one particular girl would latch onto the rhythmic footsteps within an audio-story they used quite frequently. The children required little encouragement to make rhythm in this way and so rhythm featured quite heavily where possible.

School 1 was also the same school where Intensive Interaction had been adopted as a means of encouraging and developing communication. Intensive Interaction considers that, in its broadest sense, communication is really derived from a set of characteristics that are not wholly reliant upon speech: learning to give brief attention to someone else, sharing attention, learning to concentrate on another person, developing shared attentions into activities, taking turns in changes of behaviour and so on. All against a backdrop of having fun through play. So, an Intensive Interaction session might begin by creating the opportunity by which a child might give brief attention to someone else. Someone might mimic the physical or vocal gestures a child is making such that their attention is drawn by this apparent repetition of their own actions. The child moves, the teacher moves in the same way and a connection is made. An example of this in practice occurred quite organically during the visit where a young girl had picked up a set of handbells and was shaking them rhythmically. The support-teacher joined in, copying the rhythm. When the girl swapped the bells to her other hand, the support-teacher did the same and when the girl swapped again, so did the support-teacher; game-play had occurred in a most natural sense and with no need for a set of rules.

This playful mirroring of actions went on for a substantial period of time and the key member of staff explained how the exchange that was happening was quite indicative of the way in which Intensive Interaction could work with sound, adding that the children were generally much more responsive to sound and music than they were to light and visuals. It was because of this that so many of the activities they use incorporate sound and music in some way. She also added that as the children became more responsive to these game-like exchanges they would occasionally respond to each other's rhythmic sound with one of their own. So, communication at a very fundamental yet meaningful level was being achieved with pre-speech children, perhaps simply as an indication of 'I'm here' or establishing that 'I'm part of this'. The Intensive Interaction sessions were complemented by similar sessions with a music therapist who would use musical gestures to mimic and copy the physical gestures or vocalisations of the children.

In School 8, two members of staff were creating sensory spaces within standard classrooms using a variety of technologies and materials. Though the school had dedicated sensory spaces these were seen as being timetabled areas where one-on-one work might happen whereas these sensory spaces were being designed to be group based. The environment would be set in place for perhaps a week or so at a time or until most of the children had been given the opportunity to experience it. The overall experience might be themed (the seaside, the supermarket, the forest etc.) with a multisensory approach being used to create an ambience. The staff were working from the arts-focused Touch Trust model and were using a combination of music-technology (Soundbeam etc.), lighting, projected images, fabrics, textured materials and even smells. Unlike some other uses of audio within sensory experiences there was considerable emphasis on active interaction such that there were opportunities for the children to become part of the music. As with the previous school this might involve using assistive switched-technology (BIGmack®) to trigger sound but might also involve the child making sound through continuous movement (Soundbeam).

In a similar way to School 8, staff at School 2 were also creating themed sensory environments outside of dedicated spaces, most notable in a corridor where a similar combination of technologies and materials was in use and, as with School 8, very much aimed at group-based activities. A lightweight framework was used to allow objects and lights to be held in place such that children could move in and around the environment. Similar to School 8, the audio was also being triggered using Soundbeam but here there were a number of these devices in use, each creating one part or layer of an overall musical texture. Though sound could be passive as the children interacted with the other elements of the environment they could happen across a 'beam' and produce a sound. This would change or contribute to the background ambience in some way leading to naturally occurring play with the child hearing the musical interpretation of his or her own actions. Within a context of timetabled activities this was probably quite effective though it was also clear that for those children who were making their way from one location to another it could be quite confusing. Indeed, personal experience has demonstrated that PMLD children with a visual impairment can find this sudden exposure to unexpected sound quite distressing. Ultimately, it was not made particularly clear why the corridor was regarded as being a better place to create a sensory space than a conventional classroom.

School 5 was also using resonance boards in certain activities. These are simple plywood platforms that allow a child to feel vibrations from their own interactions with the surface of the board. Very simple in construction, these are essentially a homemade resource based on an original concept by Lilli Nielson, the Danish psychologist, special education adviser and pioneer in Active Learning. Though the main function is one of transmitting vibrations to a child who is touching or possibly sitting on the board, the vibrations can originate from a sound source such that speaking onto the board or touching noise-making objects onto the surface will work well. Often, the board itself is used as the sound source with children exploring the surface by scratching or tapping to create both noise and vibration.

10.5 Technologies for Musical Play

There are numerous new and novel technologies that assist with musical improvisation and game-play and these can be placed into three broad categories: mainstream commercial technologies, specialist assistive technologies and research-based or 'novel' technologies. Mainstream commercial technologies can be thought of as those that are specifically aimed at a reasonably large market of people who wish to make music but have little or no formal training. These range from technologies that are aimed at a domestic market for home entertainment (including smart-phone apps and gaming devices) to a more professional end of the market (DJs, dance music production) where accessible but more musically sophisticated tools are in considerable demand. In the specialist assistive category, there are those musical-tools and instruments that have been designed primarily for use within special needs education and/or therapeutic settings (e.g. Soundbeam, MIDI Creator etc.). Whilst in the research/novel category, there are emerging technologies that may offer musical opportunities but are perhaps not widely available. It would be possible to discuss these technologies within the three suggested categories however there are similarities of technology and design that exist across two or more of the categories. With this in mind, some key technologies are presented in the following section grouped by virtue of the types of interactivity on offer.

10.5.1 Pressure-Sensitive

There are a range of pressure sensitive digital musical instruments (DMIs) available commercially many of which are designed to take the place of conventional hand percussion (congas, djembes, tablas, chimes etc.). There are obvious benefits of being able to replicate a wide variety of acoustic percussion instruments electronically in this way and easily switching between different and possibly quite cumbersome or heavy items is just one of these. Of the commercially available DMIs that provide pressure sensitive pads, the Korg Wavedrum and Roland Handsonic both stand out as being both accessible and versatile. The Wavedrum provides a single membrane (very much like that on an acoustic snare drum) to strike with fingers, hands or sticks. Designed to sit in a stand or on table or lap, there are sensors under the skin and within the rim that respond in different ways such that a variety of tones and timbres can be achieved. For example, tapping the rim might trigger a complementary sound to the one achieved by hitting the main skin which; the surface also lends itself to being scratched to produce different tones. Other than providing easy, expressive and responsive access to a wide variety of hand percussion sounds, it also provides an array of pitch sounds and looped samples to work with such that a single 'hit' can produce a substantial outcome.

In contrast to the Wavedrum, the Handsonic offers an array of pressure pads (10 or 15) arranged on a surface not dissimilar to the size of a conventional snare drum. The ten-pad version in particular is as easy to work with on a lap as it is on a table top or stand and the pads can be easily and effectively controlled with the fingers and/or palm of a single hand if needs be. The individual pads are particularly sensitive such that delicate fine taps can be achieved whilst not having to strike the pads that hard to reach much higher volumes. Other than offering a similar range of percussion sounds to the Wavedrum, the Handsonic also offers a number of pitched percussion patches to work with (xylophone, steel drum etc.). Both versions of the instrument have a Roland D-Beam built in that allows hand gestures to be mapped to sounds or processes (this and similar technologies are discussed in a following section). The larger, 15 pad, instrument also offers two ribbon strips for similar purpose.

10.5.2 Touch-Sensitive

Aimed primarily at the DJ market, the Korg Kaoss Pad was originally presented as an intuitive method of bringing real-time SFX processing to any audio the user might care to input (music, speech etc.). Although not functioning as an instrument as such, the processed audio can be captured as loops or single-shot sounds for further playback. In the same way that the BIGmack® can be used to quickly capture sound for playback so can devices like the Kaos Pad making them useful tools for recording key sounds and words to be used in a name-game for example. The SFX processing is controlled using a touch sensitive surface with the X and Y dimensions being mapped to different controllable parameters (volume, modulation, effect or filter depth etc.) and the housing is compact and easily held on a lap or table top. The buttons and switches are sensibly laid out with controls for key functions being immediately accessible and particularly intuitive. Much of the functionality of the device is easily accessed from the available controls.

In contrast to the Kaos Pad, the Kaossilator was designed more as an instrument. Interacting with the main pad triggers musical sounds, either individually or according to rules (e.g. arpeggios or rhythm patterns). For example, whilst using a lead instrument setting, the main pad might be mapped as pitch across the X-axis and volume or tone across the Y-axis. For users with limited dexterity, these can be incredibly powerful and expressive sound producing devices.

10.5.3 Movement-Sensitive

There are a number of DMIs that respond to non-contact body movement and the inspiration for many of these perhaps comes from the theremin. Developed by Leon Theremin in the early 1920s, the theremin is a touch-free instrument that a performer plays by moving his or her hands near to two antennae; one to control pitch and the other to control volume. The sound is continuous, responsive, highly expressive and quite ethereal in quality. Indeed the unnatural quality of the instrument's musical tones quickly led to its sound becoming synonymous with the cinematic portrayal of alien or future worlds. The instrument is very accessible as there is no requirement for individual finger dexterity and, though the instrument was designed to be used with hands, it can be accessed with any body part that is convenient. This is a feature that Magee [18] comments on, suggesting that regardless of its historic status, the instrument still has a place in mainstream Music Therapy.

There are several music controllers that are theremin like in terms of interaction. Soundbeam is almost certainly the most well known of these and is commonly used in special needs education. Using an ultrasound sensor, an invisible beam is projected across a space and, when interrupted, specific notes, sounds or events will be triggered. The length of the beam and mapping of notes within this are all adjustable such that small movements can be captured (e.g. head motion) as easily as larger whole body movement (e.g. dance). MIDI Gesture operates in a similar way, as does the D-Beam component of many Roland products though this uses infrared technology over a much shorter distance[5] than ultrasound can offer easily. Although not commercially available, the Octonic [23, 24] is a non-contact DMI that offers eight mini-beams, each of which control a different note or sound such that the player can play notes individually or by gesturing across the whole set. Originally aimed at users with mobility issues, the instrument is designed to be easy to set-up as well as play.

In a similar vein, Beamz uses a series of lasers to create a small light-harp. Breaking a beam triggers a musical phrase that is superimposed on top of a prerecorded piece and phrases are synchronised such that they fit and complement the backing music. Though the player has no direct control over the notes that are produced, the experience of triggering and releasing pre-recorded phrases across a backing is hugely entertaining.

It should also be acknowledged that popular gaming technologies (e.g. Nintendo Wiimote and Microsoft Kinect) are commonly used to create controllers for music and sound in conjunction with music software systems like MAX/MSP from Cycling '74. This is very much in the domain of the DIY and 'hacker' electronic-

---

[5] The specified distance for sensing movement with the D-Beam sensors is under 1 m though Brooks [22] has demonstrated how this can be extended greatly to around 12 m by using retroreflective microprism; a highly reflective material often found safety clothing and equipment.

musician, but the potential for creating accessible systems using such readily available and affordable systems is quite apparent. In a similar vein, it should also be acknowledged that smartphones are also being used for similar purposes, accessing the various embedded sensors to create innovative sound controllers.

10.5.4 Switch-Access

Conventional switches as commonly used in special needs education, can be included into some of the systems just described (e.g. Soundbeam and MIDI Creator[6]) where they can be used to trigger additional events. However, recent developments in music-making consumerism have led to a widespread appearance of MIDI control surfaces. Primarily aimed at providing access to recording software, such surfaces can be used to send MIDI messages to a computer where they are interpreted into actions. Using appropriate software, it is possible to use such interfaces to trigger recorded loops and samples. This could take the BIGmack® approach to playing back meaningful sounds to another level where a larger set of connected or themed sounds are available. These could be used as part of story-telling or game play in the ways described earlier but with a richer set of opportunities for interaction and exploration.

10.6 Reconsidering the Design and Use of Sensory Spaces

The original aim behind this short review of current practice was to consider how sound and music might be used most effectively within sensory spaces but in carrying out the study it quickly became apparent that there is no real consensus of opinion over the effectiveness of such spaces, their usage or their design. In some respects, this might have taken the overall aim slightly off target but there have been key themes emerging along the way that could still influence the design and use of audio-based sensory activities and ultimately the environments in which they might be employed. It seems possible that the original notion of the sensory space is becoming out-dated such that there is no real need to take an individual to a dedicated sensory room. With this in mind, it is understandable that sensory trolleys might become such a familiar feature though the equipment they contain is still quite standardised and with generally only passive applications for sound. Yet outside of this commercial sensory equipment there is evidence to suggest that individual schools and staff members are finding novel means for interacting with

---

[6] Swingler [25] provides or a comprehensive overview of the Soundbeam system and its applications in special needs education and Abbotson et al. [26] provide similar for MIDI-Creator.

sound and music on a needs-led basis. In some respects, it could be a fitting conclusion to reach by suggesting that there is quite simply a need to establish an effective network and repository for sharing observations, concepts, activities and appropriate technologies across as many special needs schools as possible. In the same way that School 4 elected to take a much more involved and prescriptive role in designing their new sensory spaces, it seems only reasonable to suggest that the same should be true for the majority of sensory spaces. After all, who is most likely to have the richer more informed understanding of the therapeutic and developmental needs of the individual than the practitioner who is already working so closely with them? So, acknowledging the effectiveness of ad-hoc, small-scale spaces built around the needs of the individual and gently moving away from the generic commercial resource may well be the way forward but where does music and sound figure within this?

In many of the commercially designed and resourced sensory spaces, music and sound were typically used in a passive sense (as an ambient backdrop to other activities) yet the consensus from the staff being interviewed was that many PMLD children respond more noticeably to sounds in place of visuals. The evidence from observing Intensive Interaction in practice showed that games could emerge quite organically where the 'rules' of the game are established during play. Rhythm, pulse and repetition were cited as having considerable influence in encouraging individual children to engage, as was the importance of meaningful sounds (names, familiar noises etc.). Empowering the individual to move freely and explore seems key within this such that a richer interaction can be achieved where a reliance on quite prescriptive cause and effect activities can perhaps be avoided. Where Intensive Interaction was observed, it was also noted that the visiting Music Therapist adopted these same approaches but by mimicking physical gesture with sound. It could be, though, that given an appropriate and enabling technology these initial gestures could have a sonic output in the first place. Specialist resources were being used in some schools to trigger sounds but these would typically involve a single action (button press) rather than being dynamically responsive (capturing gesture for example). Indeed, some of the exploratory work in one or two of the schools visited suggested that there are intuitive means for encouraging and enabling more dynamic interaction using mainstream technologies including gaming devices, smart phones and touch tablets. A quick review of other available music-technologies shows that there are numerous devices that will enable the playback of custom recorded sounds and though these are aimed at a mainstream market they are occasionally very accessible.

There is a real opportunity here to bring together a number of simple concepts to create an engaging environment for musical play within a context of sensory development. Using mainstream music technology (pads to hit, surfaces to explore, gestures to make) can already provide access to a rich variety of sounds. By itself this could be the means for enabling key aspects of basic communication but the evidence would suggest that access to personally meaningful sounds within this set could be even more effective. The early morning musical-games witnessed at School 1 used recorded sounds that 'belonged' to a specific child, as did the child-specific recorded stories where familiar names and sounds were included. It is not a huge leap beyond this to embed or attach a personalised set of such sounds to a specific device, one that can be chosen to best match the needs or preferences of the individual. Where the magic of gesturing in invisible beams works well for one child it is likely that another will perhaps respond more keenly with a tactile or haptic element. Ultimately, the musical play will exist within the concept of controlling meaningful and personalised sounds and the role of technology can be a simple one; enabling that play to take place.

References

1. Hulsegge, J., Verheul, A.: Snoezelen: Another World. Rompa, Chesterfield (1988)
2. Pagliano, P.: Using a Multisensory Environment: A Practical Guide for Teachers. David Fulton Publishers, London (2001)
3. Pagliano, P.: The Multisensory Handbook: A Guide for Children and Adults with Sensory Learning Disabilities. Routledge, London (2012)
4. Bozic, N.: Constructing the room: multi-sensory rooms in educational contexts. Eur. J. Spec. Needs 12(1), 54–70 (1997)
5. Vlaskamp, K., de Geeter, K.I., Huijsmans, L.M., Smit, I.H.: Passive activities: the effectiveness of multisensory environments on the level of activity of individuals with profound multiple disabilities. J. Appl. Res. Intellect. Disabil. 16, 135–143 (2003)
6. Stadele, N.D., Malaney, L.A.: Effects of a multisensory environment on negative behavior and functional performance on individuals with autism. J. Undergraduate Res. IV, 211–218 (2001)
7. Slevin, E., McClelland, A.: Multisensory environments: are they therapeutic? A single-subject evaluation of the clinical effectiveness of a multisensory environment. J. Clin. Nurs. 8, 48–56 (1999)
8. Hogg, J., Cavet, J., Lambe, L., Smeddle, M.: The use of Snoezelen as multisensory stimulation with people with intellectual disabilities: a review of the research. Res. Dev. Disabil. 22(5), 353–372 (2001)
9. Lotan, A., Gold, C.: Meta-analysis of the effectiveness of individual intervention in the controlled multisensory environment for individuals with intellectual disability. J. Intellect. Dev. Disabil. 34(3), 207–215 (2009)
10. Mount, H., Cavet, J.: Multisensory environments: an exploration of their potential for young people with profound and multiple learning difficulties. Br. J. Special Educ. 22(2), 52–55 (1995)
11. Bruscia, K.E.: Defining Music Therapy. Barcelona Publishers, Lower Village (1998)
12. Bailey, D.: Improvisation: Its Nature and Practice in Music. Da Capo Press, Cambridge (1992)
13. Stevens, J.: Search and Reflect. Rockschool, UK (1986)
14. Nachmanovitch, S.: Free Play: Improvisation in Life and Art. Penguin Group, New York (1990)
15. Moser, P., McKay, G.: Community Music: A Handbook. Russell House Publishing, Lyme Regis (2005)
16. Lewis, S.: Drumming, silence and making it up. In: Moser, P., McKay, G. (eds.) Community Music: A Handbook. Russell House Publishing, Lyme Regis (2005)
17. Nankivell, H.: Making new music: approaches to group composition. In: Moser, P., McKay, G. (eds.) Community Music: A Handbook, pp. 79–98. Russell House Publishing, Lyme Regis (2005)

18. Magee, W.: Electronic technologies in clinical music therapy: a survey of practice and attitudes. Technol. Disabil. IOS Press (18), 139–146 (2006)
19. Healey, R.: New technologies and music making. In: Moser, P., McKay, G. (eds.) Community Music: A Handbook, pp. 161–179. Russell House Publishing, Lyme Regis (2005)
20. Nind, M., Hewett, D.: Access to Communication: Developing Basic Communication with People Who have Severe Learning Difficulties, 2nd edn. David Fulton Publishers, London (2006)
21. Keay-Bright, W.: Designing interaction though sound and movement with children on the autistic spectrum. In: Brooks A.L. (ed.) Arts and Technology, Proceedings of ArtsIT 2011. Lecture Notes of the Institute for Computer Sciences, Social-Informatics and Telecommunications Engineering, vol. 101. Springer (2012)
22. Brooks, A.L.: Enhanced gesture capture in virtual interactive space (VIS). Digit. Creativity **16**(1), 43–53 (2005)
23. Challis, B.P., Challis, K.: Applications for proximity sensors in music and sound performance. In: Proceedings of 11th International Conference on Computers Helping People with Special Needs, ICCHP Springer Lecture Notes in Computer Science, vol. 5105, pp. 1220–1227 (2008)
24. Challis, B.: Octonic: an accessible electronic musical instrument. Digit. Creativity **22**(1), 1–12 (2011)
25. Swingler, T.: The invisible keyboard in the air: an overview of the educational, therapeutic and creative applications of the EMS Soundbeam. In: Proceedings of 2nd European Conference on Disability, Virtual Reality and Associated Technologies, pp. 253–259 (1998)
26. Abbotson, M., Abbotson, R., Kirk, P.R., Hunt, A.D., Cleaton, A.: Computer music in the service of music therapy: the MIDI grid and MIDI creator systems. Med. Eng. Phys. **16**, 253 (1994)

Part III
Technologies for Well-Being

Chapter 11
Serious Games as Positive Technologies for Individual and Group Flourishing

Luca Argenton, Stefano Triberti, Silvia Serino, Marisa Muzio and Giuseppe Riva

Abstract By fostering continuous learning experiences blended with entertaining affordances, serious games have been able to shape new virtual contexts for human psychological growth and well-being. Thus, they can be considered as Positive Technologies. Positive Technology is an emergent field based on both theoretical and applied research, whose goal is to investigate how Information and Communication Technologies (ICTs) can be used to empower the quality of personal experience. In particular, serious games can influence both individual and interpersonal experiences by nurturing positive emotions, and promoting engagement, as well as enhancing social integration and connectedness. An in-depth analysis of each of these aspects will be presented in the chapter, with the support of concrete examples. Networked flow, a specific state where social well-being is associated with group flourishing and peak creative states, will eventually be considered along with game design practices that can support its emergence.

L. Argenton (✉)
Centre for Studies in Communication Sciences—CESCOM, University of Milan-Bicocca, Building U16, Via Giolli, angolo Via Thomas Mann 20162 Milan, Italy
e-mail: argenton@campus.unimib.it

S. Triberti · G. Riva
Department of Psychology, Catholic University of Sacred Heart, Milan, Italy
e-mail: stefano.triberti@unicatt.it

G. Riva
e-mail: giuseppe.riva@unicatt.it

S. Serino · G. Riva
Applied Technology for Neuro-Psychology Lab, IRCCS Istituto Auxologico Italiano, Milan, Italy
e-mail: s.serino@auxologico.it

M. Muzio
Department of Sport Sciences, Nutrition and Health, University of Milan, Milan, Italy
e-mail: marisa.muzio@unimi.it

A. L. Brooks et al. (eds.), *Technologies of Inclusive Well-Being*,
Studies in Computational Intelligence 536, DOI: 10.1007/978-3-642-45432-5_11,
© Springer-Verlag Berlin Heidelberg 2014

Keywords Positive psychology · Positive technology · Serious games · Networked flow

11.1 Introduction

Serious applications for computer game technologies have become important resources for the actual knowledge society. Their use and effectiveness have been broadly acknowledged in different sectors, such as education, health, and business. By fostering continuous learning experiences blended with entertaining affordances, serious games have been able to shape new virtual contexts for human psychological development and growth. They have in fact supported the creation of socio-technical environments [1], where the interconnection between humans and technology encourages the emergence of innovative ways of thinking, creative practices, and networking opportunities. Further, serious games have been capable of supporting wellness and promoting happiness. That is why they can be considered as "positive technologies".

Positive Technology is an emergent field based on both theoretical and applied research, whose goal is to investigate how Information and Communication Technologies (ICTs) can be used to empower the quality of personal experience [2–4].

Based on Positive Psychology theoretical framework [5, 6], Positive Technology approach claims that technology can increase emotional, psychological and social well-being. Hence, positive technologies can influence both individual and interpersonal experiences by fostering positive emotions, and promoting engagement, as well as enhancing social integration and connectedness.

Starting from an introductory analysis of the concept of well-being as it has been framed by Positive Psychology research, this paper will reflect on the nature and the role of serious games as positive technologies. In particular, it will discuss how they can support, and train the optimal functioning of both individuals and groups, by contributing to their well-being.

Finally, it will present a practical exemplification of a serious game developed to promote networked flow, a peak creative state achieved when team members experience high levels of presence and social presence in a condition of "liminality" [7].

11.2 Positive Psychology

Inspired by the psychological and philosophical tradition that focused on individual growth and human empowerment, Martin Seligman and Mihaly Csikszentmihalyi officially announced the birth of Positive Psychology in 2000.

Within the first twenty-first century issue of the *American Psychologist*, the two authors identified an epistemological and theoretical limit for the modern psychology in the emphasis given to the study of human shortcomings, illnesses, and pathologies [6]. Therefore, psychology was perceived as "half-baked" [8], and a change in its focus from repairing deficits to cultivating human flourishing took place.

In this way, Positive Psychology emerged as the scientific study of "positive personal experience, positive individual traits, and positive institutions" [5, 6]. By focusing on human strengths, healthy processes, and fulfillment, Positive Psychology aims to improve the quality of life, as well as to increase wellness, and resilience in individuals, organizations, and societies.

Rather than representing a new formal sector or a new paradigm, Positive Psychology is a novel perspective to studying human behavior that encompasses all areas of psychological investigation [9]. Thus, the link with accurate and scientific methodological practices [10] has become the engine of interventions to study and promote the optimal expression of thought, emotions and behaviors. In particular, according to Seligman [11], three specific existential trajectories may be explored to reach such a complex goal:

- the *pleasant life*, marked by the presence of pleasure, and supported by positive feelings and emotions about the past, present and future;
- the *engaged life*, achievable through the exploitation of individual strengths and virtues, by pursuing enjoyable activities that are absorbing, and involving;
- the *meaningful life*, based on the possibility of identifying and serving of something larger than one's individual self.

Similarly, Keyes and Lopez [12] argued that positive functioning is a combination of three types of well-being: (a) high emotional well-being, (b) high psychological well-being, and (c) high social well-being. This means that Positive Psychology identifies three characteristics of our personal experience—affective quality, engagement/actualization, and connectedness—that serve to promote personal well-being. We will go into depth in each of them in the attempt to understand how they can be enhanced by technology and serious games.

11.3 The Hedonic Perspective: Fostering Positive Emotional States

The *pleasant life* and high emotional well-being have been investigated by the so-called hedonic perspective. It owes its name to the ancient greek vocabulary, where the word *edonè* means pleasure. Interpreted by Positive Psychology scholars as an essential feature of human nature, this construct encompasses a family of psychological experiences that make us feel good [13].

Although in an implicit form and with a different vocabulary, the concept of pleasure appeared several times in the psychological domain. The psychoanalytic model—in which the pleasure principle assumed a role of primary importance—or the behavioral approach—where pleasure focuses on the logic of reinforcement— are just two examples [13]. However, only thanks to the work of Kahneman et al. [14], hedonic psychology was eventually conceptualized as the study of "what makes the experience pleasant or unpleasant".

Among the different ways to evaluate pleasure in human life, a large number of studies have focused on the concept of subjective well-being (SWB), defined as "a person's cognitive and affective evaluation of his or her life as a whole" [15, 16]. Thus, SWB implies both emotional responses to life events, and cognitive judgment of personal satisfaction. At the cognitive level, opinions expressed by individuals about their life as a whole, and the level of satisfaction with specific life-domains, such as family or work, becomes fundamental. At the emotional level, SWB is indeed related to the presence of positive emotional states and the absence of negative moods.

This point is of particular interest to the hedonic perspective. Unlike negative emotions, that are essential to provide a rapid response to perceived threats, positive emotions can expand cognitive–behavioral repertoires and help to build resources that contribute to future success [17, 18]. The salience of positive emotions in increasing well-being has been recently highlighted by the "broaden-and-build" model [17, 18]. According to Fredrickson, positive emotions broaden, on the one hand, the organism's possibilities with undefined response tendencies that may lead to adaptive behaviors and mitigate the impact of negative stressors. The elicitations of positive emotions, for example, make attentional processes more holistic and gestaltic [19], stimulate a more flexible, intuitive [20], receptive and creative thinking [21].

Further, by encouraging a broadened range of actions, positive emotions build over time enduring physical, psychological, and social resources. For example, correlation with a faster recovery from cardiovascular diseases [22], an increase of immune function and lower levels of cortisol [23] have been highlighted. More-over, the presence of positive emotions is an effective predictor of the level of happiness of individuals [24] and longevity [25], triggering a virtuous circle, that implements the possible use of other positive experiences.

11.3.1 Using Technology to Foster Positive Emotional States

The hedonic side of Positive Technology analyzes the ways technologies can be used to produce positive emotional states. On the basis of Russell's model, many researchers have acknowledged the possibility to modify the affective quality of an experience by manipulating the "core affect" [26, 27]. This is a neurophysio-logical state corresponding to the combination of hedonic valence and arousal that endows individuals with a sort of "core knowledge" about the emotional features of their emotional experience. The "core affect" can be experienced as freefloating (mood) or attributed to some causes (and thereby begins an emotional episode). In this view, an emotional response is the attribution of a change in the core affect given to a specific object (affective quality).

Recent researches showed that the core affect could be manipulated by Virtual Reality (VR). In particular, Riva and Colleagues tested the potentiality of Virtual Reality (VR) in inducing specific emotional responses, including positive moods [28] and relaxing states [29, 30]. As noted by Serino et al. [4], the potential advantages of using VR technology in inducing positive emotions are essentially two:

- *Interactivity*, to motivate participants, including video and auditory feedback;
- *Manipulability*, to tailor each session in order to evaluate user's idiosyncratic characteristics and to increase task complexity as appropriate.

More recently, some studies explored the potentiality of emerging mobile devices to exploit the potential of positive emotions. For instance, Grassi et al. [31] showed that relaxing narratives supported by multimedia mobile phones were effective to enhance relaxation and reduce anxiety in a sample of commuters.

11.3.2 Can Serious Games Foster Positive Emotional States?

Serious Games and games in general are strictly connected to positive emotions, and to a wide variety of pleasant situational responses that make gameplay the direct emotional opposite of depression [32].

At first, serious games can evoke a *sensorial pleasure* throughout graphics, usability, game aesthetic, visual and narrative stimuli. This point has been analyzed by emerging trends, such as engineering aesthetics 2.0 [33] and hedonic computing [34], whose results will be able to significantly influence game design.

Secondly, serious games foster an *epistemophilic pleasure* by bridging curiosity with the desire of novelty within a protected environment where individuals can experience the complexity of their self, and developing mastery and control. In other words, they are able to recreate a "magic circle" [35], that enforces individual agency, self-confidence and self-esteem [36], by sustaining a process of acknowledgement of personal ability to perform well, solve problems, and manage with difficulties. Hence, empowered by new media affordances and possibilities, serious games can promote a dynamic equilibrium between excitement and security.

Thirdly, serious games promote the *pleasure for victory* and, by supporting virtual interactions with real people, they nurture a *social pleasure*, promoting collaborative and competitive dynamics, communication and sharing opportunities, even outside the context of the game [37].

Games have also been traditionally recognized as marked by a *cathartic pleasure* as they represent a relief valve for emotional tensions, anger and aggressiveness [38].

Finally, pleasure has a *neural* counterpart. An interesting example is that of dopamine, a neurotransmitter that affects the flow of information in the brain and that is often involved in pleasant experiences, as well as in different forms of addiction and learning. In a classic study made by Koepp et al. [39] to monitor the effects of video games on brain activity, a significant increase of dopamine (found in a quantity comparable only to that determined by taking amphetamines) was measured.

Good examples of Serious Games explicitly designed to foster positive emotion are *The Journey to Wild Divine* (http://www.shokos.com/The_Journey_to_Wild_Divine.html) and *Eye Spy: the Matrix, Wham!,* and *Grow your Chi!,* developed in Dr Baldwin's Lab at McGill University (http://selfesteemgames.mcgill.ca). In *The Journey to Wild Divine* the integration between usable biofeedback sensors and a computer software allows individuals to enhance their subjective wellbeing throughout a 3D graphic adventure. Here, wise mentors teach the skills to reduce stress, and increase physical and mental health.

Eye Spy: the Matrix, Wham!, and *Grow your Chi!* are indeed projects whose goal is to empower people with low self-esteem respectively by working on ignoring rejection information, throughout positive conditioning, or by focusing on positive social connections [40, 41].

11.4 The Eudaimonic Perspective: Promoting Individual Growth and Fulfillment

The *engaged life* is based on an eudaimonic definition of well-being. This perspective is associated with the possibility to fully realize human potential through the exercise of personal virtues in pursuit of goals that are meaningful to the individual and society [4, 9]. In the Greek tradition a *daimon* is in fact a divine spirit in charge of cherishing man, and pushing him towards happiness.

In his theory of "developmentalism", strongly influence by Aristotelian naturalism and perfectionism, Kraut [42] introduced the term "flourishing". Humans are really able to flourish when they develop, interiorize and exercise their natural capacities on a cognitive, affective, physical and social level. From a psychological point of view, this conception had a strong influence on authors like Maslow [43], Allport [44], and Rogers [45]. In the works of these and other authors, individuals, groups, and organizations are no longer perceived as passive receptors of external stimulations, but as proactive agents, fully engaged in their actualization and fulfillment.

Thus, this approach focuses on the growth of individuals as a whole, rather than merely emphasizing the pursuit to pleasure and comfort. In this case, happiness no longer coincides with a subjective form of well-being, but with a psychological one. This is based on 6 elements [46–48]:

- *Self-acceptance*, characterized by awareness and a positive attitude towards personal qualities and multiple aspects of the self, including unpleasant ones;
- *Positive relationships with others*, determined by the ability to develop and maintain social stable relationships and to cultivate empathy, collaboration and mutual trust;
- *Autonomy*, reflected by the ability of seeking self-determination, personal authority, or independence against conformism;
- *Environmental Mastery*, based on the ability to change the external environment, and to adapt it to personal needs or goals;
- *Purpose in life*, marked by the presence of meaningful goals and aims in the light of which daily decisions are taken;
- *Personal growth*, achievable throughout a continuous pursuit of opportunities for personal development.

Another author that has fully interpreted the complexity of the eudaimonic perspective is Positive Psychology pioneer Mihaly Csikszentmihalyi who formalized the concept of flow [49, 50]. The term expresses the feeling of fluidity, and continuity in concentration and action reported by most individuals in the description of this state [9]. In particular, flow, or optimal experience, is a positive, complex and highly structured state of deep involvement, absorption, and enjoyment [49]. The basic feature of this experience is a dynamic equilibrium perceived between high environmental action opportunities (challenges) and adequate personal resources in facing them (skills). Additional characteristics are deep concentration, clear rules and unambiguous feedback from the task at hand, loss of reflective self-consciousness, control of one's actions and environment, alteration of temporal experience, and intrinsic motivation.

11.4.1 Using Technologies to Promote Individual Growth and Fulfillment

Scholars in the field of human–computer interaction are starting to recognize and address the eudaimonic challenge. For example, Rogers [51] calls for a shift from "proactive computing" to "proactive people," where "technologies are designed not to do things for people but to engage them more actively in what they currently do".

Further, the theory of flow has been extensively used to study user experience with Information and Communication Technologies. It is the case of internet [52], virtual reality [53, 54] social networks [55], video-games [56–59], and serious games [60].

In fact, all these media are able to support the emergence of a flow state, as they offer an immediate opportunity for action, and the possibility to create increasingly challenging tasks, with specific rules, as well as the opportunity to calibrate an appropriate and multimodal feedback.

In addition, some researchers have drawn parallels between the experience of flow and the sense of presence, conceived as the subjective perception of "being there" in a virtual environment [61]. Both experiences have been described as absorbing states, marked by a merging of action and awareness, loss of self-consciousness, and high involvement and focused attention on the ongoing. On these premises, Riva and colleagues postulated the power of "transformation-of-flow"-based strategies [3]. They can be conceived as individuals' ability to draw upon an optimal experience induced by technology, and to use it to promote new and unexpected psychological resources and sources of involvement.

11.4.2 Can Serious Games Promote Individual Growth and Fulfillment?

Bergeron [60] defined serious games as interactive computer applications, with or without a significant hardware component, that (a) have challenging goals, (b) are fun to play with and/or engaging, (c) incorporate some concepts of scoring, (d) impart to the user skills, knowledge, or attitude that can be applied in the real world.

Interestingly, all of these aspects can be easily overlapped to Csikszentmihalyi's theory of flow. Games are in fact "flow activities" [49, 50] as they are intrinsically able to provide enjoyable experiences [32, 62], creating rules that require the learning of skills, defining goals, giving feedback, making control possible, and fostering a sense of curiosity and discovery.

In addition, the intrinsic potential of flow that characterizes serious games can be even empowered by (a) identifying an information-rich environment that contains functional real world demands; (b) using the technology to enhance the level of presence of subjects in the environment, and (c) allowing the cultivation, by linking this optimal experience to the actual experience of the subject [3]. To achieve the first two steps, it is fundamental to look at the following game design elements [58]:

- *Concentration.* Serious games should stimulate a mental focus on in-game dynamics, by providing a set of engaging, differentiated and worth-attending stimuli that limit the influence of external variables. Along with other aspects, concentration can result in hyperlearning processes that consist of the mental ability to totally focus on the task by using effective strategies aligned with personal traits [50];
- *Challenge.* As noted by Gee [63], who claims that the game experience should be "pleasantly frustrating", challenges have to match players' skills/level and to

support their improvement throughout the game. During specific stages of the game, "Fish tanks" (stripped down versions of the real game, where gameplay mechanisms are simplified) and "Sand boxes" (versions of the game where there is less likelihood for things to go wrong) can support this dynamism;

- *Player Skills.* Games must support player skills and mastery throughout game usability, and specific support systems and rewards;
- *Control.* It is fundamental for players to experience a sense of control over what they are doing, as well as over the game interface, and input devices;
- *Clear goals.* Games should provide players with specific, measurable, achievable, responsible and time-bounded goals;
- *Feedback.* Players have to be supported by feedback on the progress they are making, on their action, and the ongoing situations represented in the virtual environment;
- *Immersion.* Players should become less aware of their surroundings and emotionally involved in the game dynamics;
- *Social Interaction.* Games should create opportunities for social interaction by supporting competition, collaboration, and sharing among players.

An interesting example of an eudaimonic serious game is *Reach Out Central* (ROC), developed by ReachOut.com (http://www.reachoutpro.com). It is a Cognitive-Behaviour therapy game that encourages users to develop psychological well-being. Studied for young people aged 14–24, ROC is a single-player role play game with innovative 3D graphics and real-life scenarios and characters. Here, players can see how their decisions and reactions affect their moods, and apply skills they learn offline in their day-to-day lives. An evaluation conducted by Shandley et al. [64] found that ROC reduced psychological distress, alcohol use, and improved life satisfaction, resilience, and problem-solving abilities.

11.5 The Social Perspective: Enhancing Integration and Connectedness

Networking and participation are becoming the foundations of human performance in educational, organizational and recreational settings [65]. Here, new communities of practice [66] are being established to promote an engagement economy that will be able to foster innovation and success by sustaining collective well-being and group flourishing [32, 67]. However, the enhancement of a human capital so dynamic and heterogeneous implies a deep involvement in nurturing a social form of well-being. In particular, social well-being indicates the extent to which individuals are functioning well in their social system and it is defined on five dimensions [68]:

- *Social integration*, conceptualized as the evaluation of the quality of personal relationships with a community or a society;

- *Social contribution*, evidenced by the perception of having something important to offer to society and the world at large;
- *Social coherence*, determined by the meaning given to the quality, organization, and operations that make up the social sphere;
- *Social acceptance*, based on the belief that people proactivity and agency can foster the development of societies and culture;
- *Social actualization*, determined by the evaluation of the potential and the trajectory of society.

11.5.1 Using Technologies to Enhance Integration and Connectedness

At this level, the challenge for Positive Technology is concerned with the use of new media to support and improve the connectedness between individuals, groups, and organizations, and to create a mutual sense of awareness. This is essential to the feeling that other participants are there, and to create a strong sense of community at a distance.

Short et al. [69] introduce the term "social presence" to indicate the degree of salience of the other person in a mediated environment and the consequent salience of their interpersonal interaction. On this point, Riva and Colleagues [61] argued that an individual is present within a group if he/she is able to put his/her own intentions (presence) into practice and to understand the intentions of the other group members (social presence). Techniques to promote such a "sense of being with another" throughout a medium have a long history, going back to the first stone sculptures that evoked a sense of some other being in the mind of an ancestral observer [70].

Nowadays, social presence has been empowered by advanced ICT systems. Groupware, for example, are computing and communication technology based systems that assist groups of participants engaged in a common task, supporting communication, coordination, and collaboration through facilities such as information, discussion forums, and messaging [71, 72]. The use of groupware has been particularly effective in distributed systems, where—because of the increasing complexity of the environment—single members can access only partially the whole problem. By enhancing social presence, it thus becomes possible to overcome major shortcomings and group's inefficiencies.

Further, assembling these basic concepts with the potential of the world wide web in its most recent version (web 2.0), enterprise 2.0 was born in the business context. It implies the emerging use of social software platforms within companies to facilitate the achievement of business objectives [73]. Thus, Enterprise 2.0 allows to work on reputation, (by both monitoring the internal reality of the organization, and identifying the dynamics implemented by external stakeholders and audiences), collaboration (by developing internal communities), communication (by stimulating the development of interactive exchanges), and connectedness (by enriching the relational and logical transmission of information).

Other interesting phenomena linked to the interpersonal dimension are crowdsourcing and Collaborative Innovation Networks (COINs). The former represents an online, distributed problem-solving and production model that indicates the procurement of a set of tasks to a particularly broad and undefined group of individuals, called to collaborate through Web 2.0 tools [74]. The latter, indicates a "cyber-team of self-motivated people with a collective vision, enabled by the Web to collaborate in achieving a common goal by sharing ideas, information and works" [75].

All these technologies can promote the development of a peak collaborative state experienced by the group as a whole and known as "networked flow" [7]. Sawyer [76, 77], who referred to this state with the term of "group flow", identified several conditions that facilitate its occurrence: the presence of a common goal, close listening, complete concentration, control, blending egos, equal participation, familiarity, communication and the potential for failure. As noted by Gaggioli et al. [7], networked flow occurs when high levels of presence and social presence are matched with a state of "liminality". In particular, three pre-conditions have to be satisfied:

- group members share common goals and emotional experiences so that individual intentionality becomes a *we-intention* [78] able to inspire and guide the whole group;
- group members experience a state liminality, a state of "being about" that breaks the homeostatic equilibrium previously defined;
- group members identify in the ongoing activity the best affordances to overcome the situation of liminality.

11.5.2 Can Serious Games Enhance Integration and Connectedness?

Social presence and networked flow can be fostered by serious games as well. An interesting study realized by Cantamesse et al. [79], for example, examined the effect of playing the online game World of Warcraft (WoW), both on adolescents' social interaction and on the competence they developed on it. The in-game interactions, and in particular conversational exchanges, turn out to be a collaborative path of the joint definition of identities and social ties, with reflection on in-game processes and out-game relationships.

Another interesting example is *Mind the Game*^{TM}, developed by our research group [80] to enhance the optimal functioning of groups.

11.6 Mind the Game™: A Serous Game to Promote Networked Flow

Mind the Game™ is a multiplayer serious game developed to create a socio-technical environment [1] where the interconnection between humans and technology could encourage the emergence of networked flow.

Embedding the potential of serious gaming, *Mind the Game*™ aims to expand the range of resources that groups can access in daily contexts, allowing a greater awareness of the skills possessed both individually and as a whole, and implementing an experiential learning process that supports shared optimal experiences. As a new medium aimed at facilitating change, the serious game generates a virtual environment where groups can express their potential, dealing with a reality that constantly redefines the balance between challenges and skills. This was studied to create a virtuous circularity that promotes collective peak experiences and high levels of perceived effectiveness, both in an individual and collective sense.

11.6.1 Technology

The serious game was created with Forio Simulate™ (www.forio.com), a software that allows the development of multiplayer online simulations, based on Adobe Flash Player. We designed an interface primarily textual, enriched multimodally by clips, images, and animated graphics that make the game more interactive (Fig. 11.1).

Mind the Game™ emerged as a multiplayer game studied for small groups of 5 people, that provides the facilitator/the researcher with the ability to monitor the progress of the game.

Eventually, the serious game was embedded in a specific website (www.mindthegame.it) that consists of a welcome page, a tutorial, and a questionnaire section to evaluate the game experience.

11.6.2 Sharing Goals and Emotional Experiences: Sport as a Narrative Tool

The narrative framework—especially in technological solutions based on a textual environment—is a core element for serious game design. Narratives have to be clear, straightforward, easy to understand and memorable to capture the interest of the user [81, 82]. Therefore, the choice of plots and settings will be decisive to bring the group out of the comfort zone, nurturing the onset of spontaneous

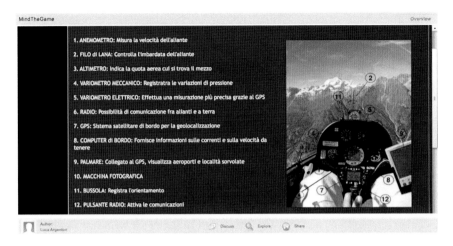

Fig. 11.1 The primarily textual interface of *Mind the Game*™ is enriched multimodally by clips, images, and animated graphics

behaviors, as well as promoting the emergence of we intentions [78], social presence and ingroup dynamics in multiplayer settings.

Moreover, the underlying potential of narratives can be amplified through the use of peculiar scenarios that have nothing to do with day by day experiences [83]. In this way, it is possible to modulate the impact of prior knowledge of users and to support common cognitive processes and knowledge sharing practices.

As a metaphor of life, sport is a powerful and effective training tool, capable of supporting learning and experiential transpositions. In particular, sport witnesses how beyond individual and team excellence there are challenges that do not end against the opponent, but in their relationship with the self.

According to the aforementioned considerations, we chose a little-known sport that could be used to promote networked flow: gliding. This is a discipline based on soaring flight, where, in the absence of the driving force of an engine, the pilot is required to take advantage of upward motions and movements of air masses [84]. In fact, thanks to the overheating of the soil and the atmospheric layers close to it, the air creates connective vertical motions, called thermals, that support the flight.

The development of the narrative plot structure on such a discipline can be particularly effective both because of an implicit and an explicit reason.

On the one hand, soaring flight embraces a deep archaic desire: the tension to the sky. Sky has represented a point of reference for a humanity that has begun to mature the dream of approaching it. Thus, before becoming the concrete possibility, outlined by the studies of Leonardo or by the efforts made by Wright and Montgolfier brothers, flight is synonym of purity and freedom, fantasy, hope, and imagination. It is the symbol of a challenge marked by a courageous and meaningful searching of the limit.

On the other hand, as a sport, gliding implies competition and collaboration. The first concept is well reflected by the Grand Prix, a race in which pilots directly compete one another. The goal is to go throughout a task—a plot delimited by specific turning points that are placed so as to form a polygon—in the shortest time. Generally, the Grand Prix is structured among several days, implying different tasks from time to time.

The choice of an individual sport to promote group creativity and of team working may instead appear paradoxical. But it is not: individual excellence is the tip of the iceberg beneath which team effort and coordination always make the difference. The collaborative dimension of gliding is present because, despite the solo flight of the pilot, his/her staff can support each step of the race from the ground. In fact, parameters to be taken into consideration are extremely numerous and they require the intervention of professionals specifically trained. In particular, according to the model described by Brigliadori and Brigliadori [85], five elements are fundamental:

- *Technical.* Managing an efficient flight and exploiting the energy available in the atmosphere in the best possible way, require specific skills: decision-making, problem-solving, control, and experience. Moreover, the maintenance of security and risk management are the foundation of successful flights.
- *Strategic.* The ability to take advantage of circumstances involves a process of decision-making able to take into account meteorological aspects, competitors, geography. The race is played on the ability to make the most from the opportunities that are revealed during the task.
- *Psychological.* Control of emotions, stress management, relaxation, high levels of concentration, resilience and self-efficacy are just some of the psychological components that may be decisive during a competition.
- *Athletic.* Pilots must take great care in athletic training, monitoring nutrition and fatigue management.
- *Meteorological.* Climate is a component whose analysis should be careful and meticulous in order to avoid unnecessary risks and make winning choices.
- *Organizational.* The athlete, together with the staff, is expected to prepare the race in every detail, monitoring equipment and logistics practices.

Thus, in the serious game users will not be asked to wear the shoes of gliding pilots: they will be the team members of an athlete that has to win the World Competition. In the effort to promote a more immersive user experience and increase the realism of the serious game we contacted a testimonial: Margherita Acquaderni, one of the best gliding pilots in the world, who holds 46 Italian records.

The sense of in-group belonging is first increased by the narrative framework, that immerses players in a collaborative environment. In this way, it is possible to encourage the emergence of a we-intention, whose genesis is the *conditio sine qua non* for the development and implementation of networked flow.

11.6.3 Creating a Space of Liminality

With the support of a suggestive graphic environment and with an emotional clip, users are introduced to a letter written directly by Margherita Acquaderni. After introducing herself, the athlete said to be in Australia where she will compete in the last gliding race of the season. Here she can realize her dream: winning for the first time the World Competition. By showing the characteristics of the race, the pilot explains that the task is going to be very technical and complicated, stressing that every detail can make an important difference.

The user begins to realize his/her role in the simulation: he/she will not be called to be an athlete or an opponent of the athlete, but a member of her team. Each player will in fact be assigned one of the following roles: team manager, strategist, technical expert, meteorologist or doctor.

After that, the players are introduced to a tutorial that illustrates the basic rules of the game, and analyzes the structure of the interface and its iconography. It is then specified how, next to a shared goal (leading the athlete to win the World Competition), each character will be motivated by personal goals, different from those of the other participants.

Finally, the team's score is defined as a result of three parameters:

- Score obtained in the race by the athlete;
- The sum individual scores;
- Time Management as each task it time bounded.

The arrangement by which each character appears to the player is the same and tends to follow the systemic model proposed by Bowman [86]. It is marked by the definition of name, age, nationality, as well as the role played within the team and the tasks he/she has to preside. The user can then discover his/her background. This is realized on three levels, indicating aspects of the past, present and future. At the same time, the user can also view the individual goals of the character.

Finally, there is a brief personality description, borrowed, even in elementary form the Jungian psychological types and the Mayers-Briggs Type Indicator System [87] that arises from them.

11.6.4 Identifying a Common Activity to Overcome the Space of Liminality

According to Steiner's model [88], it is possible to distinguish:

- *Additive* tasks, referred to situations in which the final result is determined by the sum of individual contributions;
- *Compensatory* tasks, where the result is determined as an average of the contributions made by individual subjects;

- *Conjunctive* tasks, where the success of the group depends on the success of each member;
- *Disjunctive* tasks, where each member can promote a solution of their own, knowing that the success of the group depends on a single correct alternative;
- *Complementary* tasks, that requires the sharing of knowledge, processes and methods so that the whole could exceed the sum of its single parts.

When the intention is then transformed into action, the serious game becomes functional to stimulate a synergistic collaboration. In line with the theory of Steiner [88], each task is in fact designed according to a *complementary logic*, in an attempt to involve each player. Specifically, players are called upon to deal with distributed decision-making environments in which real success can not depend on free-riding efforts, but on the emergence of group phenomena, such as social facilitation, social labouring, and team thinking.

Clearly, the effectiveness of the group will be marked by its specific characteristics, as well as on its communicative, emotional and hierarchical structure.

11.6.5 A Pilot Study

A pilot study was realized in order to gather initial and qualitative data on the impact the serious game had on the optimal functioning of different groups.

According to the taxonomy proposed by Arrow et al. [89] we organized focus groups with three different kinds of groups:

- *Task forces* (N = 4), temporary groups of business people formed to carry out a specific project, or to solve a problem that requires a multi-disciplinary approach. The task forces that participated to the study worked in four different business fields: fashion, graphic design, research, and food industry.
- *Sport teams* (N = 4), stable groups of sportsmen that work together to achieve a common goal. Focus groups were conducted with the members of a football team divided according to their position (goalkeepers, defenders, midfielders, forwards).
- *Social groups* (N = 4), groups of peers where members are linked by informal relations.

A total of 60 subjects (age = 25.7, d.s: 7.15) participated to focus groups. Sessions were chaired by a facilitator, who began by introducing *Mind the Game*^{TM} and explaining the reasons of the study.

Anagraphic and socio-demographic data, including age, gender, marital status, occupation, and education, were collected. Computer experience and skills were considered too. Then, groups played the serious game.

At the end of the game, each group was asked to identify ten factors that *Mind The Game*^{TM} was able to stimulate. Results are shown in Figs. 11.2, 11.3 and 11.4.

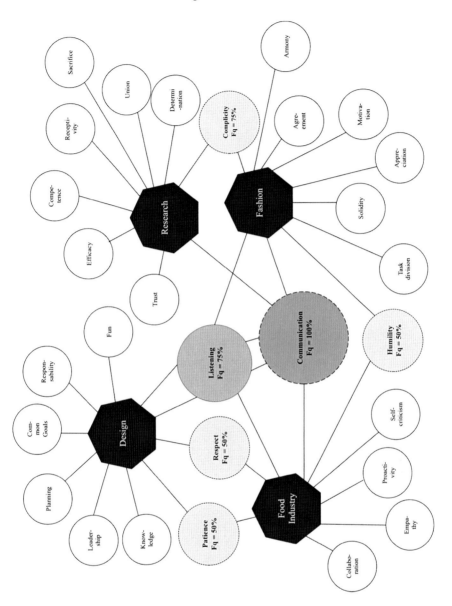

Fig. 11.2 After the homogenization of similar lexical forms, factors mentioned with a percentage of occurrence greater than 60 % were communication ($Fq = 100$ %), and listening ($Fq = 75$ %). Overall, task force members focus on personal characteristics (humility, patience, listening), as well as on their explication in a relational framework (communication, collaboration and respect)

All groups reported that the serious game had a strong impact on communication and collaboration, fostering active listening and collective efficacy.

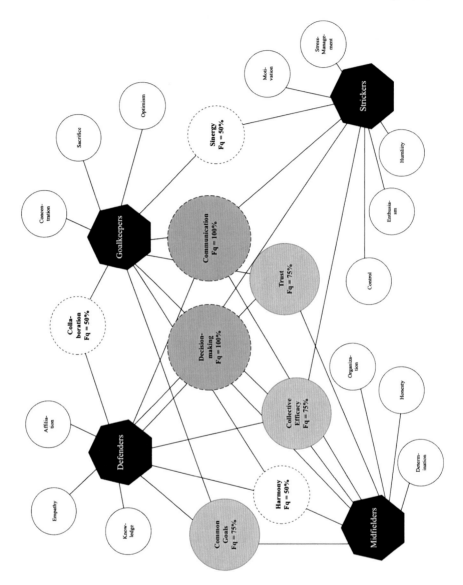

Fig. 11.3 Within the football team, communication and decision-making were mentioned by all groups (Fq = 100 %). Trust, collaboration and common goals had a high percentage of occurrence too (*Fq* = 75 %)

However, further research has to be done in order to gather empirical data on the topic.

Moreover, the identification of the ten factors supported the development of a detailed discussion that focused on strengths and weaknesses of the game. Overall,

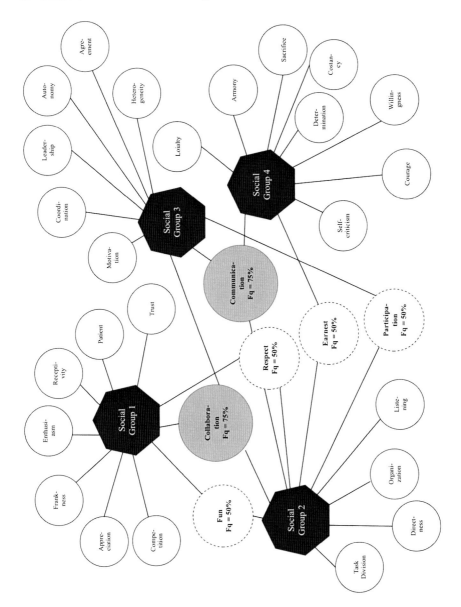

Fig. 11.4 Within social groups, the factors mentioned with the highest frequency were communication and collaboration ($Fq = 75$ %), followed by fun, respect, participation, and earnest (50 %)

participants reported that *Mind the Game*[TM] has the potentiality to become an effective positive technology to empower team working and networked flow.

On the one hand, *Mind the Game*[TM] might be considered as a tool to both train and assess individual and social skills. Team and individual measures may be considered along with outcome and process measures. Moreover, within an assessment perspective, the SG could be considered as an assessment tool itself, allowing an on-line evaluation of human behaviours, or it can be integrated with other assessment instruments.

On the other hand, it can be used to maintain high levels of ecological validity and experimental control, giving the researcher the possibility to manipulate specific variables in everyday life environments.

11.7 Conclusion

In this chapter we discussed the role of serious games as positive technologies. According to Positive Psychology theoretical framework and Positive Technology approach, we demonstrated that these applications are able to promote the three life trajectories identified by Seligman [11]: *the pleasant life, the engaged life, and the meaningful life.*

First of all, serious games can foster positive emotional states by enhancing the different forms of pleasure they are intrinsically made of. In particular, we discussed the importance of sensorial, epistemophilic, social, cathartic and neural pleasure.

Secondly, serious applications for computer game technologies can be associated with flow experiences and, thus, with psychological well-being. Throughout high level of presence and flow, technologies can, in fact, promote optimal experiences marked by absorption, engagement, and enjoyment.

Finally, serious games are able to increase connectedness and integration. To achieve such a complex goal they have to work on a mutual sense of awareness, as well as social presence and situations of liminality. In this way, groups can access peak creative states, known as networked flow optimal experiences, that are based on shared goals and emotions, collective intentions, and proactive behaviors. We eventually presented an empirical exemplification of how all these three aspects may be implemented by psychology-based user experience design.

References

1. Fisher, G., Giaccardi, E., Eden, H., Sugimoto, M., Ye, Y.: Beyond binary choices: integrating individual and social creativity. Int. J. Hum. Comput. Stud. **12**, 428–512 (2005)
2. Botella, C., Riva, G., Gaggioli, A., Wiederhold, B.K., Alcaniz, M., Banos, R.M.: The present and future of positive technologies. Cyberpsychol. Behav. Soc. Networking **15**, 78–84 (2012)
3. Riva, G., Banos, R.M., Botella, C., Wiederhold, B.K., Gaggioli, A.: Positive technology: using interactive technologies to promote positive functioning. Cyberpsychol. Behav. Soc. Networking **15**, 69–77 (2012)
4. Serino, S., Cipresso, P., Gaggioli, A., Riva, G.: The Potential of Pervasive Sensors and Computing for Positive Technology. In: Mukhopadhyay, S.C., Postolache, O.A. (eds.)

Pervasive and Mobile Sensing and Computing for Healthcare. Smart Sensors, Measurement and Instrumentation. Springer, New York (2013)

5. Seligman, M.E.P.: Positive psychology: fundamental assumptions. Psychol. **16**, 26–27 (2003)
6. Seligman, M.E.P., Csikszentmihalyi, M.: Positive psychology. An introduction. Am. Psychol. **55**, 5–14 (2000)
7. Gaggioli, A., Riva, G., Milani, L., Mazzoni, E.: Networked Flow: Towards an Understanding of Creative Networks. Springer, New York (2013)
8. Lopez, S.J., Snyder, C.R.: The Oxford Handbook of Positive Psychology. Oxford University Press, New York (2011)
9. Delle Fave, A., Massimini, F., Bassi, M.: Psychological Selection and Optimal Experience across Culture. Social Empowerment Through Personal Growth. Springer, London (2011)
10. Seligman, M.E.P., Steen, T.A., Park, N., Peterson, C.: Positive psychology progress: empirical validation of interventions. Am. Psychol. **60**, 410–421 (2005)
11. Seligman, M.E.P.: Authentic Happiness. Using the New Positive Psychology to Realize your Potential for Lasting Fulfillment. Free Press, New York (2002)
12. Keyes, C.L.M., Lopez, S.J.: Toward a science of mental health: positive direction in diagnosis and interventions. In: Snyder, C.R., Lopez, S.J. (eds.) Handbook of Positive Psychology. Oxford University Press, New York (2002)
13. Peterson, C., Park, N., Seligman, M.E.P.: Orientations to happiness and life satisfaction: the full life versus the empty life. J. Happiness Stud. **6**, 25–41 (2005)
14. Kahneman, D., Diener, E., Schwarz, N.: Well-Being: The Foundations of Hedonic Psychology. Sage, New York (2004)
15. Diener, E.: Subjective well-being: the science of happiness and a proposal for a national index. Am. Psychol. **55**, 34–43 (2000)
16. Diener, E., Diener, M., Diener, C.: Factors predicting the subjective well-being of nations. J. Pers. Soc. Psychol. **69**, 851–864 (1995)
17. Fredrickson, B.L.: What good are positive emotions? Rev. Gen. Psychol. **2**, 3000–3019 (1998)
18. Fredrickson, B.L.: The role of positive emotions in positive psychology: the broaden-and-build theory of positive emotions. Am. Psychol. **56**, 222–252 (2001)
19. Fredrickson, B.L., Branigan, C.A.: Positive emotions broaden the scope of attention and thought–action repertoires. Cogn. Emot. **19**, 313–332 (2001)
20. Bolte, A., Goschkey, T., Kuhl, J.: Emotion and intuition: effects of positive and negative mood on implicit judgments of semantic coherence. Psychol. Sci. **14**, 416–421 (2003)
21. Isen, A.M., Daubman, K.A., Nowicki, G.P.: Positive affect facilitates creative problem solving. J. Pers. Soc. Psychol. **52**, 1122–1131 (1987)
22. Fredrickson, B.L., Mancuso, R.A., Branigan, C., Tugade, M.M.: The undoing effect of positive emotions. Motiv. Emot. **24**, 237–258 (2000)
23. Steptoe, A., Wardle, J., Marmot, M.: Positive affect and health-related neuroendocrine, cardiovascular, and inflammatory responses. Proc. Natl. Acad. Sci. USA **102**, 6508–6512 (2005)
24. Fredrickson, B.L., Joiner, T.: Positive emotions trigger upward spirals toward emotional well-being. Psychol. Sci. **13**, 172–175 (2002)
25. Moskowitz, J.T.: Positive affect predicts lower risk of AIDS mortality. Psychosom. Med. **65**, 620–626 (2003)
26. Russell, J.A.: Core affect and the psychological construction of emotion. Psychol. Rev. **110**, 145–172 (2003)
27. Russell, J.A.: Emotion, core affect, and psychological construction. Cogn. Emot. **23**, 1259–1283 (2009)
28. Riva, G., Mantovani, F., Capideville, C.S., Preziosa, A., Morganti, F., Villani, D., Gaggioli, A., Botella, C., Alcaniz, M.: Affective interactions using virtual reality: the link between presence and emotions. Cyberpsychol. Behav. **10**, 45–56 (2007)
29. Villani, D., Lucchetta, M., Preziosa, A., Riva, G.: The role of interactive media features on the affective response: a virtual reality study. Int. J. Hum. Comput. Interact. **1**, 1–21 (2009)

30. Villani, D., Riva, F., Riva, G.: New technologies for relaxation: the role of presence. Int. J. Stress Manage. **14**, 260–274 (2007)
31. Grassi, A., Gaggioli, A., Riva, G.: The green valley: the use of mobile narratives for reducing stress in commuters. Cyberpsychol. Behav. **12**, 155–161 (2009)
32. McGonigal, J.: Reality is Broken. The Penguin Press, New York (2010)
33. Liu, Y.: Engineering aesthetics and aesthetic ergonomics: theoretical foundations and a dual-process research methodology. Ergonomics **46**, 1273–1292 (2003)
34. Wakefield, R.L., Whitten, D.: Mobile computing: a user study on hedonic/utilitarian mobile device usage. Eur. J. Inf. Syst. **15**, 292–300 (2002)
35. Huizinga, J.: Homo Ludens: A Study of the Play Element in Culture. Routledge, London (1944)
36. Oatley, K., Jenkins, J.M.: Understanding Emotion. Wiley, New York (1996)
37. Reeves; B. Read, J.L.: Total Engagement: How Games and Virtual Worlds Are Changing the Way People Work and Businesses Compete. Harvard Business School Publishing, Boston (2009)
38. Bruner, J.S.: On Knowing. Essays for the Left Hand. Belknap, Cambridge (1964)
39. Koepp, M.J., Gunn, R.N., Lawrence, A.D., Cunningham, V.J., Dagher, A., Jones, T., Brooks, D.J., Bench, C.J., Grasby, P.M.: Evidence for striatal dopamine release during a video game. Nature **393**, 266–268 (1998)
40. Baccus, J.R., Baldwin, M.W., Packer, D.J.: Increasing implicit self-esteem through classical conditioning. Psychol. Sci. **15**, 498–502 (2004)
41. Dandeneau, S.D., Baldwin, M.W.: The inhibition of socially rejecting information among people with high versus low self-esteem: the role of attentional bias and the effects of bias reduction training. J. Soc. Clin. Psychol. **23**, 584–603 (2004)
42. Kraut, R.: What is Good and Why: The Ethics of Well-Being. Harvard University Press, Boston (2007)
43. Maslow, A.H.: Motivation and Personality. Routledge, New York (1954)
44. Allport, G.W.: Pattern and Growth in Personality. Holt, Rinehart & Winston, New York (1961)
45. Rogers, C.: On Becoming a Person. Houghton Mifflin, Boston (1961)
46. Ryff, C.D.: Happiness is everything, or is it? Explorations on the meaning of psychological well-being. J. Pers. Soc. Psychol. **57**, 1069–1081 (1989)
47. Ryff, C.D., Singer, B.: The contours of positive human health. Psychol. Inq. **9**, 1–28 (1998)
48. Ryff, C.D., Singer, B.: Ironies of the human condition: well-being and health on the way to mortality. In: Aspinwall, L.G., Staudinger, U.M. (eds.) A Psychology of Human Strengths: Fundamental Questions and Future Directions for a Positive Psychology. American Psychological Association, Washington (2003)
49. Csikszentmihalyi, M.: Beyond Boredom and Anxiety. Jossey-Bass, San Francisco (1975)
50. Csikszentmihalyi, M.: Flow. The Psychology of Optimal Experience. Harper & Row, New York (1990)
51. Rogers, Y.: Moving on from Weiser's vision of calm computing: engaging UbiComp experiences. In: Dourish, P., Friday, A. (eds.) UbiComp 2006 Proceedings. Springer, Heidelberg (1990)
52. Chen, H.: Exploring Web Users' On-line Optimal Flow Experiences. Syracuse University, New York (2000)
53. Gaggioli, A., Bassi, M., Delle Fave, A.: Quality of experience in virtual environments. In: Riva, G., Ijsselsteijn, W., Davide, F. (eds.) Being There: Concepts Effects and Measuerement of User Presence in Syntetic Environments. IOS Press, Amsterdam (2003)
54. Riva, G., Castelnuovo, G., Mantovani, F.: Transformation of flow in rehabilitation: the role of advanced communication technologies. Behav. Res. Methods **38**, 237–244 (2006)
55. Mauri, M., Cipresso, P., Balgera, A., Villamira, M., Riva, G.: Why is Facebook so successful? Psychophysiological measures describe a core flow state while using Facebook. Cyberpsychol. Behav. Soc. Networking **14**, 723–731 (2011)

56. Jegers, K.: Pervasive game flow: understanding player enjoyment in pervasive gaming. Comput. Entertainment **5**, 9 (2007)
57. Sherry, J.L.: Flow and media enjoyment. Commun. Theory **14**, 328–347 (2004)
58. Sweetser, P., Wyeth, P.: GameFlow: a model for evaluating player enjoyment in games. ACM Comput. Entertain. **3**, 1–24 (2005)
59. Wang, L., Chen, M.: The effects of game strategy and preference-matching on flow experience and programming performance in game-based learning. Innov. Educ. Teach. Int. **47**, 39–52 (2010)
60. Bergeron, B.P.: Developing Serious Games. Charles River Media, Hingham (2006)
61. Riva, G., Waterworth, J.A., Waterworth, E.L.: The layers of presence: a bio-cultural approach to understanding presence in natural and mediated environments. Cyberpsychol. Behav. **7**, 402–416 (2004)
62. Michael, D.R., Chen, S.: Serious Games: Games that Educate, Train and Inform. Thompson, Boston (2006)
63. Gee, J.P.: What Video Games have to Teach us About Learning and Literacy. Palgrave MacMillan, New York (2004)
64. Shandley, K., Austin, D., Klein, B., Kyrios, M.: An evaluation of reach out central: an online therapeutic gaming program for supporting the mental health of young people. Health Educ. Res. **15**, 563–574 (2010)
65. Barabasi, A.L.: Linked: The New Science of Networks. Perseus Publishing, Cambridge (2002)
66. Wenger, E.: Communities of Practice: Learning, Meaning, and Identity. Cambridge University Press, Cambridge (1999)
67. Richardson, J., West, M.: Teamwork and engagement. In: Albrecht, S.L. (ed.) Handbook of Employee Engagement. Edward Elgar, Cheltenham (2012)
68. Keyes, C.L.M.: Social well-being. Soc. Psychol. Q. **61**, 121–140 (1998)
69. Short, J., Williams, E., Christie, B.: The Social Psychology of Telecommunications. Wiley, New York (1976)
70. Biocca, F., Harms, C., Burgoon, J.: Towards A More Robust Theory and Measure of Social Presence: Review and suggested criteria. Presence-Teleop. Virt. Environ. **12**: 456–480 (2003)
71. Borghoff, U.M., Schlichter, J.H.: Computer-Supported Cooperative Work: Introduction to Distributed Applications. Springer, New York (2000)
72. Ellis, C.A., Gibbs, S.J., Rein, G.L.: Groupware: some issues and experiences. Commun. ACM **34**, 39–58 (1991)
73. McAfee, A.: Enterprise 2.0: New Collaborative Tools For Your Organization's Toughest Challenges. Harvard Business Press, Boston (2009)
74. Estellés-Arolas, E., González-Ladrón-de-Guevara, F.: Towards an integrated crowdsourcing definition. J. Inf. Sci. **38**, 189–200 (2012)
75. Gloor, P.: Swarm Creativity, Competitive Advantage Throughout Collaborative Innovation Networks. Oxford University Press, New York (2006)
76. Sawyer, K.R.: Group Creativity: Music, Theatre, Collaboration. Basic Books, New York (2003)
77. Sawyer, K.R.: Group Genius: The Creative Power of Collaboration. Oxford University Press, New York (2008)
78. Searle, J.: Intentionality. Cambridge University Press, Cambridge (1983)
79. Cantamesse, M., Galimberti, C., Giacoma, G.: Interweaving interactions in virtual worlds: a case study. Stud. Health. Technol. Inform. **167**, 189–193 (2011)
80. Muzio, M., Riva, G., Argenton, L.: Flow, benessere e prestazione eccellente. Dai modelli teorici alle applicazioni nello sport e in azienda. Franco Angeli, Milano (2012)
81. Bateman, C.M.: Game Writing: Narrative Skills for Video Games. Charles River Media, Independence (2007)
82. McQuiggan, S.W., Rowe, J.P., Lee, S., Lester, J.C.: Story-based learning: the impact of narrative on learning experiences and outcomes. Intelligent tutoring system. Lect. Notes Comput. Sci. **5091**, 530–539 (2008)

83. Edery, D., Mollick, E.: Changing the Game: How Video Games are Transforming the Future of Business. FT Press, Upper Saddle River (2009)
84. Piggott, D.: Gliding: A Handbook on Soaring Flight. A&C Black, London (2002)
85. Brigliadori, L., Brigliadori, R.: Competing in Gliders. Winning with Your Mind. Pivetta Partners, Vedano al Lambro (2011)
86. Bowman, S.L.: The functions of role-playing games. How Participants Create Community, Solve Problems and Explore Identity. McFarland & Copany Inc., London (2010)
87. Myers, I.B.: Introduction to Type: A Description of the Theory and Applications of the Myers-Briggs Type Indicator. Consulting Psychologists Press, Palo Alto (1987)
88. Steiner, I.D.: Group Process and Productivity. Academic Press, New York (1972)
89. Arrow, H., Mccrath, I.E., Berdhal, J.L.: Small Groups as Complex Systems: Fomation, Coordination, Devlopment and Adaptation. Sage, Thousand Oaks (2000)

Chapter 12
Spontaneous Interventions for Health: How Digital Games May Supplement Urban Design Projects

Martin Knöll, Magnus Moar, Stephen Boyd Davis and Mike Saunders

Abstract Health games seem to provide for attractive play experiences and promise increased effects on health-related learning, motivation and behavior change. This chapter discusses the further possibility of mobile games acting as a springboard for communication on health and its correlations to the built environment. First, it introduces the notion of spontaneous interventions, which has been used to characterize co-design projects in which citizens seek to improve infrastructure, green and public spaces, and recreational facilities of their local neighborhoods by adding temporary objects and installations to the built environment. Focusing on interventions, which aim to stimulate physical activity, this chapter identifies potentials and challenges to increase their impact from an ICT perspective. Second, the chapter gives an overview into current research and best practice of health games which seek to enable interaction with urban spaces through mobile and context-sensitive technologies. Specifically, it highlights "self reflective" games in which players seem to adjust their behavior in response to interacting with real time bio-physiological and position data. Observing how mapping technology enables users to relate objective data to subjective context, the chapter identifies how health games may supplement future urban research and design in the following aspects: Raising attention to new complexes, stimulating

M. Knöll (✉)
Department of Architecture, Technische Universität Darmstadt, Darmstadt,
Hessen, Germany
e-mail: knoell@stadt.tu-darmstadt.de

M. Moar
Department of Media and Performance Arts, Middlesex University, London, UK
e-mail: M.Moar@mdx.ac.uk

S. Boyd Davis
Royal College of Art, School of Design, London, UK
e-mail: stephen.boyd-davis@rca.ac.uk

M. Saunders
Open City Labs, London, UK
e-mail: mike_saunders@blueyonder.co.uk

A. L. Brooks et al. (eds.), *Technologies of Inclusive Well-Being*,
Studies in Computational Intelligence 536, DOI: 10.1007/978-3-642-45432-5_12,
© Springer-Verlag Berlin Heidelberg 2014

participation, identifying locales for potential improvement and evaluating impact. The chapter concludes with an outline of future research directions to facilitate serious games supplementing health-related urban design interventions.

Keywords Mobile games · Context-sensitive games · Health games · Urban design · Urban planning · Active design · Playful intervention

12.1 Introduction: Morning Stroller Clubs and an iPhone Stair Climbing Game

At first sight, it seems astonishing how disparate disciplines such as urban design, preventive healthcare and serious gaming begin to form overlapping research interests. In the face of today's health challenges, such as obesity and type-2 diabetes, multi-disciplinary strategies are highly sought after. While health games so far principally focus on internal aspects such as motivation, learning and behavior change [1], public health experts also stress the importance of external factors to tackle obesity. For example, James and colleagues provide an overview into environmental aspects such as whole grocery and fresh food supply, proximity of recreational facilities and walkability of our cities [2]. The general public, however, may be hardly aware of the link between urban design and health. In this chapter, we will be speculating on how serious games may mediate between internal and external aspects of promoting health. We introduce this argument with an illustration of how intertwined personal healthcare and urban planning have been in history.

There is much scholarship describing how doctors, sanitarians and town planners teamed up to build housing, streets and parks from as early as the late nineteenth century [3]. It is important to note that developing design standards for food, drinking water, sidewalks, streets, parks, housing or furniture went hand in hand with a growing interest in personal regimes to regain and sustain one's health. Around 1860, alternative healing advocate Herrmann Klencke observed that in various German cities so called "Morning Stroller clubs" had been founded. Their members met in the morning, regardless of the weather to go for one-hour walks in the natural landscape. Dropping out of sessions without giving notice to fellow club members was punished with a substantial monetary fine. "A necessary disciplinary regime", finds Klencke, "for the modern man would be easily overwhelmed and remain all day inactive in his office [4]." Goeckenjan observes how nineteenth century alternative medicine started off as an emancipatory movement, in which dissident doctors and lay persons explored vegetarianism, sunbathing or physical activity. However, for Goeckenjan, stroller clubs signify a certain level of stagnation in that movement. They would be merely inventing tools to reinforce already established doctrines [5]. It is important to note that these nineteenth century stroller clubs seem innovative in organizing new ways to motivate physical activity in face of changing working and living conditions of the industrial city. However, they did not intend stimulating members to also explore, question or re-shape their urban environment.

Today, mobile health games seek to motivate for increased daily physical activity, too. The iPhone game *Monumental* invites players to climb iconic monuments such as the Eiffel Tower or Empire State Building on their smartphone screens, while tracking players' movement in the real world through the phone's built-in sensors. Its designers hope that players would begin using the stairs more frequently, aim to beat personal high scores and compete with friends via Facebook [6]. As Knöll and Moar have noted in earlier articles, mobile and context-sensitive health games seem to interact with their topographic [7], social [8] and cultural [9] context in various ways. To us, today's health games too often appear like morning stroller clubs: Despite their manifold potentials in interactive storytelling, digital games so far hardly intend to make users aware of the link between inner aspects of personal health and wider causes for health outcomes such the built environment. In this article, we aim to further discuss this argument and will focus on how play experiences may stimulate interest and even inform health-related urban design projects.

12.2 Active Design Guidelines and Community Games

In this section, we will introduce urban design projects in which citizens seek to stimulate physical activity by re-shaping their local neighborhoods. It is important to note that there is a broad spectrum of possibilities to stimulate physical movement through architecture, urban planning and urban design. In 2010, the New York City public planning departments have provided a comprehensive set of guidelines for architects and urban planners based on most recent research and best practices. Focusing on the question how to design so that people would more often prefer using stairs over the elevator, possibilities already vary widely. The Active Design guidelines highlight position, accessibility and visibility of staircases within a facility, to actual design features such as stair dimensions, provision of light, colors, and using valuable finishing and materials [10]. At the same time architects Janson and Tigges emphasize the positive, enhancing experience of climbing and descending well-designed staircases in palaces, operas and some public buildings. As well as the affective pleasures of being gently guided by spatial features, in which one would enjoy changing position in space and touching fine materials, they note the influence of stairs, which may become a stage on which to enjoy one's own and observe other people's movement [11]. Designers have only started to explore how such potentials for health resulting from the physical qualities of the built environment may be stimulated, augmented and supported with playful ICT applications and interventions.

Coenen and Laureyssens have recently noted that urban practices seeking to engage citizens into urban design interventions do not yet involve ICT on a large scale [12]. They refer to projects such as parents joining up to organize a pedestrianized "play street" in their neighborhood. We can confirm this impression for the field of health-related urban interventions. Borden has pointed to self-organized building processes of skate parks and mini ramps [13], but we may also take runners creating new paths in their local park by putting wood chipping on the green to provide for more comfortable running. Coenen and Laureyssens see one possible approach to research how more "social cohesion" in local neighborhoods can be activated and sustained through playful technology. They have presented their ongoing work on Sustainable Community Games (SCOGA), which will be tested in Ghent in February 2013, at the time we are writing this chapter. In the Ghent community game, two neighborhoods will contest each other in various web and location-based features to win an audience with the city mayor to discuss how to realize a project in their neighborhood. Next to extrinsic game factors such as scoring and competing, Coenen and Laureyssens emphasize the intrinsic motivators to do something good for one's neighborhood and to enjoy the more "ludic" aspects such as interacting with augmented objects [12]. We welcome more research into this direction to highlight interventions that have an informal, temporary and co-design character.

12.3 Spontaneous Interventions for Health

The US Pavilion for the thirteenth Architectural Biennale in Venice 2012 has collected over 100 projects in which New Yorkers have been seeking to improve traffic, infrastructure, and recreation facilities of their neighborhoods. The curators have framed the title "Spontaneous Interventions" to characterize their provisional, temporary and co-design character, as opposed to long-term planning schemes [14]. In this section, we discuss two such spontaneous interventions and review how they explore the boundaries between urban design and mobile technologies.

First, a project from Alison Uljee and Sierra Seip, from *Design That Moves You* (DTMY) may illustrate how spontaneous installations in the public realm can be realized with a low budget and small resources. Their most engaging example may be "Stairway Stories", for which they adjusted stickers on to a two flight stairway leading up to Manhattans' Highline—a recreational space having been developed along a retired train track. A hint on the elevator doors points towards the experience of reading an "entertaining, sexy story" while taking the stairs close by for which the authors promise to make peoples' "gorgeous faces" glow [15]. Second, the Stair Piano is an example for an intervention that blends built environment and ICT technology seamlessly. Their designers attached sensor pads to a flight of stairs leading down to a Stockholm underground station and connected each step to a piano sound being amplified through a speaker system. Being installed for not more than a day, they observed 66 % more people taking the steps than the next door escalator—enjoying the sounds while they climb the stairs [16].

In the following, we will take both projects as a starting point to discuss aspects that we feel may be worth improving in such spontaneous interventions. We will argue in terms of user experience, evaluating actual impact, participation and their potential to be transformed into sustainable, scalable projects.

12.3.1 Lack of Expertise to Explore ICT in Urban Design

The Stairway Story project shows how even a small budget (90 USD production cost, with two people involved, adjusting the stickers in less than 30 min) may have a startling impact [17]. However, with increasing advertisement and information being attached to objects in public spaces, such low budget interventions will have to compete for city travellers' attention with ever more projects. This is why designers may want to explore a bigger variety of design solutions including more subtle, more sophisticated and more technically advanced ways of enchanting physical space. Rogers observes how through affordable "plug and play" technologies, tools and materials such as Arduino, prototyping interactive ICT applications now seems to require less technical expertise [18]. Indeed, increasingly research and design projects try to balance research on existing conditions and starting to intervene with temporary installations [19]. We observe that many architects and urban designers begin to become more sensitive towards the potentials of mobile and context sensitive technologies for design practice and education curricula [20]. However, we may note that the design and development of systems consisting of built and IT interventions still seems too complex to be integrated in more mainstream urban design and research practice. The broad mass of designers and citizens are not able to explore such a broad range of possibilities to develop interactive and site-specific media.

12.3.2 Lack of Data Evaluating Medical and Social Impact

We may also question if and for how long spontaneous interventions may have an impact on peoples' daily behavior. Hansen makes an interesting observation regarding the lack of sound research on the long-term impact of projects such as piano stairs. Rather than just setting up a camera over a day or two, he points to standards in sociology and behavioral science that future projects would have to meet. He wonders how people would feel about the sounds at their second, third or fourth visit. Hansen questions the acceptance by local residents, who are using the underground on a daily basis and would have to cope with a noisy installation for a whole week [21]. The Active Design guidelines, introduced above, present studies and early experience of best practice in the field of stimulating physical activity through the built environment. However, most studies gain their results by comparing existing circumstances over different time periods and user groups retrospectively. Those best practices featured in guidelines—that seem to us most fruitful as they are set up in contemporary environments—have hardly been investigated yet [10]. This lack of data on the impact of temporary urban interventions and recently built architecture seems remarkable. By contrast, research on the relation between environments and behavior in Human Computer Interaction (HCI) has moved on considerably over the last couple of years. Rogers observes a paradigmatic change from merely observing ordinary behavior and consequentially suggesting design implications, to creating and evaluating new technologies at the same time, out "in the wild" and not in the lab [18:58]. Many more interdisciplinary research projects begin to explicitly address urban design and development on different scales and in various collaborations. One of its latest editions, Intel's joint venture with University College London and Imperial College London, investigates urban conditions by analyzing users' movement patterns through London's underground. By using everyday technologies such as travel cards, they hope to also find more cost-efficient and flexible ways to research the city [22]. Despite the insights to be gained out of projects on such a scale, we may note that such research cooperation models have not been adopted to smaller and local interventions yet. To our knowledge, there is little research that combines the possibilities of prototyping ICT "in the wild" with the approach of spontaneous interventions.

12.3.3 Not Stimulating User Co-design

The example of Stairway Stories provokes a discussion on user participation in urban design and planning. Its low costs, and few requirements for an ICT expertise would potentially enable many lay persons to get involved either by following up this concept or adapting it to comparable interventions. This is not the place to discuss the pros and cons of experts in city planning. However, we want to point to the merits of ICT supporting communication between users and experts. A popular example is the website SeeClickFix.Com on which citizens can report non-emergency failings of pavements, streets, benches, etc. As Mergel points out, more research needs to be done to investigate their success, with several more local governments embracing the service in order to provide users with sufficient feedback to their postings [23]. Social platforms, that let users also participate in developing constructive ideas and visualizations of their suggestions are currently mushrooming [24]. So far, there seems no service that seeks input from citizens in that respect and focuses on aspects of health and wellbeing.

12.3.4 Lack of Cooperation and Business Models for Further Development

Ho characterizes spontaneous interventions as being about rolling up one's sleeves, "personally bankrolling or finding creative sources of funding [25]." Earlier in this section, we have pointed to potentials and difficulties that low quality interventions in the public realm might be causing. The overall budget of the Staircase Piano project, which features considerable ICT skills and expertise, is hard to estimate. The installation was developed with the support of The Fun Theory initiative, which, supported by a big car company, aims to show that joyful, entertaining and engaging interaction would be key to stimulate behavior change [26]. The project emphasizes philanthropic motives at the expense of more detailed research into its actual impact. So far, it is hard to find business models aiming to bridge the gap between business and research institutions, activists and bottom-up approaches in the field of health-oriented interventions.

In the remaining sections, we will discuss how serious games may contribute to tackle challenges that we have identified above.

12.4 Context-Sensitive Games as a Spring Board for Communication

Debra Lieberman showed as early as the 1990s how video games can motivate better therapy management in young diabetes and asthma patients. She coined the idea of health games acting as a "springboard" for communication between patients, family members, friends and health experts [27]. More recently, she has pointed to the new potentials for learning that result from new game technologies such as motion tracking and collaborative learning in multiplayer games (MUDs) [28]. Lieberman's scholarship serves us as a starting point to speculate how mobile and context-sensitive health games may act as a springboard for communication on health and the built environment. In order to better understand how the latter are being played in their urban environment, Knöll has suggested that we should distinguish current design practice into "collaborative, expressive and reflective" health games [29]. In this section, we will use his analysis to discuss which kind of games may be best suited to supplement spontaneous interventions.

Knöll has observed how players collaborate in Jane McGonigal's Mixed Reality Game *CryptoZoo* [30] with the support of real world locales. On the one hand, players develop running styles in response to objects and topography such as stairs, benches, parking lots or squares. On the other hand, CryptoZoo seems to form a temporary and provisional stage, where players meet to perform runs, observe and communicate with fellow players. CryptoZoo therefore augments the built environment with videos, pictures and maps showing users' activities. Knöll observes that while collaborative play seems to motivate for physical activities as part of the game sessions, CryptoZoo hardly intends to stimulate discussions on serious matters outside of game situations [29]. Like the morning stroller clubs mentioned earlier in this article, games such as CryptoZoo waste a lot of potential for learning, if they just want to re-use the city as a playground consisting of objects and topographies. From our perspective, it could be precisely peoples' experiences and their ability to articulate and share them, that can become useful for health-oriented urban design.

Game theorist Ian Bogost has described "persuasive games" as videogames that are confronting users' more established views on everyday situations with a most expressive game rhetoric. By interacting with the world view as expressed by game designers in game rules, conducts and behaviors, Bogost hopes a "simulation fever" is being stimulated. Artistic techniques such as defamilarization would make users rethink deeply inscribed behaviors in the real world [31]. In his article on mobile persuasive games, Bogost points to the importance of carefully choosing real world locales to support digital game play. For Bogost, his "airport game", which criticizes non-transparent security policies at US airports, needs to be played precisely while standing in line and waiting for one's bags to be checked [32]. Knöll shows how health games use real world locations as "expressive" backgrounds to more effectively articulate their critique of health-related behavior. In future, choosing the right site for a health game may unfold potential in making players aware of environmental causes for healthy behavior. By design, he finds, such expressive health games may hardly be suitable to explore new correlations [29:155–158]. We may specify that site-specific health games help to make users aware of how the built environment influences our daily routines—for example the use of an elevator over that of stairs. Further research needs to elaborate how the combination of mobile persuasive games and spontaneous interventions may also help to indicate new correlations and new locales for improvement.

Elsewhere, Boyd Davis and colleagues have observed how digital games related to promoting physical activities would often instill guilt in players. They present established answers on what is good or bad for players' health [33]. With their multidisciplinary project *Ere Be Dragons*, Boyd Davis and colleagues have pointed to a further way of playing health games in the city and in response to interacting with the urban environment. They describe the game experience as follows: In its simplest form, the player wears a portable heart rate monitor and inputs her age into a smart phone device featuring a GPS module. On the basis of the player's age, the game calculates an optimal heart rate according to a standard formula. Players then proceed to walk along wherever and however they wish. During the walk an on-screen landscape is being built up that corresponds to the player's position and movement in the real world. On the other hand, the virtual landscape also corresponds to the measured heart rate. If the player does well, e.g. is exercising adequately, the landscape flourishes. The heart rates are split into five ranges on the basis of each player's optimal heart rate. His territory is represented in isometric tiles with distinctive colors and landscape features. Insufficient exertion causes the current landscape to impoverish, while over-exertion leads to the growth of a dark, forbidding forest [33:200–201]. Elsewhere, Boyd Davis has borrowed the term "reflection-in-action" to describe this kind of gameplay, in which players seem to adjust their behaviors—for instance moving in the city—in response to a visualization of their position and pulse data [34]. Players set out with a goal such as to keep their pulse in a certain area, but are free to explore how to achieve this.

We agree that more research is required to underline that more creative, "free" ways of playing as opposed to competitive play can result in long term effects on health related behavior change. In the remaining paragraph, we wish to elaborate further why we highlight reflective health games as most promising to supplement and inform spontaneous interventions. Elsewhere, Boyd Davis has pointed to Nold's Bio Maps as a parallel approach to Ere Be Dragons. Whereas players of Ere Be Dragons would reflect in action, Nold's visualizations allow users to reflect on their journeys—on different routes taken and body data gathered—in retrospect [34:48]. Nold emphasizes the need of users interpreting "objective" data such as stress levels and positioning and relating them to "subjective" data. He has explored various technologies to provoke user input—personal anecdotes, context, biographical notes and feelings—in order to gain a set of data that are more meaningful [35]. Nold's projects illustrate the need for users taking an active part in interpreting data for themselves, but especially when the data is being gained to inform policy making or planning legislations. Knöll has pointed out how combining these two approaches—reflecting-in-action and reflecting-on-action—is one crucial challenge to build serious games that inform health-oriented urban design [29:199–211]. A main part from the perspective of game design is to implement the serious aspect of co design into the gameplay of location-sensitive mobile apps.

More recently, Waterson and Saunders have presented their insights into a mobile service they developed for Kew Gardens in London. Based on extensive surveys on visitors' expectations and information needs, which among others stated that users would like unguided exploration and serendipitous discoveries, they developed the idea of helping users "to get delightfully lost." The resulting app provides no pre-planned tours, but help users find information where they want them. This may include browsing user generated content and news, but also interacting with QR-Codes and geo-referenced Augmented Reality functions. They found that the app succeeded in directing attention to some of the more "off the beaten track" attractions. Even though it was not primarily meant to be a guide, the app was perceived as an "expert companion" aimed at stimulating "information hunger" in the gardens and its plants [36]. In our view, such unobtrusive, flexible applications are most suitable to articulate how we perceive positive and negative impacts of the built environment on our health in a playful way. It is in a parallel step that users articulate their experiences and thus inform potential projects.

Fig. 12.1 This diagram lists potential synergies between mobile context-sensitive health games and spontaneous interventions in public space

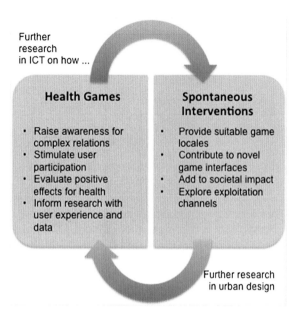

12.5 Roadmap Towards ICT Supported Spontaneous Interventions

Earlier in this article, we have identified four main challenges of spontaneous interventions for health to overcome from an ICT perspective: First, the lack of interdisciplinary expertise, second the few strategies to evaluate positive effects for health, third, the lack of specific co-design platforms and fourth the little research on business and cooperation models. Having moved on to show what mobile and context sensitive health games may contribute to overcome these challenges, we will speculate on future research and design directions to further elaborate potential synergies (See Fig. 12.1).

In the following, we will list four steps to pursue these goals. Based on what we have found to be challenges of spontaneous interventions earlier in this chapter, we have structured this outline with a view to increasing user experience, evaluation methods, stimulating participation and more sustainable concepts.

12.5.1 Encouraging Interdisciplinary Educational and Research Projects

As touched upon above, designing and developing health games involves several research disciplines. Mehm and colleagues provide a comprehensive list of experts and roles [37]. We are convinced that involving urban designers and architects can help to extend the specific positive effects, which mobile and context sensitive health games aim for to improve users' health and wellbeing. Specifically, action is needed to develop multi-disciplinary teaching formats dealing with serious games. An interdisciplinary discourse on design of ICT applications may help to integrate health games into users' everyday lives. In turn, architects and urban designers, who have been trained in basic concepts of location-based technologies will become more sensitive towards the potential contributions of serious games for urban design in raising user participation and promoting health prevention.[1] We consider such an increasing scientific exchange as one important step to also bring the knowledge and skills of agile prototyping to a wider audience including those citizens already interested in spontaneous interventions.

12.5.2 Developing New Strategies to Evaluate the Effects of Mobile Health Games in the Wild

Göbel and others have presented first attempts to integrate adaptive game play scenarios to static and dynamic data by deriving information from various sensor technologies. They have been focused on evaluating the medical impact of exer-games mainly with indoor games so far [38]. As we have shown above to evaluate spontaneous interventions for health and their positive effects in their environment is key to their further development. We therefore would like to see further research that combines Rogers' approach on prototyping and evaluating "in the wild" that we have described earlier in this article, with research groups that have an expertise in Serious Games for Health and/or urban design.

12.5.3 Improve Access to Co-design Projects with Playful Approaches

We have described a burgeoning field of social websites inviting users to partic-ipate in urban planning and design projects. We are convinced that playful applications, such as what we have described here as self-reflective health games, can supplement social platforms such as SeeClickFi.com or Betaville. Further-more, projects such as the community game for Ghent begin to extend the field to location-based features. Precisely in interactive objects being placed in public

---

[1] Since February 2013, TU Darmstadt has established a University Industry Collaborative (UNICO) research group on "Urban Health Games", which is situated at the architecture department and strongly interacts with the Multimedia Communications Lab (KOM). Further information on research projects and teaching formats that are currently being developed are available at http://www.stadtspiele.tu-darmstadt.de

places or distributed to facades, Coenen and Laureyssens see the biggest potential for more ludic activities and as a result to reach out to a wider audience [12]. We also consider development of augmented objects, which become a part of game play activities as crucial to include people with and without access to smartphones. Thus, integrating urban design can contribute to develop new game play interfaces, which may also benefit from the research in urban planning that has been presented here as Active Design. We welcome projects that develop such entertaining interventions in closer relation to the urban environment it aims to stimulate awareness for. In our view, involving urban designers and planners, who are sensitive towards location-based technologies, will support these co-design aspects of spontaneous interventions.

12.5.4 Working on New Business Models that also Target Local Neighborhoods

In order to conduct interdisciplinary research, design and implementation of spontaneous interventions, more sustainable business models need to be developed. As we have stated earlier, so far collaborations between research and corporate institutions focus on large scale projects. On the other hand, there are short-term interventions that are either low budget and/or hardly involve ICT. Site specific interventions such as the piano staircase project have been able to feature more sophisticated technology. Their collaboration between artists and corporate funds has so far focussed on short term promotion and less on long-term studies. We have stated that spontaneous interventions enriched with mobile health games may become a very powerful tool to gain more insights into this new complex and multidisciplinary field. We therefore hope to see a continuation of the approaches we have sketched in this chapter, which seek to bring together experts from urban and interaction design with corporate institutions and private interest groups. On basis of such research and design projects, strategies can be developed to convince stakeholders, improve design quality and implement spontaneous interventions on a bigger and more sustainable scale.

12.6 Conclusion and Outlook

In this chapter we have described our use of the term "spontaneous interventions for health" for an ongoing phenomenon which seems to merge context sensitive and mobile health games with urban design. We have pointed to four aspects that can be improved as (1) the lack of interdisciplinary research and educational formats, (2) the lack of studies that provide data on positive effects on users' health, (3) the lack of social platforms for co-design that focuses on health and wellbeing and (4) the lack of research on business models and cooperation between research, design and user groups. Second, we have discussed recent examples of health game practice with respects to their possible contribution to health-related spontaneous interventions. We have highlighted self-reflective games as a springboard for discussions that will inform and possibly stimulate spontaneous interventions. Finally, we have outlined necessary future research directions highlighting the need for (1) more multidisciplinary training, (2) new ways to evaluate the impact of temporary installations "in the wild", (3) improve access to co-design projects through novel interfaces and (4) developing business and cooperation models that will allow to develop spontaneous interventions for health into sustainable and scalable projects.

References

1. Health game research, http://www.healthgamesresearch.org/about-us. Accessed 10 Feb 2010
2. Philip, W., James, T., Jackson-Leach, R., Rigby, N.: An international perspective on obesity and obesogenic environments. In: Lake, A., Townshend, T.G., Alvanides, S. (eds.) Obesogenic Environments: Complexities, Perceptions, and Objective Measures, pp. 1–10. Blackwell, Oxford (2010)
3. Benevolo, L.: The Origins of Modern Town Planning. Routledge, London (1967)
4. Klencke, H.: Die physische Lebenskunst oder praktische Anwendung der Naturwissenschaften auf Förderung des persönlichen Daseins. Eduard Kummer, Leipzig (1864)
5. Göckenjan, Gerd: Kurieren und Staat machen—Gesundheit und Medizin in der bürgerlichen Welt. Suhrkamp, Frankfurt am Main (1985)
6. Me You Health, http://www.meyouhealth.com/monumental/. Accessed 9 July 2013
7. Knöll, M., Moar, M.: On the importance of locations in therapeutic serious games. In: 5th International ICST Conference on Pervasive Computing Technologies for Healthcare, pp. 538–45. Dublin: University College, Dublin, (2011)
8. Knöll, M., On the top of high towers—discussing locations in a mobile health game for diabetics. In: Blashki, K. (ed.) Proceedings of the IADIS International Conference Game and Entertainment Technologies, pp. 61-82010, IADIS Press, Freiburg (2010)
9. Knöll, M, Moar, M: The space of digital health games. Int. J. Comput. Sci. Sport. **11** (2012)
10. Burney, D., Farley, T., Sadik-Khan, J., Burden., A.: Active Design Guidlines—Promoting Phyiscal Activity and Health in Design. City of New York, New York, p. 70–9 (2010)
11. Janson, A., Tickes, F.: Fundamental Concepts of Architecture: The Vocabulary of Spatial Situations, pp. 330–335. Birkhäuser, Basel (2013)
12. Coenen, T, Laureyssens, T: Ludic Citizen Engagement. Presentation given at The Ludic City: Urban practices through a playful Lens, iMinds-SMIT, VUB, DiGRA, Brussels, Nov 2012
13. Borden, I.: Skateboarding, Space and the City—Architecture and the Body. Berg, New York (2001)
14. Ho, C.L.: Institute for Urban Design: Spontaneous Interventions: Design Actions for the Common Good. Exhibition for the U.S. Pavilion at the 13th Venice Architecture Biennale 2012
15. Design that moves you, 'stairway stories'. In: Spontaneous Interventions. http://www.spontaneousinterventions.org/project/stairway-stories. Accessed 10 Jan 2013
16. 'Piano staircase'. In: The Fun Theory. http://www.thefuntheory.com/piano-staircase. Accessed 7 Jan 2013

17. Spontaneous   interventions.   www.spontaneousinterventions.org/project/stairway-stories. Accessed 10 Jan 2013
18. Rogers, Y.: Interaction design gone wild: Striving for wild theory. Interact. **18**, 58–62 (2011)
19. See for instance The Bartlett, in Centre for Advanced Spatial Analysis, http://www.bartlett. ucl.ac.uk/casa. Accessed 10 June 2013 or Massachusetts Institute of Technology (MIT). SENSEable City Laboratory, http://senseable.mit.edu/. Accessed 10 June 2013
20. Personal conversation with Dr. Katherine Willis, who co-runs a design course for architecture students dealing with interaction design at University of Plymouth, UK, www. inhabitingtheinbetween.wordpress.com/about/. Accessed 22 July 2012
21. Hansen, P.G.: The piano stairs of fun theory—short run fun and not a nudge!. In: I Nudge You,   http://www.inudgeyou.com/the-piano-stairs-of-fun-theory-short-run-fun-and-not-a-nudge/. Accessed 6 Jan 2013
22. Smith, C., Quercia, D., Capra, L.: Finger on the Pulse: Identifying Deprivation using Transit Flow Analysis. In: Proceedings of the 16th ACM Conference on Computer Supported Cooperative Work and Social Computing (2013)
23. Mergel, I.A.: Distributed Democracy: SeeClickFix.Com. for Crowdsourced Issue Reporting, Social Science Research Network (SSRN), 2012
24. See for instance Brooklyn Experimental Media Center at Polytechnic Institute of New York University, 'Betaville: A collaborative online platform for proposals on urban design', http:// betaville.net/" http://betaville.net/. Accecced 30 Jan 2013
25. Ho, C.L.: Statements. In: Spontaneous Interventions—Design Actions for the Common Good, http://www.spontaneousinterventions.org/statements. Accessed 24 Jan 2013
26. Volkswagen. The fun theory http://www.thefuntheory.com/. Accessed 23 Jan 2013
27. Lieberman, D.A.: Interactive video games for health promotion: Effects on knowledge, self-efficacy social support, and health. In: Street Jr, R.L., Gold, W.R., Manning, T. (eds.) Health Promotion and Interactive Technology: Theoretical Applications and Future Directions, New Jersey & London: Mahwah, pp. 103–20 (1997)
28. Lieberman, D.A.: Designing serious games for learning and health in informal and Formal Settings. In: Ritterfeld, U., Cody, M., Vorderer, P. Serious Games—Mechanisms and Effects, New York; London: Routledge, pp. 117–30, (2009)
29. Knöll, M.: Urban Health Games. Collaborative, Expressive & Reflective. (PhD dissertation, Universität Stuttgart, 2012) http://elib.uni-stuttgart.de/opus/volltexte/2012/7782/
30. McGonigal, J.: Who invented CryptoZoo, and why?—CryptoZoo. In: CryptoZoo—a Secret World of Strange and Fast-moving Creatures, http://cryptozoo.ning.com/profiles/blogs/ who-invented-cryptozoo-and-why. Accessed 10 March 2010
31. Bogost, I.: Persuasive Games: The Expressive Power of Videogames. MIT Press, Cambridge (2007)
32. Bogost, I.: Persuasive games on mobile devices. In: Fogg, B.J., Eckles, D. (eds.) Mobile Persuasion: 20 Perspectives on the Future of Behavior Change, pp. 29–37. Palo Alto : Stanford University (2007b)
33. Boyd Davis, S., Moar, M., Jacobs, R., Watkins, M., Shackford, R., Capra, M., Oppermann, L.: Mapping inside out. In: Magerkurth, C., Röcker, C. (eds.) Pervasive Gaming Applications—A Reader for Pervasive Gaming Research, vol. 2, pp. 199–226. Shaker, Aachen (2007)
34. Boyd Davis, S.: Mapping the unseen: Making sense of the subjective image. In: Nold, C. (ed.) Emotional Cartography: Technologies of the Self, pp. 39–51. Creative Commons, London (2009)
35. Nold, C.: Introduction: Emotional cartography—technologies of the self. In: Nold, C. (ed.) Emotional Cartography: Technologies of the Self, pp. 3–14. Creative Commons, London (2009)
36. Waterson, N, Saunders, M.: Delightfully lost: A new kind of wayfinding at Kew. In: Museums and the Web 2012: The International Conference for Culture and Heritage On-line, San Diego, USA, Archives and Museums Informatics (2012)

37. Mehm, F., Hardy, S., Göbel, S., Steinmetz, R.: Collaborative Authoring of Serious Games for Health. In: Proceedings of the 19th ACM International Conference on Multimedia (New York: ACM, 2011), **IXX**, pp. 807–808
38. Göbel, S., Hardy, S., Wendel, V., Mehm, F., Steinmetz, R.: Serious games for health—personalized exergames. In: Proceedings ACM Multimedia 2010, Fiorenze, pp. 1663–1666

Chapter 13
Using Virtual Environments to Test the Effects of Lifelike Architecture on People

Mohamad Nadim Adi and David J. Roberts

Abstract While traditionally associated with stability, sturdiness and anchoring, architecture is more than a container protecting from the elements. It is a place that influences state of mind and productivity of those within it. On the doorstep of adaptive architecture that exhibits life like qualities, we use virtual reality to investigate if it might be a pleasant and productive place to be; without incurring the expense of building. Thus this work has a methodological contribution of investigating the use of aspects of virtual reality to answer this question and the substantive contribution of providing initial answers. It is motivated by juxtaposing (1) responsive architecture (2) simulation in architectural design (3) adaptive computer mediated environments, and (4) use of VR to study user responses to both architecture and interactive scenarios. We define lifelike architecture as that which gives the appearance of being alive through movement and potentially responds to occupants. Our hypothesis is that a life like building could aid the state of consciousness known as flow by providing stimuli that removes the feeling of being alone while not being overly distracting. However our concern is that it might fail to do this because of appearing uncanny. To test this we hypothesise that occupying a simulation of a life like building will measurably improve task performance, feelings of wellbeing, and willingness to return. Our four experiments investigate if people feel more at ease and concentrate better on task and others when the walls around them appear to organically move, are happy for the walls to help them, and prefer to come back to a building that reacts to them.

Keywords Architectural design · Construction · Experiment methods · Simulation and behavior · Social virtual environments

M. N. Adi (✉)
University of Alberta, 4-110E NREF Building, NE of 116 St and 91 Ave,
Edmonton, AB T6G 2W2, Canada
e-mail: m.n.h.adi@edu.salford.ac.uk

D. J. Roberts
University of Salford, Newton Building, Manchester M5 4WT, UK
e-mail: d.j.roberts@salford.ac.uk

A. L. Brooks et al. (eds.), *Technologies of Inclusive Well-Being*,
Studies in Computational Intelligence 536, DOI: 10.1007/978-3-642-45432-5_13,
© Springer-Verlag Berlin Heidelberg 2014

13.1 Introduction

The use of interactive elements in buildings is on the rise [1] and buildings that possess them are proving to be popular in both private and public projects [2]. However, the temporary nature of most examples does not lend them to studying long-term effects such as social behaviour [3]. It seems that popularity relies on a balance between design, content and novelty. Despite this growing popularity, interactive architecture is widely viewed as novelty. We postulate that a convergence of interactive architecture and adaptive computer mediated environments might allow form and content to be responsive to the activity and mood of occupants, thus replacing novelty with usefulness. A more comprehensive picture is set by convergence between and within: animated, reactive and organic and intelligent architecture; and socially intelligent, empathic and virtual environments. It is hoped that this amalgamation would allow buildings to be able to interpret social intentions and needs, evaluate the impact of actions in a group and steer a group of users towards a common goal based on their actions and needs. Yet even without this union, the base question of how a building that seems to move as a living entity could affect its users is still very useful and interesting and this is what is being focused on. Also no direct comparison has been made between a conventional building and an interactive one to assess [4] the effect of interactivity in a building.

With regard to appropriate methods for studying adaptive architecture, the literature falls into one of two broad categories: Interactive or Intelligent Architecture and the appropriateness of virtual reality to test our hypothesis that life like architecture will have a positive effect on their users. Subcategories in interactive and intelligent architecture include architectural theory, building materials and building examples. In the use of virtual reality as a test medium subcategories include presence, virtual environments and the use of virtual reality in fields including psychotherapy. We also consider the potential of adaptive computer mediated environments.

Responsive Architecture may be categorised as that which responds to the user, either through the design or more interactively during use.

The first major category is architecture for which the design has responded to people. This is less relevant to our work, that focuses on buildings designed to respond interactively. However, describing them not only helps to set the scene but also explains the inspiration for the organic appearance of our later experiments.

Firstly we consider response to cultural or religious needs or events. Examples include: The Water Temple by Tadao Ando [5], The Umayyad Mosque in Damascus [6], The Jewish museum by Daniel Libeskind [7], The Berlin Memorial by Eisenman [8]. All these projects focus on giving visitors unique experiences that attempt to respond to their needs and provide them with insight to cultural events that led to the construction of these buildings.

Secondly we consider animation-based design. Here the final shape of the building is a response to the recorded and then analyzed movement of people or objects in a similar place over a set amount of time. The designer records the movement of users, anticipates it then designs buildings based on the shapes resulting from that movement pattern. Examples include The Endless House by Kiesler [9], and the works of Berkel [10], Lynn [11, 12] and Spuybroek [13]. Most used flowing organic shapes that through their naturalness are intended to give the feeling of comfort. We use this as inspiration within the design of our lab-based experiments that will be described later.

What Keisler, Lynn and Berkel [9–12, 14] have in common is that the organic flowing shapes, which architects produced in these examples, create and encourage comfort as they mimic natural shapes. While this is vital in forming a more thorough understanding of user movement, it does not cover real time change in the building after it is finished, which is the main focus in interactive design. The main focus of these projects is that the final form of the building does not need to be moving or changing colour in order to be responsive.

The second major category is architecture that responds interactively to the occupant. Reflexive architecture was introduced as a concept by Neil Spiller a decade ago. At that time he felt that little research had been done in the area. He described reflexive architecture as "architecture that is highly responsive and intelligent, able to translate and connect to its contextual environmental surroundings at a new level, while also operating in three or more spaces simultaneously". Although the need for at least three spaces is unclear.

The responses between interactive architecture and its occupants may include animated behavior that is responsive within seconds or minuets to the user. This is a rapidly emerging area for experimental architectural design. Examples include: Kas Oosterhuis (the trans_PORTs project) [15, 16], Nox (the H_2O Expo) [13, 17], Paul Sermon Tele presence [18, 19], Frazer [20], Decoi (Aegis Hyporsurface) [21], Usman Haque [22, 23] and David Fisher, The rotating towers in Dubai and Moscow [23, 24]. In the past decade architects, designers and artists have increasingly worked in the field of interactive structures. In many instances these projects were temporary installations rather than being permanent ones [3, 25], An example of a more permanent interactive building is the rotating towers projects by David Fisher [23, 24]. When designing temporary structures or installations an architect often has more freedom to experiment. This is not least as the building typically has fewer functions to support.

Architecture can be considered as a type of language in which the architect and the user engage in speech through the building [26]. Palassma proposes that interactive architecture allows a livelier conversation. Thanks to emerging technologies and materials, structures have become able to interact with users instantaneously [17, 23]. The movement and behaviour of its visitors trigger structures to change any combination of shape, appearance, sound or smell. Arguably in these projects architects arrange rather than design and allow the users to have conversation directly with the architecture itself.

From this analysis it seems that responsiveness of space is linked to two main factors. First, the materiality of the project and how it is constructed. Second, the sociological dimension of space and how that can be controlled and used to further focus the relationship between architecture and its users. These new ways of interacting with architecture are helping to form a whole new understanding of space that such space becomes part of the user and interaction. Space becomes looked at as a medium of communication. This is apparent in the works of the artist Paul Sermon [18, 19], in which he links users in different locations via live chroma-keying and video conferencing equipment to explore user behaviour and interaction within such telespace. This creates some awkward responses from the users at the beginning and causes them to re-evaluate their environment in a new understanding which depends on space as well as advanced technology.

All the previous architects and artists designed interactive buildings that can change colour, shape, sound, layout or all of these elements combined to create an enriched user experience. There are a lot of different types of buildings here ranging from installations [13, 18, 19, 21, 22] to full buildings [15–17, 23, 24]. A recurring theme here, however, is that all these changes are predetermined by the architect at the time of design. It is noticed that there is a great deal of theoretical investigation being done by the designers on the role of such architecture in its environment and its relation to the user's experience [17, 20, 23]. When I interviewed Usman Haque in 2005 he mentioned that the role of the architect or the designer should be more like the role of an operating system in that he sets the rules in which the building operates but the users are the ones who determine the final outcome of the design, this is also supported by Kronenberg [17]. Usman also states that for a building to be interactive there has to be information exchange to and from the building otherwise it's just a reactive system [27]. Other design theories include the works of Joye [28], Thomsen [29], Pallasmaa [26], Forty [30], Zellner [31], Holl [32] and Hillier [33, 34] all of whom explore the relation between users and space, and how that can affect the experience gained. In particular Joye [28] looks at how organic and natural elements in architecture can help in reducing stress levels in users. This is an important theme of our work. Buildings here have a somewhat open ended design which builds on the users actions. Haque [23, 27], Thomsen [29] and Spiller [35] view architecture as a changing user interface, while Pallasmaa [26], Forty [30], Holl [32] and Hillier [33, 34] view it as an unchangeable setting for user interaction (which is the classical view of architecture). The term intelligent building or architecture also has many definitions as well, sometimes overlapping with interactive architecture. However, the unifying characteristic in all definitions looked at is that an intelligent building utilizes sensors to acquire user data so it can automatically provide services depending on user behaviour [36–44].

"Human life is interactive life in which architecture has long set the stage" [45]. Humans are flexible creatures capable of manipulating a wide verity of objects and living in varying environments [17] as such it is no surprise that humans want to reflect that flexibility in our buildings as our genes are expressed through our environment [46]. It appears that users are becoming captivated by interactive or intelligent structures as they are seeing their effects on the structure directly and clearly [47]. Their movement patterns and behaviour in general change in accordance to the reactions they see appearing before their eyes as architecture and its users become parts of each other [48]. The possibility of having a more lifelike or living building is becoming more feasible as more artists and architects are experimenting with it [49, 50]. It can be argued that users of such buildings are learning to see themselves in a new light, looking at their new reflections in these unorthodox 'mirrors'. Such new visual reflections can produce new behaviours of the users in accordance to these reflections. Also, with such buildings attracting more audience and gaining in popularity [51], it indicates that they can have a distinct advantage over none interactive buildings especially in public and retail spaces where a high amount of visitors is preferred.

In all the previous projects the materials used were as vital as the designs themselves as the desired level of instantaneous reaction by the building cannot be achieved by traditional materials. Combinations of sensors, processors and changeable components (e.g. scent releasing systems, mechanical or hydraulic movement systems, colour changing LEDs...) were used. Materials like Aero Gel [17], OLED [17], Litracon [17, 52] and smartwrap [53] all push the limits of architectural boundaries and help in creating better models for interactive and intelligent Architecture. Research from the MIT, shows materials that can change shape in real-time [54]. This opens up new possibilities for designing such spaces [54]. Research from Cornell University shows robots that have a limited ability to self reproduce [55]. When incorporated in designs this has the potential to produce self sustaining buildings. While some materials are still in concept mode they are very important in helping to create better designs and conceptual models for interactive or intelligent buildings.

13.1.1 Using Virtual Reality for Testing and Evaluating

This section examines different test mediums that are available in virtual reality assessing each one to determine which is more suitable for testing interactive buildings. It highlights available test environments, how people react in virtual reality environments and how close such reactions are to real life ones.

In architecture, the architect depends on his experience when trying to design and produce a building. In essence knowledge of materials, space and building methods is gained through observation and experience gained from sometimes disastrous trial and error [1]. Apart from architectural drawings and images of 3D models the client is usually buying a product that they cannot see fully until it is constructed in real life [56–60]. Even when using 3D reconstruction video the angles and areas the client sees are limited [56]. It is commonly known that this method has some major draw backs. First it is very time consuming to construct a building, second it is very expensive to do so. Third and perhaps most importantly is that the client will not know what they are exactly getting until the building is completed. This means that there is no room for error and that amending or adjusting a project would be very difficult. Experimental buildings are not unheard of in architecture, for example: Sky Ear [22], Scents of Space [16, 23, 61], H_2O

Expo [13, 17]... etc., but making these buildings is very expensive and time consuming. Amending or adjusting them would also be very difficult for the same aforementioned reasons.

A potential test environment is the use of an online social environment. Online social environments are becoming accepted as credible tool for social studies [62, 63]. In particular a number of researchers have indicated that people tend to behave in a very similar way when in Second Life as they would in real life [63, 64]. Users seem to react naturally to social space even though such environments are not immersive [63]. Online environments like Second Life have some distinct advantages. It can be argued that creating content in such environments is quick and cheap. All that is needed is making a 3D model of the building and placing it in the virtual environment, so no time is wasted on construction issues that might arise in real life models. This also means that adjusting a model can be done quickly as well. Research implies that online (even forum like) test environments function as a more generalized simulation than that of a mathematical simulation tool [64]. Equipment wise, all a user needs to use the programme is a midrange computer and an internet connection. This coupled with the fact that the usage of the software is free means that a lot of users will be able to access such an environment with ease. This can increase the number of visitors as every day millions of users spend an average of 22 h a week interacting with each other thorough avatars [65–67]. Also since visitors can come and go as they please, their behaviour can be closer in quality to real life behaviour. Lastly placing a test building for prolonged periods of time there is relatively cheap.

One issue in such an environment is the lack of realism which can lead to users feeling detached from the environment they are experiencing, another issue is the quality of immersion, since users have to see the environment through a computer screen, they will not feel as immersed as the other two methods we are going to discuss (Immersive environments and head mounted displays) and as such this might cause their reactions to be less realistic [68]. However, previous research shows that online virtual environment users behave in a very similar fashion to real life [62]. Researchers are using Second Life as a viable tool to evaluate social trends [64] and as such we feel that this medium would be good for tests that might require longer periods of time and a larger sample size.

Another method examined is immersive virtual environments IVEs. This technology first existed in 1965 as a lab-based idea [69, 70]. It is usually a room with graphics being projected on its surfaces, usually referred to as a CAVE (square shaped room). The number of surfaces used can vary from three upwards. The advantages of using such a method is that it provides a high level of realism as test subjects literally step into the virtual model being tested giving it an advantage over desktop-based methods [71]. Test subjects are highly immersed in the virtual environment and as such might react to it in a more realistic manner. A wealth of presence research indicates that people in such environments react exactly as they would in real life [70, 72, 73] even in low-fidelity scenarios [74, 75]. Mel Slater points out that in an immersive environment almost all test subjects avoid colliding with virtual objects even though they know that they are not there [72].

Also participants usually respond in a realistic manner to events shown to them in IVEs [70]. Such environments have been used as an effective tool for social studies [74, 76, 77] particularly in spatial cognition [78], education [79, 80] and psychotherapy [81–83]. In numerous cases they were successfully used to treat social phobias [84–93] and post traumatic stress disorder [94, 95]. Research also shows that IVEs can evoke real emotions and mental activity as a real situation would [96, 97]. Any adjustments to the model can be made quickly and easily. Also, since such an environment is a lab-based one, all the environmental factors can be easily controlled (such as lighting, time limits, etc.) enabling researchers to specifically focus on the variables or elements that they want to study.

Since a CAVE or OCTAVE is a lab-based environment it has some disadvantages. The amount of people that you can have there at once is limited due to the size of the room. Also people cannot stay there for long periods of time. Since the equipment is delicate a researcher has to be present with the test subject at all time which means that reactions of test subjects might not be as natural as hoped. Perhaps the biggest disadvantage is that people cannot come and go at will. Even with such disadvantages it is still the preferred test method for short term experiments. It provides a high level of immersion that is comparable to that of constructing a life-sized model of the environment while at the same time having the flexibility, controllability and repeatability of using 3D simulations.

The final immersive display method examined is head mounted displays; it has the same advantages as the immersive environments mentioned before [68]. It allows for a very high level of immersion and content can be created and edited quickly as well [68]. Our main reservation about this medium is that the person becomes disembodied (as they cannot see their body in the virtual environment) and thus the experience is less lifelike. Also the equipment a test subject has to wear is heavy and it can cause discomfort. The low field of view it provides has been linked to motion sickness and a lower sense of presence which is likely to impact the awareness, attention and action of users [68]. This will be a major issue as people tend to lose focus and interest when they are fatigued. Also issues might arise if the experiments incorporate the use of real objects rather than virtual ones.

13.1.2   Crossover and Relations

This section examines possible relations between the two main research areas investigated previously and how they relate to each other. One project that is relevant here is an interactive entertainment space built for the Swiss national exhibition Expo in 2002. It was investigating if users can associate buildings with life in what is called the ADA project [23, 51, 98–100]. This project consisted of a room where the floor was covered with pressure sensitive plates that changed colours and collectively displayed different colour patterns as you stepped over them. If one followed them fast enough the room rewarded the user with a special pattern. It turns out that the majority of users enjoyed being in this responsive space and a good percentage of them were convinced that the space could be considered as a living organism. The public reaction to the project was overwhelmingly positive [101]. ADA was the most popular IT related exhibit at the expo [51]. This suggests that with the right level of interaction people can actually start to view buildings as active participants in a group rather than just passive spectators. It also shows how popular and crowd attracting interactive buildings can be, although no direct comparison have been made between an interactive building and a non-interactive one to understand the explicit effect of interactivity on visitor numbers. The ADA project was an entertainment space; it would be interesting to see whether similar success could be achieved in a work oriented environment. Also this project demonstrated how human response can be inferred from observing behaviour as well as the potential to use ADA to automatically deduce group attitudes opening the door to possibly influence their behaviour [99]. It is noticeable that ADA had only short term reasoning as it responded to users directly but had no mid or long-term goals. Even though users still enjoyed being in this space which is encouraging. Theoretically, this would suggest that a building with two layers of reasoning would provide a better and more enjoyable environment for its users and visitors. This project is the only project that we found that combined research in virtual environments and architecture. It demonstrates that users can have meaningful experiences with an interactive environment, and as mentioned before also highlights the popularity of this type of space [51, 101].

The use of virtual reality or rather virtual reality visualising methods is common in architecture. What is meant here is that architects often use virtual 3D models of buildings to produce rendered images or fly through videos that they show to clients and some potential users [56–58]. The use of virtual reality or virtual environments in design focuses more on the construction methods rather than design elements in most cases [56]. The main issue with this approach is that architects tend to show clients what they, as designers, want them to see without giving the client the ability to navigate and experience the space on their own [58]. It can also be argued that the virtual models shown to clients and users tend to be at the final stages of design when most decisions have been taken. Models placed in virtual environments such as the Second Life, CAVE or OCTAVE like environments tend to give the client (users of the building) a high level of freedom enabling them to experience every part of the building and interact with it however they want [56].

Virtual environments have been used earlier to effectively treat phobias and other mental issues such as trauma. The papers looked at in the virtual reality section earlier indicate that VEs are useful tools that can produce meaningful results because people react in them in the same manner as they would in real life [70, 73, 102]. Research done in this field also suggests that, when measured in similar scenarios, the human brain exhibits the same level of neural activity in both virtual and real scenarios [81–86, 88, 89, 91–95].

An issue that remains to be seen is if users were interacting with the environment or with objects in that environment. In the case of shell shock it is almost certain that users were reacting to the environment as the experiment procedure focuses on having users go through similar war-like scenarios. Although in these cases it can be argued that users are interacting with either the environment (shell shock) [94, 95] or objects in the environment like avatars (anxiety and public speaking phobia) [84, 85, 87, 88] or the cause of phobia itself (heights, flying, etc.) [91–93, 96]. But the main interaction is guided through the presence of a therapist who uses and controls these objects as they see fit during the treatment procedure [72, 91, 103–105]. What this research wants to see is the effect of the building itself on the user with minimal interference from the designer or owner isolating and studying what effect interactive or intelligent buildings might have on their users.

In general papers reviewed on the matter of the use of virtual environments in the treatment of phobias, regardless of phobia type, establish two things. First, Molinari [88] concluded that virtual environments are as good if not better in treating phobias than real-life environments. They also have the advantage of allowing high control and to tailor fit the required virtual environment to the exact needs of the individual's treatment [88] also there are over a hundred research papers within that reference that agrees and reaches the same conclusion.

Second, when using a VE even through a desktop system it was recorded that subjects reacted in the same manner as they would in real life [70, 73–75, 102]. One major issue arises here is the lack of physical interaction between the subjects and their environment [84, 85, 87–96]. This can be attributed to the nature of phobia treatment. Phobia treatment, in real or virtual environments is based on exposing the patient gradually to their fear [72, 91, 103–105]. This means that the person treating them is the one who interacts with the patient and controls how the treatment session goes. Even in the case of treating shellshock where the patient goes through a premade scenario there is no interaction with the environment just exposure to different conditions [94, 95]. In these cases it can be argued that users are interacting with either the environment (shell shock, post traumatic stress disorder, etc.) [94, 95] or objects in the environment like avatars (anxiety and public speaking phobia) [84, 85, 87, 88] or the cause of phobia itself (heights, etc.) [91–93, 96] the main interaction is guided through the presence of a therapist who uses these objects as they see fit and there is no interaction happening in from the environment to the user. This leads us to believe that there might be a possible lack of literature on how might having an interactive/intelligent environment affect its users.

13.2 Research Direction

The previous sections reviewed papers in relation to the fields of interactive and intelligent architecture and the use of virtual environments as an evaluation method. That main aim was to propose a definition of interactive, intelligent or lifelike architecture, assess the usefulness of virtual environments as a test medium and see if there are any research projects that have attempted to study the effect of interactive or intelligent buildings on their users.

Architecture is becoming so intertwined with a network of other disciplines that a new hybrid form of practice and architecture itself is emerging [106]. While the sheer amount of new interactive architecture is apparent from previous examples, we attempted to go past the momentary popularity and ask if such projects are useful to their users. It was found that first, there are multiple definitions of interactive architecture and through reviewing a selection of projects we defined interactive architecture as architecture that can react to its users and change its properties (colour, shape, sound) in real time, intelligent architecture furthers that by adding a level of reasoning with which the building analyzes input from its users to achieve a set of goals. Based on these definitions derived from the literature, we further define lifelike architecture as interactive architecture that has some resemblance of being alive. This might be through moving or interacting as if alive and or through exhibiting intelligence. Interactive projects have proved to be popular and potentially valuable when it comes to attracting visitors. Second, papers reviewed indicated that virtual environments are a viable test medium with users reacting in similar or almost the same way in a virtual environment as they would in real life to the same input or scenario. Papers reviewed in the medical field indicate that brain activity in virtual environments is the same as brain activity in real life, given that the scenario is the same. It is concluded that both online social environments and immersive virtual environments are suitable test environments that can be used for long-term and lab-based experiments respectively. They both provide sufficient levels of freedom and immersion (semi immersion in the case of online environments) and they are both accepted test mediums for social interaction as people tend to react very naturally in them, to social or environmental stimuli. Third, it is apparent that every method of measurement in virtual environments has some advantages and disadvantages [107]. Methods used in measuring presence in virtual environments are varied and no single method is universally accepted [107]. It is possible that a combination of evaluation methods would provide better data for experiments, as mixed methods can assist and complement each other, eliminating or minimizing the disadvantages that can occur from using a single method. From that it can be concluded that a mixture of evaluation methods should be used in experiments to produce good high quality data from experiments.

13.3 Experiments

Each experiment involves having a test subject complete different yet similar tasks in a series of virtual environments. Test subjects can be completely alone or accompanied by an examiner depending on the individual experiment and the research question it is trying to address. Apart from the last experiment all our experiments are undertaken within a surround display system called the OCTAVE. The OCTAVE is an immersive large screen projection system. It projects computer graphics images on 8 surround walls and the floor (Fig. 13.1).

Fig. 13.1 The Octave, the environment and the author

This test environment surrounds the subject in a life size simulation, and thus can give the impression of being within a simulated room better than looking into that room through a desktop display, thus providing a better sense of being there (presence) to test subjects and making the experience seem more natural. In this work immersive stereo was not used, so the participants did not have to wear stereo glasses. The last experiment (experiment 4) was made using an online social virtual environment called Second Life.

The first experiment designed was a pilot experiment that explored the second research question: How might being inside a room with walls that appear to come to life impact on a person's feeling and performance? Doing a quick pilot experiment at the beginning of the research helped in understanding practical limits and issues that might rise in later experiments. Issues that were made clearer through the pilot experiment were the number of participants needed, how long should the experiment last and what task should be used. For instance, in the pilot experiment we used a task that required the user to play a game on a laptop while in the OCTAVE. We noticed that test subjects were focusing more on the laptop and taking no notice of their surroundings, also they were repeating the same game in all test conditions which made any results unusable (their score improved because of repetition). That is why we opted for a simpler task in later experiments that can be placed directly in the test environment without the need for interface equipment (laptop or 3D glasses) and can be varied with ease to eliminate the problem of repetition. Feedback from the experiment using questionnaires and interviews also helped in highlighting any potential issues that might have not been covered in the literature survey. All experiments can potentially help in answering unclear points identified by the literature review and some problems raised via the methodology.

13.3.1 Experiment 1

How might being inside a room with walls that appear to come to life impact on a person's feeling and performance?

Task: Solving a jigsaw puzzle alone in an Octave, while walls are blank and static or display moving patterns.

To be able to assess the impact of an animated environment on an individual we had to conduct the following experiment. Test subjects had to complete a task in different environment settings. Their performance in each task was measured and compared. Post experiment interviews and questionnaires were also used to assess the appeal and attractiveness of animated environments.

The task that test subjects had to do was completing a jigsaw puzzle. We felt that this is a simple task that most test subjects should be familiar with and that we can vary with ease. Another reason jigsaws were chosen was that they require both concentration and allow performance to be easily quantified. Tasks were completed in an immersive display system called the OCTAVE. Two test environment settings were used, one with blank walls and one with moving lifelike walls, which simulate that they are moving around the test subject. The performance of test subjects was measured by the amount of puzzle pieces they assembled in each setting.

A within subjects design was used for this experiment, this method provides more statistically sound results and is more practical as less resources are required in terms of number of test subjects and time required [108–110].

To ensure that participants performance was improved because they were familiar with the task, we varied the order of puzzles and conditions across subjects. This meant that some began with puzzle A and some with B; and 50 % were inside a simulation of moving walls first while the remainder where surrounded by blank display walls first. Twenty Test subjects were used.

13.3.1.1 Findings

The results of this experiment have shown that when a person is placed in a small room both their comfort and performance when doing a jigsaw puzzle are increased if the walls appear to move. 90 % of subjects performed better when the walls appeared to move around them and upon analysis results were highly significant with a probability value less than 0.0001 this meant that the chances of having the same results would be nearly 100 % if the experiment was repeated. No subjects reported disliking the moving walls and more often than not test subjects reported feeling more comfortable, and better able to concentrate and do the work.

Many reported the feeling of something missing when being in the blank environment while specifically mentioning being calmed and more relaxed when the walls moved around them.

13.3.2 Experiment 2

How can seemingly lifelike architecture impact people's ability to follow instructions from a teacher?

Task: Solving a puzzle in an octave by depending solely on verbal instructions from a tutor, while different projections are displayed on its walls.

To answer the above research question we conducted the following experiment. This time, instead of working alone, the participant had the part of a learner in a simple teacher learner scenario following a set of standardized spoken instructions given to them by an instructor to guide them in completing a task in three different conditions. The flow of data was one way from instructor to subject. This was done to standardize the amount of information given to test subjects. The performance of test subjects for each task was measured and compared. Questionnaires and interviews were used to assess the appeal and comfort of animated environments.

The task remained similar to the previous experiment so comparisons can be drawn allowing the effect of the teacher learner experience to be isolated from that of working alone. An instructor (confederate) explained the procedure of the test and what the test subject should do before the experiment started. Every experiment involved both the test subject and confederate entering a series of conditions. Tasks performed in each condition were split into a series of stages. Test subjects had to follow a series of spoken verbal instructions given to them in key stages during the test by the instructor in order to complete the task, while being aware that they could not communicate with the instructor during the test. The task test subjects had to do was completing a jigsaw puzzle while relying only on the oral instructions given to them by the instructor. Tasks were completed in the Octave. The performance was measured by calculating the percentage of completion for each puzzle in each setting. Thirty test subjects were used here to provide better statistical data than the previous experiment.

13.3.2.1 Findings

In this teacher learner setting the performance of 93 % people was improved when they were in the animated environment the overall average improvement was 38.77 % compared to the blank environment and 21.25 % when compared to the patterned environment. When analyzed, the results proved to be highly significant with P values less than 0.0001. Previously people universally preferred the experience of seemingly moving walls and some reported finding it comforting. Questionnaires and interviews made in this experiment indicate that most people prefer a room with moving walls stating that it improved their concentration levels which was reflected in the improvement of their scores. This suggests that surround projection or physically moving walls are likely to be beneficial in the classroom setting and unlikely to be detrimental. In particular they seem to complement the presence of a teacher especially when giving instructions. Using a combination of virtual models and surround projection gave clear results. The ability to transfer such results to the real world is yet to be tested, but results further imply that real environment tests would be worth doing in the future.

13.3.3 Experiment 3

How can seemingly intelligent lifelike architecture impact people's ability to follow instructions from a teacher?

Task: Solving a jigsaw puzzle in an octave with verbal instructions from a tutor, while different projections are displayed on its walls. In one of the settings one of the walls provides visual hints to aid the test subject.

Here we expand the previous experiment by convincing the participant that the environment around them is intelligent and that it can understand what is happening inside it and try to assist them in their task (Fig. 13.2). The same teacher learner setting was used here as in the previous experiment. The participant had the role of a learner in a simple teacher learner experience following a set of standardized spoken instructions given to them by an instructor to guide them in completing a task in three different conditions. The flow of data was one way from instructor to subject. This was done to standardize the amount of information given to test subjects. The Performance of test subjects in each task was measured and compared. Questionnaires and interviews were used to assess the appeal and comfort of animated environments.

In the last condition a wizard of OZ approach was used to make the room appear intelligent through the use of an additional confederate. The task in this experiment is the same as the one in the previous experiment so that a direct comparison can be drawn allowing the effect of having a helpful environment to be separated and compared to the effect of an animated one. An instructor (confederate) explained the procedure of the test and what the test subject should do before the experiment started. Every experiment involved both the test subject and confederate entering a series of conditions. Tasks performed in each condition were split into a series of stages. Test subjects had to follow a series of spoken verbal instructions given to them in key stages during the test by the instructor in order to complete the task, while being aware that they could not communicate with the instructor during the test. The task test subjects had to do was completing a jigsaw puzzle while relying only on the oral instructions given to them by the instructor except in the final condition. Tasks were again completed within an immersive display system called the OCTAVE. The performance was measured by calculating the percentage of completion for each puzzle in each setting.

Fig. 13.2 The position of the test subject within the environment in accordance to the helpful wall. The *arrow* shows the flashing area where the assembled puzzle pieces should be placed

A within subjects design was used for the experiment. The order of puzzles and environments was randomized across subjects to avoid any effects that puzzle order or environment order might have had on our findings. We split the participants into two equal groups. The first group environment order was blank, static patterned, intelligent and the second group environment order was intelligent, static patterned, blank. The puzzles used in each condition was randomized as well. Thirty test subjects were also used here.

13.3.3.1 Findings

Building on a previous experiment where we improved the performance and moods of people doing jigsaw puzzles by surrounding them with projections of walls that appeared to move, a teacher and informative graphics were added to these walls to begin to study the potential impact on a teacher learner interaction. The previous experiment [111] improved the performance of 90 % of participants and the average improvement was 14 %. In this teacher learner setting the performance of 93.33 % people was improved and the average improvement was 62.84 % in comparison with the average score of test subjects in the blank environment and 42.07 % in comparison to the average score of test subjects in the patterned environment. When analyzed the difference was statistically significant with a P value of 0.001. Questionnaires and interviews conducted after the experiment indicate that most people prefer a room with moving helpful walls. Analyses of the video recordings also indicate that people were more engaged with the helpful environment than any other condition regularly looking at the helpful wall. In addition to the results transferring to concentration on instructions from a teacher, people liked the helpful graphics in front of the moving walls. This suggests that surround projection or physically moving interactive helpful walls are expected to have a positive effect in a classroom setting and unlikely to have a hindering one. Specifically interactive buildings seem to complement the presence of a teacher and other information on the walls. The combination of using virtual models and surround projection has once more yielded clear results. The transferability of these results to the real world is remains to be tested, but the results indicate that physical, real environment tests are a very viable option.

13.3.4 Experiment 4

In an online virtual environment, would people be in favour of visiting, returning and staying in an interactive building over a static counterpart?

Users did not have to do a task here, instead the number of visitors and the amount of time spent in each virtual building was recorded and compared.

Our main concern about previous experiments was that they were made in a lab setting and they were short time wise. In this experiment we wanted to test the appeal and attractiveness of an interactive environment over an extended period of time while trying to provide as much freedom to visitors as possible. To do that, two virtual building models were constructed in an online social virtual environment (Second Life). The only difference between them was that one was interactive (Fig. 13.3) and the other was not. They were placed in Second Life for a total of 6 months, during which we took various measurements to do with the number; length and properties of visits (e.g. number of groups and number of return visits). Feedback on each building was collected from a separate group of 20 participants as an extra measure.

The online social virtual environment was used to provide greater levels of freedom for participants as there were no examiners present and there were no restrictions on access times. This means that reactions and actions of test subjects should mirror what might happen in a real life scenario more accurately and conclusions drawn from this experiment could be applied to real life situations more rapidly. The reason for conducting the test for 6 months was that we felt it was a sufficient period to produce meaningful results, it was also due to this research time restrictions. In addition to measuring visitor numbers and the qualities of their visits to each building feedback from a separate focus group was collected. It was calculated that using 20 people for feedback would provide meaningful results for this type of experiment.

13.3.4.1 Findings

With the rising popularity of using interactive elements in architecture, we asked: Would people more willingly return and bring their friends to a building if it had interactive elements? Second life, an online social environment was used to

Fig. 13.3 Interactive building. The ground tiles illuminate as visitors walk on them and the walls move away from the visitor when they get close to them

answer this question, testing the appeal of interactive architecture in comparison to a non-interactive equivalent.

The aforementioned research question was divided to two sub-questions. The first sub- question was "could an interactive building produce more visitors, return visits, group visits and be a more sociable place when compared to a static building?". Results demonstrated that people revisited our interactive online gallery more often, stayed there longer and were more likely to bring friends when its walls and floor where interactive. The interactive building managed to generate 408 visits compared to 92 visits generated by the static building. There was also more return visits to the interactive building as well with nearly nine times as much as the static building. The interactive building also had more unique visitors (102 visitors) than the static building (57 visitors) and a bigger percentage of its visitors chose to return to the interactive building over its static counterpart. (68.627 % compared to 26.316 %). Another finding is that interactive building had more group visits than the static building with the static building having only one group visit throughout the period of study to 22 group visits that were made to the interactive one. The main and perhaps the most significant difference however, was the total amount of time spent in each building. Visitors to the interactive building spent a significantly longer period of time in comparison to the visitors of the static building with the average of 50.322 min spent in the interactive building for each minute spent in the static counterpart.

The second sub-question was "would the presence of interactivity within a building create a significant boost in the above mentioned qualities?". Our analysis proved results to be significant with all our P values below 0.0001 which is statistically highly significant. Questionnaires indicated that the vast majority of people (85–90 %) favoured the interactive building and found it more appealing, with the ability of interacting with the building being a major attraction factor. No dislikes to the interactive building were stated and the majority of visitors favoured socializing within it.

This implies that interactive buildings are more appealing to be in and are a more attractive place to socialize within. Results also showed that interactivity can generate a sustainable interest in a building thus increasing its projected life span and revenue, especially in buildings that depend on visitors like exhibition halls, museums and public buildings. The online virtual environment Second Life proved a valuable tool for this study. In particular it addressed the issue of allowing regular and freedom of access to participants over a period of weeks or longer. It also provided anonymity and as such made test subjects more comfortable and willing to take part in such an experiment.

13.4 Discussion

In all experiments the interactive or seemingly intelligent environments proved to have an advantage over normal (non intelligent, interactive or moving) environments. In the three lab-based experiments people performed better and stated that they were more comfortable and preferred interactive or seemingly intelligent environments over other environments which were blank or patterned environments. From observation and interviews people seemed to behave similarly in both normal (where the walls were either blank or with static patterns) and seemingly interactive or intelligent settings. There was no mention of distraction or feelings of alienation during interviews. This was also reflected in the scores as the majority of participants scored better in the interactive and intelligent settings. In the third experiment test subjects were interacting with the intelligent environment as it gave them visual hints on how to complete their task and their scores improved in the intelligent environment. Subjects stated that they were more comfortable in interactive and seemingly intelligent environments.

Results from experiment four indicate that over longer periods of time it seemed that the interactive building generated more visitors and appeared to have more appeal than a static building. This agrees with papers relating to the ADA project [2, 51, 101] which indicate that interactive buildings appear to be more popular and attract more visitors than none interactive buildings. The interactive projects reviewed in the literature did not have none interactive counterparts to be compared to. This means that their popularity could be contributed to other factors than interactivity (e.g. design, materials or novelty). The results from this experiment suggest that interactivity has a strong positive effect on the popularity of a building. This is also supported by Haque [47] as he states that interactive or intelligent buildings captivate their users as they see their effects on them. They also confirm the views of Delbrück and Bäbler [51] that interactive or intelligent buildings attract more audience and are more popular than non-interactive ones.

   In lab-based experiments a within subject design was used and subjects had to do different tasks in test environments in one go. Care was taken to ensure that test subjects were not bored or tired out by the end of the experiment, mainly by changing the order in which they experienced the environments. It can still be argued that having the test subjects do multiple visits and perform different tasks on different visits might produce better results, although upon analysis the results were highly significant. There is also a risk in having test subjects perform multiple visits, as they could practice tasks at home thus corrupting the data. To counter both, tasks must be sufficiently different or complex, both of these things would making analysing data difficult or even make data sets incomparable due to the difference. Another potential issue of having complicated or different tasks is that there would be a higher learning curve involved. This means that experiment have to be longer to counter that which risks people becoming tired or losing interest, it would also make recruiting test subjects more challenging.

   Test subjects recruited for the experiment were mostly students and employees of the University of Salford. Also there was imbalance in the sample between male/female or nationality or profession (most were research students from the computer science department). It could be argued that the test subjects sample is not representative nether in gender nor in profession. However, since we are looking for non gender/profession specific findings that can be overlooked. Further, as these experiments are possibly the first to investigate this field we felt that a random sample of people would be sufficient. Since performance and the possibility of improving it can be linked to work or study environments, which in many cases might not be varied or balanced, performing experiments on a random group of test subjects also made sense. From the experiments looked at in the literature review a number of 20 participants was adequate in most cases. The number of participants aimed for was 30 people for each lab experiment, this was done to provide better data and produce more significant results. The first experiment only had 20 participants but produced statistically significant results.

   In the last experiment, made in an online environment, the main issue was that the experiment was performed in an area with low traffic (visitors). Also the low amount of content in both buildings might have deterred people from visiting or revisiting the experiment area. However, it can be argued that the simplicity of the buildings helped in isolating and clarifying the effect interactive components had on visitors. The results of the experiment were highly significant and even with low traffic the experiment managed to attract 159 visitors in total. This increases the confidence in our results and supports what the literature indicated that interactive buildings tend to be popular. It also indicates that the popularity of such projects can be attributed to their interactivity.

   While the results of the experiments all confirm that simulated interactive or intelligent architecture can have a positive effect on its users, it is not known if such results would be transferable to real life. However, literature reviewed indicates that virtual reality is a credible tool for studying how people might react to a real life situation [62–64, 70, 73–75, 102]. Ultimately, the only way to be sure is to perform similar tests to our experiments using real life buildings. Results from experiments made in this research indicate that doing them, although requiring a lot of time and resources, is a credible future option that can yield meaningful results.

13.5 Conclusion

We set out to determine from analysis of the results if lifelike architecture is likely to be a pleasant and productive place to be.

Experiment one concluded that being surrounded by walls that appear to come life and move has a positive effect on an individual's performance. 90 % of test subjects performed better when the walls moved around them. Results also show that 60 % of individuals preferred it to blank walls as well as 65 % of them felt more comfortable in it.

Experiment two concluded that being inside animated lifelike architecture improves the ability of people to follow instructions from a tutor. 93.33 % of subjects performed better in the animated environment. Data shows that 54.33 % of test subjects prefer it to similar environments with blank or patterned walls and 60 % of them felt more comfortable in it.

Experiment three furthers the results of the previous experiment by introducing a seemingly intelligent to test environment. It concluded that people's performance increases in a seemingly intelligent environment in a task that requires following instructions from a teacher. 93.33 % of subjects had better scores in the seemingly intelligent environment. 63 % of test subjects were more comfortable there and 60 % of them preferred it. Comparison with experiment two revels that 90 % of people performed better in a seemingly intelligent environment compared to an animated one.

Experiment four conducted online concluded that over a long period of time an interactive building significantly generates more visitors and social activity than a non-interactive counterpart. The interactive building had, on average 4.5 visits to each visit that was made to the static building. As for return visits the ratio was 8.743 to 1 in favour of the interactive building. Visitors also spent more time there with an average of 50 min spent in the interactive building for each minute spent in the static counterpart. Data from questionnaires indicate 85 % of people preferred the interactive building, wanted to spend time there and 90 % wanted to promote it and socialize within it.

All of this suggests that the presence of interactive or intelligent elements within a building is likely to have positive effects on its users, increasing the productivity and comfort of its users. Experiment 4 also suggests that interactive buildings can generate more visitors and that people tend to socialise more within them. The transferability of these results to experiments made using real buildings or models is yet to be tested. Experiments made here coupled with the increasing popularity of interactive or intelligent building projects suggest that conducting real life experiments are a viable option for future experimentation.

There was strong correlation between the measures. An excellent example of this is that during experiment one a participant was heard humming to herself when the walls were static but not when they were moving. Her task performance was significantly better with seemingly moving walls around her. She reported in questionnaires having an improved experience when the walls appeared to move. In post interview she volunteered that she had felt lonely in the static wall condition and so hummed to herself but had not felt lonely or hummed when the walls appeared to move.

References

1. Addington, D.M, Schodek, D.L.: Smart Materials and New Technologies: For the Architecture and Design Professions. Elsevier, Oxford (2005)
2. Bullivant, L.: 4D space: interactive architecture introduction. Archit. Des. **75**(1), 5–7 (2005)
3. Garcia, M.: Otherwise engaged: new projects in interactive design. Archit. Des. **77**(4), 44–53 (2007)
4. Bauman, Z.: Liquid arts. Theo. Culture Soc. **24**(1), 117–127 (2007)
5. Galinsky People Enjoying Buildings Worldwide: Water Temple (Shingonshu Honpukuji). Available from: http://www.galinsky.com/buildings/watertemple/index.htm (2006). Cited 15 Dec 2011
6. Stierlin, H. (ed.): Islam Volume Early Architecture From Baghdad to Cordoba. Taschen, Italy (1996)
7. Schneider, B., et al.: Jewish Museum Berlin: Between the Lines. Prestel/NAL Pressmark, Munich/New York (1999)
8. Eisenman, P. (ed.): Blurred Zones: Investigations of the Interstitial. Monacelli Press, New York (2002)
9. Kiesler, F.J. (ed.): Endless Space Los Angeles. MAK Center for Art and Architecture, Los angeles (2001)
10. Van Berkel, B., Bos, C. (eds.): Techniques Network Spin Amsterdam. UN Studio and Goose Press, Amsterdam (1999)
11. Lynn, G.: Predator. Archit. Des. **72**(1), 64–71 (2002)
12. Lynn, G. (ed.): Animate Form. Princeton Architectural Press, New York (1999)
13. Spuybroek, L. (ed.): NOX: Machining Architecture. Thames and Hudson, London (2004)
14. Van Berkel, B. (ed.): Mobile Forces. Ernst and Sohn, Berlin (1994)
15. Structures of Other Projects, Available from: http://www.azw.at/otherprojects/soft_structures/oosterhuis/trans_PORTs.htm (n.d). Cited 25 Aug 2008
16. Bouman, O.: Architecture, liquid and gas. Archit. Des **75**(1), 14–22 (2005)
17. Kronenburge, R. (ed.): Flexible Architecture that Responds to Change. Laurence King Publishing, London (2007)
18. Net, M.A.: Telematic Dreaming. Available from: http://www.medienkunstnetz.de/works/telematic-dreaming/ (1992). Cited 15 Dec 2011
19. Net, M.A.: Telematic Vision. Available from: http://www.medienkunstnetz.de/works/telematic-vision/ (1993). Cited 15 Dec 2011
20. Frazer, J. (ed.): An Evolutionary Architecture. Architectural Association, London (1995)
21. Burry, M.: Between surface and substance. Archit. Des. **73**(2), 8–19 (2003)
22. Bullivant, L.: Sky ear, Usman Haque. Archit. Des. **75**(1), 8–11 (2005)
23. Fox, M., Kemp, M. (eds.): Interactive Architecture. Princeton Architectural Press, New York (2010)
24. Dynamic Architecture. Available from: http://www.dynamicarchitecture.net/ (2011). Cited 15 Dec 2011

25. Croci, V.: Relational interactive architecture. Archit. Des. **80**(3), 122–125 (2010)
26. Pallasmaa, J. (ed.): The Eyes of The Skin : Architecture and The Senses. Academy Group Ltd, London (1996)
27. Haque, U.: Architecture, interactions, systems. Arquitetura & Urbanismo **149**, 68–71 (2006)
28. Joye, Y.: Cognitive and evolutionary speculations for biomorphic architecture. Leonardo **39**(2), 145–152 (2006)
29. Thomsen, M.R.: C.O.R. Development. Available from: http://artsresearch.brighton.ac.uk/research/academic/ramsgard_thomsen (2008). Cited 25 Aug 2008
30. Forty, A. (ed.): Words and Buildings a Vocabulary of Modern Architecture. Thames and Hudson, London (2000)
31. Zellner, P. (ed.): Hybrid Space: New Forms in Digital Architecture. Thames and Hudson, London (2000)
32. Holl, S., Pallasmaa, J., Perez-Gomez, A.: Questions of Perception Phenomenology of Architecture and Urbanism (A + U). A + U Publishing Co., Ltd., Tokyo (1994)
33. Hillier, B. (ed.): The Social Logic of Space. Cambridge University Press, Cambridge (1989)
34. Hillier, B. (ed.): Space is the Machine: A Configurational Theory of Architecture. Cambridge University Press, Cambridge (1996)
35. Spiller, N.: Reflexive architecture. Archit. Des. **72**(3), 88–93 (2002)
36. Yamahra, H., Takada, H., Shimakawa, H.: An individual behavioural pattern to provide ubiquitous service in intelligent space. WSEAS Trans. Syst. **6**(3), 562–569 (2007)
37. Aoki, S., et al.: Detection of a Solitude Senior's Irregular States Based on Learning and Recognizing of Behavioral. IEEJ Trans. Sens. Micromach. **125**(E6), 259–265 (2005)
38. Hara, K., Omori, T., Ueno, R.: Detection of Unusual Human Behavior in Intelligent House. In: Paper presented at the Neural Networks for Signal Processing XII, IEEE Signal Processing Society Workshop. pp. 697–706 (2002)
39. Mori, T., et al.: Sensing Room: Distributed Sensor Environment for Measurement of Human Daily Behavior. In: Paper presented at the 1st International Workshop on Networked Sensing Systems (INSS2004). pp. 40–43 (2004)
40. Nakauchi, Y., et al.: Vivid Room: Human Intention Detection and Activity Support Environment for Ubiquitous Autonomy. In: Paper Presented at the 2003 IEEE/RSJ International Conference on Intelligent Robots and Systems. pp. 773–778 (2003)
41. Isoda, Y., Kurakake, S., Nakano, H.: Ubiquitous Sensors Based Human Behavior Modeling and Recognition Using a Spatio-Temporal Representation of User States. In: Proceedings of the 18th International Conference on Advanced Information Networking and Applications. p. 512 (2004)
42. Kidd, C.D., et al.: The Aware Home: A Living Laboratory for Ubiquitous Computing Research. In: Paper presented at the Second International Workshop on Cooperative Buildings, Integrating Information, Organization, and Architecture. pp. 191–198 (1999)
43. Matsuoka, K.: Smart House Understanding Human Behaviors: Who did what, where, and when. In: Paper presented at the 8th World Multi-Conference on Systems, Cybernetics, and Informatics. pp. 181–185 (2004)
44. Sherbini, K.: Overview of Intelligent Architecture. In: 1st ASCAAD International Conference, e-Design in Architecture KFUPM, Dhahran, Saudi Arabia (2004)
45. McCullough, M. (ed.): Digital Ground. MIT Press, Cambridge (2004)
46. Anderson, J.: Manifesto upgrade:from comfort to happy, flourishing super monkeys. Urban Scrawl **3**, 16–19 (2009)
47. Haque, U.: Distinguishing concepts: lexicons of interactive art and architecture. Archit. Des. **77**(4), 24–31 (2007)
48. Cruz, M.: Cyborgian interfaces. Archit. Des. **78**(6), 56–59 (2008)
49. Catts, O., Zurr, I.: Growing semi-living structures concepts and practices for the use of tissue technologies for non-medical purposes. Archit. Des. **78**(6), 30–35 (2008)
50. Armstrong, R.: Artificial evolution a hands-off approach for architects. Archit. Des. **78**(6), 82–85 (2008)

51. Delbrück, T., Eng, K., Bäbler, A.: ADA a Playful Interactive Space. In: Interactions, IFIP 2003, vol. 1. p. 4 (2003)
52. Litracon Light-Transmitting Cincrete: Walls Pavements Design Art. Available from: http:// www.litracon.hu/aboutus.php (2011). Cited 15 Dec 2011
53. Design Boom. Smartwrap: The Mass Customizable Print Facade. Available from: http:// www.designboom.com/eng/funclub/smartwrap.html (2010). Cited 15 Dec 2011
54. Regine. A Makes Move Toward Vehicles that Morph. Available from: http://www. we-make-money-not-art.com/archives/2006/03/mit-engineers-r.php (2011). Cited 15 Dec 2011
55. Zykov, V., et al.: Robotics: self-reproducing machines. Nature **435**, 163–164 (2005)
56. Patel, N.K., Campion, S.P., Fernando, T.: Evaluating the Use of Virtual Reality as a Tool for Briefing Clients in Architecture. In: Sixth International Conference on Information Visualisation (IV'02). London, England (2002)
57. Aouad, G., et al.: An IT map for a generic design and construction process protocol. J Constr. Procure. **4**(1), 1–14 (1998)
58. Barrett, P., Stanley, C. (eds.): Better Construction Briefing. Blackwell Science, Cornwall (1999)
59. Bucolo, S., Impey, P., Hayes, J.: Client Expectations of Virtual Environments for Urban Design Development. In: Information Visualisation, Fifth International Conference. London, UK, pp. 690–694 (2001)
60. Frost, P., Warren, P.: Virtual Reality Used in a Collaborative Architectural Design Process in Information Visualization, IEEE International Conference London, UK (2000)
61. Haque, U.: Scents of Space: An Interactive Smell System. In: SIGGRAPH '04 ACM SIGGRAPH 2004 Sketches. New York, USA (2004)
62. Friedman, D., Steed, A., Slater, M.: Spatial social behavior in second life. Lect. Notes Comput. Sci. **4722**, 252–263 (2007)
63. Heldal, L., et al.: Immersiveness and Symmetry in Copresent Scenarios in Virtual Reality, IEEE, (2005)
64. Yee, N.: The unbearable likeness of being digital: the persistence of nonverbal social norms in online virtual environments. Cyberpsychology Behav. **10**(1), 115–121 (2007)
65. Yee, N.: The demographics, motivations and derived experiences of users of massively-multiuser online graphical environments. PRESENCE: teleoperators and virtual environments. Cyberpsychology Behav. **9**(6), 772–775 (2006)
66. Yee, N.: The labor of fun: how video games blur the boundaries of work and play. Games Culture **1**(1), 68–71 (2006)
67. Griffiths, M., Davies, O., Chappel, D.: Breaking the stereotype: the case of online gaming. Cyberpsychology Behav. **6**(1), 81–91 (2003)
68. Wolff, R., et al.: A review of telecollaboration technologies with respect to closely coupled collaboration. Int. J. Comput. Appl. Technol. **29**(1), 11–26 (2007)
69. Sutherland, I.E.: The Ultimate Display. In: IFIPS Congress. New York (1965)
70. Slater, M.: Place illusion and plausibility can lead to realistic behaviour in immersive virtual environments. Royal Soc. Biol. Sci. **364**(1535), 3549–3557 (2009)
71. Mizell, D.W., et al.: Comparing Immersive Virtual Reality With Other Display Modes for Visualising Complex 3D Geometry. University College London, technical report (2002)
72. Marks, I.M., Gelder, M.G.: A controlled retrospective study of behavior therapy in phobic patients. Br. J. Psychiatry **111**, 571–573 (1965)
73. Slater, M.: Measuring presence: a response to the Witmer and singer presence questionnaire. Presence **8**(5), 560–565 (1999)
74. Pertaub, D.P., Slater, M., Barker, C.: An experiment on public speaking anxiety in response to three different types of virtual audience. Presence: Teleoper. Virtual Environ. **11**(1), 68–78 (2001)
75. Roberts, D., et al.: Factors influencing flow of object focussed collaboration in collaborative virtual environments. Virtual Reality **10**(2), 116–133 (2006)

76. Blascovich, J., et al.: Immersive virtual environment technology as a methodological tool for social psychology. Psychol. Inq. **13**(2), 103–124 (2002)
77. Jang, D.P., et al.: Analysis of physiological response to two virtual environments: driving and flying simulation. Cyberpsychol. Behav. **5**(1), 11–18 (2002)
78. Peruch, P., Gaunet, F.: Virtual environments as a promising tool for investigating human spatial cognition. Current Psychol. Cogn. **17**(4/5), 881–899 (1998)
79. Roussos, M., et al.: Learning and Building Together in an Immersive Virtual World. Presence: Teleoper. Virtual Environ. **8**(3), 247–263 (1999)
80. Salzman, M.C., et al.: A model for understanding how virtual reality aids complex conceptual learning. Presence: Teleoper. Virtual Environ. **8**(3), 293–316 (1999)
81. Rothbaum, B.O., et al.: A controlled virtual reality exposure therapy for the fear of flying. Consult. Clin. Psychol. **68**, 1020–1026 (2000)
82. Vincelli, F.: From imagination to virtual reality: the future of clinical psychology. Cyber Psychol. Behav. **2**(3), 214–248 (1999)
83. Rothbaum, B.O., Hodges, L.F., Kooper, R.: Virtual reality exposure therapy. Psychother. Pract. Res. **6**, 291–296 (1997)
84. Slater, M., et al.: An experimental study on fear of public speaking using a virtual environment. Cogn. Behav. Pract. **10**, 240–247 (2003)
85. Anderson, P., Rothbaum, B.O., Hodges, L.F.: Virtual reality in the treatment of social anxiety: two case reports. Cogn. Behav. Pract. **10**, 240–247 (2003)
86. Harris, S.R., Kemmerling, R.L., North, M.M.: Brief virtual reality therapy for public speaking anxiety. CyberPsychol. Behav. **5**(6), 543–550 (2002)
87. Roy, S., et al.: Definition of a VR-based protocol to treat social phobia. CyberPsychol. Behav. **6**(4), 411–420 (2003)
88. Molinari, E., Riva, G., Wiederhold, B.K. (eds.): Virtual Environments in Clinical Psychology and Neuroscience. IOS Press, London (1998)
89. Stanney, K.M. (ed.): Handbook of Virtual Environments: Design, Implementation, and Applications. Human Factors and Ergonomics, pp. 12–32. Lawrence Erlbaum Associates Publishers, Mahwah (2002)
90. Rizzo, A., et al.: Virtual Environment Applications in Clinical Neuropsychology in IEEE Virtual Reality Conference 2000 (VR 2000). New Brunswick, New Jersey (2000)
91. Hodges, L.F., Kooper, R., Meyer, T.C.: Virtual environments for treating the fear of heights. Computer **28**(7), 27–34 (1995)
92. Wiederhold, B.K., Gevirtz, R., Wiederhold, M.D.: Fear of flying: a case report using virtual reality therapy with physiological monitoring. Cyberpsychol. Behav. **1**(2), 97–103 (1998)
93. Huang, M.P., et al.: Comparing virtual and real worlds for acrophobia treatment. Stud. Health Technol. Inf. **50**, 175–179 (1998)
94. Hodges, L.F., et al.: A virtual environment for the treatment of chronic combat-related post-traumatic stress disorder. Cyberpsychol. Behav. **2**(1), 7–14 (1999)
95. Rothbaum, B.O., et al.: Virtual reality exposure therapy for PTSD Vietnam veterans: a case study. Trauma. Stress **12**(2), 263–272 (1999)
96. Regenbrecht, H.T., Schubert, T.W., Friedmann, F.: Measuring the sense of presence and its relation to fear of heights in virtual environments. Int. J. Human Comput. Interact. **10**(3), 233–250 (1998)
97. Brogni, A., et al.: Touching Sharp Virtual Objects Produces a Haptic Illusion Virtual and Mixed Reality—New Trends. Springer, Berlin. pp. 234–242 (2011)
98. Eng, K., et al.: An investigation of collective human behavior in large-scale mixed reality spaces. Teleoper. Virtual Environ. **15**(4), 403–418 (2006)
99. Eng, K., Mintz, M., Verschure, P.F.M.J.: Collective Human Behavior in Interactive Spaces. In: IEEE International Conference on Robotics and Automation (ICRA05) (2005)
100. Eng, K.A., Bäbler, A.: Design for a brain revisited: the neuromorphic design and functionality of the interactive space. Rev. Neurosci. **14**(1–2), 145–180 (2003)
101. Bullivant, L.: Ada the intelligent room. Archit. Des. **75**(1), 86–90 (2005)
102. Slater, M.: Depth of presence in virtual environments. Presence **3**, 130–144 (1994)

103. Kaplan, H.I., Sadock, B.J., Grebb, J.A.: Synopsis of psychiatry: behavioral sciences, clinical psychiatry, 6th edn. Williams and Wilkins, Baltimore (1991)
104. Crowe, M.J., et al.: Time-limited desensitisation, implosion and shaping for phobic patients: a crossover study. Behav. Res. Ther. **10**(4), 319–328 (1972)
105. Tran, C., et al.: 2011 Toronto Notes Clinical Handbook. Toronto Notes for Medical Students Inc, Toronto (2011)
106. Guallart, V., Diaz, M.: Hyper Habitat: reprogramming the world. In: Proceedings of the 11th International Architecture exhibition: Out There: Architecture beyond Building (2008)
107. Van Baren, J., IJsselsteijn, W.: Measuring presence: a guide to current measurement approaches (2004)
108. Psych Connections: Within-Subjects Design. Available from: http://web.mst.edu/ ~psyworld/experimental/within_subjects.html (2011). Cited 15 Dec 2011
109. Experiment-Resources: Within Subject Design. Available from: http://www. experiment-resources.com/within-subject-design.html (2011). Cited 15 Dec 2011
110. MacKenzie, I.S.: Within-Subjects Designs. Available from: http://www.yorku.ca/mack/ RN-Counterbalancing.html (2011). Cited 15 Dec 2011
111. Adi, M.N., Roberts, D.: Can you help me concentrate room?. In: IEEE ACM Virtual Reality. Waltham, MA, pp. 131–134 (2010)

Part IV
Technologies for Education and Education for Rehabilitative Technologies

Chapter 14
An Overview of Virtual Simulation and Serious Gaming for Surgical Education and Training

Bill Kapralos, Fuad Moussa and Adam Dubrowski

Abstract The rising popularity of video games has seen a recent push towards the application of serious games to medical education and training. With their ability to engage players/learners for a specific purpose, serious games provide an opportunity to acquire cognitive and technical surgical skills outside the operating room thereby optimizing operating room exposure with live patients. However, before the application of serious games for surgical education and training becomes more widespread, there are a number of open questions and issues that must be addressed including the relationship between fidelity, multi-modal cue interaction, immersion, and knowledge transfer and retention. In this chapter we begin with a brief overview of alternative medical/surgical educational methods, followed by a discussion of serious games and their application to surgical education, fidelity, multi-modal cue interaction and their role within a virtual simulations/serious games. The chapter ends with a description of the serious games surgical cognitive education and training framework (SCETF) and concluding remarks.

Keywords Serious games · Virtual simulation · Surgical education

B. Kapralos (✉)
Faculty of Business and Information Technology, Health Education Technology Research Unit (HETRU), University of Ontario Institute of Technology,
Oshawa, ON L1H 7K4, Canada
e-mail: bill.kapralos@uoit.ca

F. Moussa
Division of Cardiac and Vascular Surgery, Schulich Heart Centre, Sunnybrook
Health Sciences Centre, Toronto, ON, Canada
e-mail: Fuad.Moussa@sunnybrook.ca

A. Dubrowski
Disciplines of Emergency Medicine and Pediatrics, Memorial University of Newfoundland,
St. John's, Newfoundland, Canada
e-mail: adam.dubrowski@med.mun.ca

A. L. Brooks et al. (eds.), *Technologies of Inclusive Well-Being*,
Studies in Computational Intelligence 536, DOI: 10.1007/978-3-642-45432-5_14,
© Springer-Verlag Berlin Heidelberg 2014

14.1 Introduction

The acquisition of medical skills in general and surgical skills (both cognitive and technical) in particular, has historically been based on Halsted's apprenticeship model whereby the resident (trainee) acquires the required skills and knowledge in the operating room [1]. However, the present era brings with it stresses on the apprenticeship model for surgical training, leading to increased resource consumption (e.g., monetary, faculty time, and time in the operating room), and increased costs [2]. For example, a study conducted by Lavernia et al. [3] examined the cost of performing a total knee arthroplasty surgical procedure (replacement of the painful arthritic knee joint surfaces with metal and polyethylene components that serve to function in the way that bone and cartilage previously had), in teaching hospitals vs. non-teaching hospitals. They found that patients who underwent surgery at a teaching hospital had higher associated charges (\$30,311.00 ± \$3,325.00) and longer surgeries (190 ± 19 min) as opposed to those who underwent similar surgery in a non-teaching hospital (\$23,116.00 ± \$3,341.00, 145 ± 29 min). They attribute this increase in resource consumption to the hands-on approach required to train residents.

Compounding the problem is the growing trend towards decreasing resident work hours in North America and globally [4]. Thus the available training time in the operating room and consequently operative exposure, teaching, and feedback are continuously shrinking [5]. Therefore, the available operative time must be maximized in order to maintain a high level of surgical training. Although the amount of repetition necessary to obtain the surgical competence required of residents is still unclear, medical literature suggests that technical expertise is acquired through years of practice [6] and indicates a positive correlation between volume and patient outcome [7]. As a result, traditional educational methods in surgery have come under increasing scrutiny [8] and as Murphy et al. [9] describe, the current situation calls for a reassessment of medical education practises. It is evident that given the increasing time constraints, trainees are under great pressure to acquire complex surgical cognitive and technical skills. Therefore, efforts must be made to optimize operative room exposure by devising training opportunities using artificial settings before exposure to patients.

14.1.1 Alternative Educational Models

In accordance with new educational models, such as competency based curricular approaches [10], new teaching modalities and technologies are necessary to augment the traditional teaching practices in light of the shrinking operative time. Other available alternative methods for surgical training include the use of animals, cadavers, or plastic models; each option with its share of problems [11]. More specifically, the use of animals for medical education may be prohibited in some countries, animal anatomy can vary greatly from humans, and there are ethical concerns with the use of animals in medical research and education [12]. Cadavers cannot be used multiple times, and plastic models don't necessarily provide realistic visual and haptic feedback [11]. In addition to availability issues, and changed tissue behavior, cadavers are expensive (up to $5,000 each), require specialized facilities, and disposal arrangements [12, 13]. Furthermore, animals and cadavers can be cost prohibitive when considering extensive training [14], and ethical issues are being raised with respect to practising surgical procedures on anesthetized humans and animals [15]. Simulation (and virtual simulation in particular) offers a viable alternative to practice in an actual operating room, offering residents the opportunity to train until they reach a specific competency level. Unlike working with live patients (and animals), virtual simulation allows trainees to intentionally make and correct mistakes [16]. In addition, according to Reiner and Harders [12], additional advantages of virtual reality-based technologies include: allowing students to practice independent of busy operating room schedules and patients, allow for a large number of diverse anatomies and pathologies in a small period of time, allow for the training of rare yet dangerous complications that a trainee may not otherwise have the opportunity to experience, allow for objective assessment, and can lead to reduced costs.

The rising popularity of video games has seen a recent push towards the application of video game-based technologies to teaching and learning. Serious games (that is, video games that are used for training, advertising, simulation, or education [17]), provide a high level of interactivity not easily captured in traditional teaching/learning environments. In contrast to traditional teaching environments where the teacher controls the learning (e.g., *teacher-centered*), serious games and virtual simulations present a *learner-centered* approach to education, so that the player controls the learning through interactivity thus allowing the player to learn via an active, critical learning approach [18]. Game-based technologies have also been used for many years as training simulators for vehicle control (e.g., flight simulators) and are growing in popularity within medical education including surgery. Through game constructs, realistic situations can be simulated to provide valuable experience to support discovery and exploration in a fun, engaging, and cost-effective manner.

Although (virtual) simulations and serious games are similar (according to Becker and Parker [19], all serious games (or game simulations as they refer to them) are games, and all games are simulations [19]; see Fig. 14.1), and can employ identical technologies (hardware and software). Being a video game, serious games should strive to be fun and include some of the primary aspects of games including challenge, risk, reward, and loss or more formally, as defined by Thiagarajin and Stolovich [20], serious games should include the following five characteristics:

1. Conflict: can be described as challenge.
2. Constraints on a player's behaviors: rules.
3. Closure: the game must come to an end.
4. Contrivance: all games are contrived situations.
5. Correspondence: designed to respond to some selected aspects of reality.

Fig. 14.1 The relationship between simulation, games, and serious games (or simulation games as referred to by Becker and Parker [19]). After Becker and Parker [19]

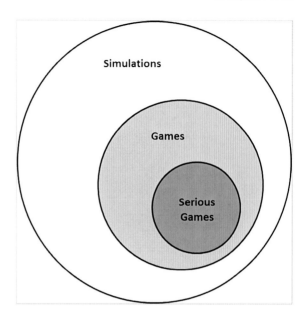

14.1.2 Open Problems with Virtual Simulation and Serious Games

Despite the benefits of virtual simulation and serious games, there are a number of open, fundamental issues that must be addressed before they become more widespread. Tashiro and Dunlap [21] developed a typology of serious games for healthcare education and explored the strengths and limitations of serious games for improving clinical judgment. They identified seven areas that require research and improvements for the effective development of serious games: (1) disposition to engage in learning, (2) impact of realism/fidelity on learning, (3) threshold for learning, (4) process of cognitive development during knowledge gain, (5) stability of knowledge gain (retention), (6) capacity for knowledge transfer to related problems, and (7) disposition toward sensible action within clinical settings. Our own work is focused on the impact of realism/fidelity and multi-modal interactions on learning and for the scope of this paper, we will therefore focus on this problem specifically.

Fidelity and Multi-Modal Interactions In the context of a simulation (and a serious game), fidelity denotes the extent to which the appearance and/or behavior of the simulation matches the appearance and behavior of the real system [22, 23]. Fidelity can be divided into two components: (1) psychological fidelity, and (2) physical fidelity [24]. Psychological fidelity denotes the degree that the skills inherent in the real task being simulated are captured within the simulation [24] and may also include the degree of reality perceived by the user of the simulation (the trainee) [25]. Physical fidelity covers the degree of similarity between the training situation and the operational situation which is simulated [23, 24]. Physical fidelity can be further divided into equipment fidelity that denotes the degree that the simulation replicates reality and environmental fidelity that denotes the degree that the simulation replicates the sensory cues [24, 25].

Knowledge transfer can be defined as the application of knowledge, skills, and attitudes acquired during training to the environment in which they are normally used [26]. Ker and Bradley [24] describe twelve factors that most effectively promote learning transfer. With respect to clinical training, one of these factors includes the re-creation of the real clinical environment, which both aids in the suspension of disbelief and in the transfer of competence to performance. This effect is also consistent with a situated learning approach whereby the learning environment is modeled in the context that the knowledge is expected to be applied [27, 28]. Therefore, a serious game/virtual simulation should attempt to remain faithful in its representation of the real environment (where multiple senses are engaged simultaneously at any time). However, it is unclear just how close this relationship needs to be. In other words, (1) how much fidelity is actually needed to maximize transfer and retention? and (2) what effect do multi-modal interactions have on knowledge transfer and retention? These questions have a number of implications when considering that any training device-be it a virtual or physical simulator-will never be able to completely replicate the real world, and in virtual worlds we have (typically) eliminated one sensory modality-smell, and reduced or restricted the haptic senses (touch and movement). Therefore, complete (perfect) multi-sensory fidelity appears to be impossible to achieve, at least with our current technology. Furthermore, it remains unclear if such a high level of fidelity is actually needed for either enjoyment or knowledge transfer and retention, and striving to reach higher levels of fidelity can also lead to increased computational requirements (processing time). Moreover, despite the great computing hardware advances, particularly with respect to graphics rendering, real-time high fidelity audio and visual rendering particularly of complex environments, is still not feasible [29]. In addition, striving for such high fidelity environments increases the probability of lag and subsequent discomfort and simulator sickness [30]. Finally, here fidelity has been explicitly defined (see above) but as Cook et al. [31] describe, fidelity itself is complex, and encompasses various aspects of a simulation activity including the sensory modalities (visual, auditory, olfactory and haptics), learning objectives and task demands and therefore, referring to a simulation as high fidelity can convey diverse meanings.

14.1.3 Paper Overview

In the remainder of this paper we present an overview of serious games with an emphasis on medical/surgical education. We focus specifically on fidelity and multi-modal interactions, a problem that we believe can have significant implications in the widespread use of serious games. We also outline some of our current work in the serious games domain applied to surgical education and training. More specifically, in Sect. 14.2, an overview of serious games and virtual simulations in a surgical learning context is provided. Section 14.3 focuses on fidelity, and multi-modal interactions while Sect. 14.4 introduces the serious game surgical cognitive education and training framework (SCETF) being developed both as a learning tool for surgical cognitive education and training and as a research tool to examine the role of fidelity and multi-modal interactions on learning are also provided. Finally, concluding remarks are provided in Sect. 14.5.

14.2 Virtual Simulation and Gaming in Surgical Education

Simulations range from de-contextualized bench models and virtual reality (VR)—based environments, to high fidelity recreations of actual operating rooms [32] whose fidelity is high enough to allow training that is equal to and at times better than traditional methods [15]. Virtual reality has been widely and successfully used for training and education in a variety of industries including nuclear, aviation, military, and surgery for many years [15, 33]. One of the prevailing arguments for using simulation in the learning process of trainees is their ability to engage the trainee in the active accumulation of knowledge by doing. Focusing on laparoscopic surgery education and training, according to Smith [15], the literature supports four hypotheses with respect to the impact and acceptance of virtual reality and game-based technologies:

1. Training in laparoscopic surgery can be accomplished at a lower cost using virtual reality and game-based technologies than existing methods of training.
2. Virtual reality and game-based training environments provide better access to representative patient symptoms and allow more repetitive practise that existing training methods and approaches.
3. Virtual reality and game-based training environments can reduce the training time required to achieve proficiency in laparoscopic procedures.
4. Virtual reality and game-based training can reduce the number of medical errors caused by residents and surgeons learning to perform laparoscopic procedures.

Smith [15] focused on laparoscopic surgery due to the similarities between this type of surgery and virtual reality systems, and the surgical interface used during laparoscopic surgery lends itself to virtual reality and gaming. However, this is not to say that virtual reality and game-based training is only applicable to laparo-scopic surgery. Rather, virtual reality and game-based technologies are becoming more widely accepted methods of training within medical education curriculums in general and the technological problems plaguing such systems in earlier years are quickly being overcome [15]. The use of virtual reality in medical education in general is one part of a change over in medical education that also includes the use of mannequins, task trainers, and online learning modules [34]. In fact, Smith [15] proposed a model of medical education where virtual reality and game-based learning technologies are the next major transformation of the medical education curriculum.

To date, simulation in the surgical domain has been primarily developed and studied as an educational tool for the development of foundational skills (i.e., basic technical skills and appropriate use of instruments). In addition, simulators can be rather complex and very costly. For example, the vascular intervention system training (VIST) simulator allows surgeons to practice laparoscopic surgical skills while providing the context of a surgical procedure. The VIST simulator displays photorealistic organs that can be cut and sutured and includes the simulation of cauterization. However, the VIST simulator costs approximately $300,000 US making it prohibitively expensive for many institutions [35].

14.2.1 Transfer of Skills to the Operating Room

Competent surgical performance requires mastery of not only technical skills but cognitive skills as well (i.e., the capability of responding and adapting to the wide range of contextual variations that may require adjustments to the standard approach) [36]. As with the development of technical skills, cognitive judgment takes practice to develop [6]. It has been suggested that cognitive skills training could accelerate the understanding and planning of a particular procedure, leads to a reduction in the training time required to become proficient with the procedure, may provide greater meaning to the actions being practiced, and creates more effective learning while making more efficient use of resources [6, 36]. It has also been suggested that certain non-technical aspects of performance can enhance or, if lacking, contribute to deterioration of surgeons' technical performance [37]. The addition of cognitive skills training to residents early on provides the opportunity for residents to detect errors and this ultimately helps prevent errors in the oper-ating room; cognitive skills training helps surgeons judge the correctness of their own actions [36]. Finally, is has been suggested that cognitive skills training may help accelerate the understanding and planning of a particular surgical procedure, providing the surgeon with greater meaning to the actions being practised, and reduce the overall training time required to become competent both cognitively and technically [6].

According to Fitts and Posner [38], learning is a sequential process and we move through three distinct phases when learning a new skill. The three stages are as follows:

1. Cognitive phase: Identification and development of the component parts of the skill—involves formation of a mental picture of the skill.
2. Associative phase: Linking the component parts into a smooth action—involves practicing the skill and using feedback to perfect the skill.
3. Autonomous phase: Developing the learned skill so that it becomes automatic—involves little or no conscious thought or attention whilst performing the skill—not all performers reach this stage.

Serious games and virtual simulations particularly lend themselves to the cognitive phase, allowing trainees to focus on the understanding and planning of a particular surgical procedure. Focusing on the cognitive aspects of a procedure can also be cost-effective given that potentially specialized (and costly) equipment (e.g., haptic devices), typical in technical skills training are not generally required. Furthermore, such virtual simulations/serious games can be developed to run on common computing and mobile platforms ensuring that the applications are available to the trainees/residents outside of the regular training/educational setting, at all times.

Serious games provide an opportunity to acquire cognitive and technical surgical skills outside the operating room thereby optimizing operating room exposure with live patients, and this is particularly so when considering novice trainees in the cognitive stage of motor skills acquisition where the majority of errors occur. In addition to their use in training and as part of a curriculum, virtual simulations and serious games can be used by surgeons to rehearse/practise an operation (particularly complex cases) that they are about to perform using data that simulates the patient that they will be operating on [34]. For example, a research study (by researchers with Beth Israel and the National Institute on Media and the Family at Iowa State University), that investigated whether good video game skills translate into surgical skills demonstrated that laparoscopic surgeons who played video games at least 3 h each week made approximately 37 % less mistakes in laparoscopic surgery and performed the task 27 % faster than their counterparts who did not play video games [39]. Further work remains, but in addition to the use of serious games, video game playing may also be an integral part of the training program of future surgeons. As Dr. Paul J. Lynch, a Beth Israel anesthesiologist who has studied the effects of video games for years comments [40]: "The study landmarks the arrival of Generation X into medicine. We grow up with computers, with PDAs, with video games systems, with the Internet, with handheld video games, with cable TV, with remote controls. We've grown up saturated in this technology era that we are in and now we are bringing these skills into the medical profession." Finally, according to Professor Pamela Andreatta, the Director of the Clinical Simulation Center at the University of Michigan Medical School in Ann Arbor [34], although there seems to be obvious patient safety benefits to training outside of the patient setting and preliminary data suggests that simulation is effective in the early learning phase of clinical skills acquisition, a general conclusion stating that simulation (and serious games) has a significant impact on clinical applications cannot be made given the lack of data to date further research is required.

14.3 Serious Games: Fidelity and Multi-Modal Interactions

In the real world, our senses are constantly exposed to stimuli from multiple sensory modalities (visual, auditory, vestibular, olfactory, and haptic), and although the process is not exactly understood, we are able to integrate/process this multisensory and acquire knowledge of multisensory objects [41]. As described previously, it is currently beyond our capability to faithfully account for all of the human senses within a virtual simulation/serious game, and although the emphasis of (virtual) simulations in general (including serious games) is on the visuals/graphics [42], visuals within such environments are rarely presented in silence but rather, include sound of some type. In the real world, visuals and auditory stimuli influence one another.

Various studies have examined the perceptual aspects of audio-visual cue interaction, and it has been shown that sound can potentially attract part of the user's attention away from the visual stimuli and lead to a reduced cognitive processing of the visual cues [43]. Bonneel et al. [44] examined the influence of the level of detail of auditory and visual stimuli in the perception of audiovisual material rendering quality. In each trial of their experiment, participants were presented with two sequences, each sequence of an object falling on a table, bouncing twice and producing audible bounce sounds. One of the sequences was a reference (highest quality with respect to sound and graphics) while for the other sequence, auditory and visual levels of detail varied. Auditory level of detail was defined with respect to modal synthesis (a physical-based model of a vibrating object), while visual level of detail was defined with respect to the bidirectional reflection distribution function (BRDF) which describes the reflection, absorption, and transmission of light at the surface of a material [45]. The participants' task for each trial was to rate, on a scale of 0–100, the similarity of the falling objects in the two sequences. The authors observed significant interactions between visual and auditory level of details and the perceived material quality. In other words, visual level of detail was perceived to be higher as the auditory level of detail was increased.

Mastoropoulou et al. [43] examined the influence of sound effects on the perception of motion smoothness within an animation and more specifically, on the perception of frame-rate. Their study involved forty participants that viewed pairs of computer generated walkthrough animations at five different frame-rates. The visuals were consistent across the animation pairs although the pairs differed with respect to sound; one contained sound effects while the other did not (it was silent). The participants' task was to choose which animation had a smoother motion.

There was a significant effect of sound on perceived smoothness and more specifically, sound attracted a viewer's attention from the visuals leading to a greater difficulty in distinguishing smoothness variations between animations containing sound cues displayed at different rates, than between silent animations [43]. It was inferred that sound stimuli attract part of the viewer's attention away from any visual defects inherent in low frame-rates [43]. Similarly, Hulusic et al. [46] examined the interaction of sound with visuals for the purpose of reducing computational requirements of visual rendering with the use of motion-related sound effects. They found that such sound effects allowed slow animations to be perceived as smoother than fast animations and that the addition of footstep sound effects to walking (visual) animations increased the animation smoothness perception. Hulusic et al. [46] conclude that for certain conditions, the rendering rate can be reduced by incorporating the appropriate sound effects, leading to a reduction in the required computation without the viewer being aware of this reduction.

Greater details regarding the influence of sound over visual rendering is provided by Hulusic et al. [47] while an overview of "crossmodal influences on visual perception" is provided by Shams and Kim [48].

14.4 The Serious Games Surgical Cognitive Education and Training Framework

Given the strain on resources associated with the current master-apprenticeship surgical training model, and the importance of cognitive skills training, development of a multi-modal, serious game surgical cognitive education and training framework (SCETF) has recently begun. Domain-specific surgical modules can then be built on top of the existing framework, utilizing common simulation elements/assets and ultimately reducing development costs. The SCETF focus is on the cognitive components of a surgical procedure and more specifically, the proper identification of the sequence of steps comprising a procedure, the instruments and anatomical/physiological knowledge required for performing each step, and the ability to respond to unexpected events while carrying out the procedure. By clearly understanding the steps of a procedure and the surgical knowledge that goes along with each step, trainees are able to focus solely on the technical aspect of the procedure. In other words, with respect to the three stages to skills acquisition described by Fitts and Posner [38], trainees can transition from the cognitive through the integrative, and perhaps into the automatic stage) in higher fidelity models or in the operating room thus making more efficient use of the limited available resources. The SCETF is also being developed as a research tool where various simulation parameters (e.g., levels of audio/visual fidelity) can be easily adjusted allowing for the controlled testing of such factors on knowledge transfer and retention and this will ultimately lead to more effective serious games. The SCETF consists of graphical and spatial sound rendering engines and various other

Fig. 14.2 Sample SCETF
screenshot

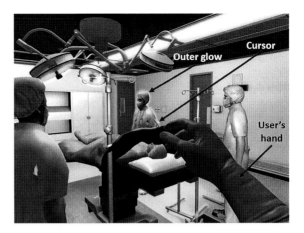

components that are common (generic) to all serious games including a scenario
editor (currently being developed) that allows users of the module (educators/
instructors) to create and/or modify/edit specific scenarios using a graphical-based
user interface. The scenario editor that allows a scenario to be easily developed by
clicking and dragging various interface components in a "what you see is what you
get" (WSYWIG), manner.

Within a given module, trainees take on the role of the surgeon, viewing the
environment through their avatar in a first-person perspective and therefore, only
their hand is visible (see Fig. 14.2). In each module, the task of the trainee is to
complete the surgical procedure following the appropriate steps and choosing the
correct tools for each step. Along the way, complications can arise that will require
some action from the trainee. These complications will appear in the form of visual
or auditory cues adapted according to predefined features of the simulated surgical
scenario. Several other non-player characters (NPCs) also appear in the scene
including the patient (lying on a bed), assistants, and nurses. The SCETF includes
networking capabilities to allow these NPCs to be controlled by other users and
provide an entire surgical team the opportunity to practice remotely and allow for
interprofessional education. The trainee (user) can move and rotate the "camera"
using the mouse in a first-person manner thus allowing them to move within the
scene. A cursor appears on the screen and the trainee can use this cursor to point at
specific objects and locations in the scene.

A SCETF module for the off-pump coronary artery bypass (OPCAB) grafting
cardiac surgical procedure is currently being developed [49]. The OPCAB procedure itself is complex and technically challenging and it has been suggested that
appropriate training be provided before being performed on patients [50]. We
hypothesize that by learning the OPCAB procedure in a first-person-shooter
gaming environment, trainees will have a much better understanding of the procedure than by traditional learning modalities and therefore, we anticipate that a
serious game for OPCAB training will be a beneficial educational tool.

14.4.1 Graphical Rendering

The 3D graphics rendering engine was developed completely in-house and is based on the C++ programming language and the OpenGL 3D graphics API. Real-time rendering is accomplished using the graphics processing unit (GPU) via the OpenGL shading language (GLSL). The SCETF utilizes GPU-based effects such as outer glow (used to indicate a selectable object), reflection mapping, bloom filtering (to provide more realistic lighting effects particularly when considering objects in front of a light source), and various effects to generate realistic metal effects given the widespread use of metal (stainless steel) in an operating room. An example of the outer glow effect is provided in Fig. 14.2. In this example, the outer glow surrounds the outline of the operating room nurse after the end-user (taking on the role of the surgeon in a first-person perspective), moves the cursor over the nurse to indicate that they are able to interact with the nurse. Interaction is initiated by clicking the left mouse button. Finally, the SCETF also supports stereoscopic 3D viewing using active stereo technologies were the user wears a pair of LCD shutter glasses that are synchronized with the display refresh rate. Stereoscopic 3D viewing has been linked to increased player engagement in the gameplay [51], and increased engagement in an educational setting has been linked to higher academic achievement. With respect to medical education and training, it has been suggested that the use of stereoscopic 3D can (1) lead to improved understanding of anatomical relationships and pathology, (2) improve the quality of the student's learning experience, and (3) create more life-like training simulations [52]. Furthermore, stereoscopic 3D provides the ability to establish foreground and background information [53], which can be useful in a variety of training situations. However, the technology can also be problematic, as it may lead to users hyperfocusing on the foreground while ignoring potentially important information within the periphery [53]. Therefore, the usefulness of stereoscopic 3D for training simulations and serious games needs further study.

14.4.2 Sound Rendering

Sound plays an important role in the operating room. For example, with respect to the OPCAB procedure, the surgeon listens to the auditory component of the electrocardiogram; they can hear the onset of changes such as bradycardia (slowing of the heart rate), tachycardia (speeding up of the heart rate), decreasing blood oxygen saturation, and hemodynamic changes. As a result, appropriate sound modeling should be included within a virtual operating room intended for education and training. The addition of realistic spatial sound (that is, the simulation of realistic "spatial" auditory cues within a virtual environment such that it allows users to perceive the position of a sound source at an arbitrary position in three-dimensional space; see [54]), within virtual environments in general is beneficial for a number of reasons. More specifically, spatial sound can add a new layer of realism [55], contributes to a greater sense of presence, or immersion [56], can improve task performance [57], convey information that would otherwise be difficult to convey using other modalities (e.g., vision) [57], and improve navigation speed and accuracy [58]. In addition, when applied to virtual environments for medical training, sound can play an important role by aiding in the acquisition of enhanced skills and dexterity, and lead to increased effectiveness [12]. In addition, although greater studies are required, it has been suggested that sound can influence higher-level cognitive processes [59].

The SCETF supports high fidelity sound to allow for the inclusion of spatial sounds such as those described above. The SCETF also supports spatial (3D) sound rendering including reverberation and occlusion/diffraction modeling using novel GPU-based methods that approximate such effects at interactive rates [60, 61]. The system also supports head-related transfer functions (HRTFs) which describe the individualized location dependent filtering of a sound by the listener's head, shoulders, upper torso, and most notably, the pinna [42]. HRTF filtering is accomplished using GPU-based convolution ensuring interactive frame-rates [62]. Spatial sound by default is not activated; it is an option that can be chosen during start-up. By default, sounds are non-spatialized and output in a traditional stereo format. Although each module has its corresponding sounds (e.g., background sound, sound effects, and dialogue), the user is also provided the opportunity to provide their own sounds during start-up (e.g., the user can provide their own background sound or can choose to have no sound output at all).

14.4.3 Multi-cue Interaction and Cue Fidelity

As previously described, the SCETF is being developed as a research tool to enable the investigation of the effect of multi-modal cue interaction on knowledge transfer and retention. Currently, the SCETF supports the alteration of audio and visual fidelity and the interaction of audio and visual cues particularly if they are incongruent and mis-matched (i.e., high quality audio and poor quality visuals and vice versa). With respect to audio fidelity currently, the following options are supported: (1) spatial sound vs. non-spatial sound, (2) no sound at all (background sound and/or all sound effects are turned off), (3) adjustable quantization levels, (4) addition of white noise to background sounds and/or sound effects, and (5) adjustment of loudness and dynamic range. Such effects (and their corresponding settings) are chosen during start-up and cannot be adjusted dynamically. Visual (graphical) fidelity ranges from high to low quality, defined with respect to polygon count, and resolution (both texture resolution and overall resolution). These particular fidelity measures cannot be adjusted dynamically but rather, their settings must be specified during an initialization phase at start-up. The SCETF also provides for the dynamic adjustment of visual fidelity through various graphical filtering effects implemented using the graphics processing unit (GPU).

The degree of filtering introduced by each of these effects can be dynamically adjusted via a slider control and these effects can also be combined dynamically. For example, the scene can be blurred using the blurring filter and noise can also be added with the noise filer.

In addition to its use as a training tool the OPCAB module that is currently being developed will also serve as a testbed to allow for the methodical investigation of the effect that varying audio/visual simulation fidelity and the interaction of audio/visual cues have on knowledge/skills/behaviours (KSB) transfer and retention. Furthermore, the OPCAB module will be used to measure the feasibility/usefulness of using serious games as a tool for assessing/screening cognitive skills (e.g., surgical steps and knowledge) in surgical training.

14.5 Conclusions

Surgical training has predominantly taken place in operating rooms placing a drain on the limited available operating room resources. Simulations, both physical and virtual, have been effectively used to complement residents' training and education. Serious games, or the use of video game-based technologies for applications whose primary purpose is other than entertainment, are becoming very popular in a variety of applications including medical education in general. This popularity stems in part from the current generation of tech-savvy learners who play games. In addition to promoting learning via interaction, serious games allow users to experience situations that are difficult (even impossible), to achieve in reality and they support the development of various skills including analytical and spatial, strategic, recollection, and psychomotor skills as well as visual selective attention. Despite the growing popularity of serious games and their inherent benefits, before their use becomes more widespread, a number of open problems must be addressed. Here, in this paper, we focused on the problem multi-modal interaction and fidelity on learning. Although great progress has been with respect to both of these problems, many open issues remain and plenty of work remains. In this paper we have also described the serious games surgical cognitive education and training framework that is currently being developed specifically for cognitive surgical skills training. Domain-specific surgical modules can then be built on top of the existing framework utilizing common simulation elements and assets and ultimately reducing development time and costs. The SCETF is also being developed as a research tool where various simulation parameters such as levels of audio and visual fidelity, can be easily adjusted allowing for the controlled testing of these factors on knowledge transfer and retention. In addition to examining the effect multi-modal interactions may have on knowledge transfer and retention, the SCETF also allows us to methodically investigate perceptual-based rendering and more specifically, the role of sound with respect to visual quality perception which may ultimately lead to reduced rendering (computational) requirements and ultimately allow for more effective virtual simulations and serious games.

As serious gaming becomes more widespread, care must be taken to ensure that they are properly designed to meet their intended goals. With traditional entertainment game development, designers/developers start fresh with a blank slate, and are primarily concerned with creating an engaging gameplay experience that will keep the players playing the game; to improve gameplay and engagement, they are free to modify the design of the game throughout the entire design and development process [19, 63]. However, serious games designers/developers are not afforded this luxury but rather, must strictly adhere to the content/knowledge base while ensuring that their end product is not only fun and engaging, but is also an effective teaching tool [19, 63]. In addition to knowledge and expertise in game design and development, serious games designers/developers must therefore also be knowledgeable in the specific content area covered by the serious game and possess some knowledge in teaching methods and instructional design in particular. In other words, the development of effective serious games is not a trivial task and knowledge in game design solely is not sufficient to develop an effective serious game. Serious games development is an interdisciplinary process, bringing together experts from a variety of fields including game design and development and although serious games designers are not expected to be experts in instructional design and the specific content area, possessing some knowledge in these areas will, at the very least, promote effective communication between the interdisciplinary team members.

Acknowledgments The financial support of the *Social Sciences and Humanities Research Council of Canada* (SSHRC), in support of the iMMERSE project, and the *Canadian Network of Centres of Excellence* (NCE) in support of the *Graphics, Animation, and New Media* (GRAND) initiative is gratefully acknowledged.

References

1. Halsted, W.S.: The training of the surgeon. Johns Hopkins Hosp. Bull. **15**(162), 267–275 (1904)
2. Reznick, R.K.: Teaching and testing technical skills. Am. J. Surg. **165**(3), 358–361 (1993)
3. Lavernia, C.J., Sierra, R.J., Hernandez, R.A.: The cost of teaching total knee arthroplasty surgery to orthopaedic surgery residents. Clin. Orthop. Relat. Res. **380**, 99–107 (2000)
4. Zuckerman, J., Kubiak, E., Immerman, I., DiCesare, P.: The early effects of code 405 work rules on the attitudes of orthopaedic residents and attending surgeons. J. Bone Joint Surg. **87**(A(4)), 903–908 (2005)
5. Weatherby, B.A., Rudd, J.N., Ervin, T.B., Staff, P.R.: The effect of resident work hour regulations on orthopaedic surgical education. J. Surg. Orthop. Adv. **16**(1), 19–22 (2007)
6. Ericsson, K.A., Krampe, R., Tesch-Romer, C.: The role of deliberate practice in the acquisition of expert performance. Psychol. Rev. **100**(3), 363–406 (1993)
7. Halm, E.A., Lee, C., Chassin, M.R.: Is volume related to outcome in healthcare? A systematic review and methodological critique of the literature. Ann. Intern. Med. **137**(6), 511–520 (2002)
8. Gorman, P.J., Meier, A.H., Rawn, C., Krummel, T.M.: The future of medical education is no longer blood and guts, it is bits and bytes. Am. J. Surg. **180**(5), 353–356 (2000)

9. Murphy, J.G., Torsherand, L.C., Dunn, W.F.: Simulation medicine in intensive care and coronary care education. J. Crit. Care **22**(1), 51–55 (2007)
10. Sonnadara, R.R., Garbedian, S., Safir, O., Nousiainen, M., Alman, B., Ferguson, P., Kraemer, W., Reznick, R.: Orthopaedic boot camp ii: examining the retention rates of an intensive surgical skills course. Surgery **151**(6), 803–807 (2012)
11. Heng, P.A., Cheng, C.Y., Wong, T.T., Xu, Y., Chui, Y.P., Tso, S.K.: A virtual reality training system for knee arthroscopic surgery. IEEE Trans. Inf. Technol. Biomed. **8**(2), 217–227 (2004)
12. Riener, R., Harders, M.: Psychology Applied to Work. Springer, London (2012)
13. Meyerson, S.L., LoCascio, F., Baldersonand, S.S., D'Amico, T.A.: An inexpensive, reproducible tissue simulator for teaching thoracoscopic lobectomy. Ann. Thorac. Surg. **89**(2), 594–597 (2010)
14. Bridges, M., Diamond, D.L.: The financial impact of teaching surgical residents in the operating room. Am. J. Surg. **177**(1), 28–32 (1999)
15. Smith, R.: Game Technology in Medical Education: An Inquiry into the Effectiveness of New Tools. Modelbenders LLC, USA (2009)
16. Ziv, A., Ben-David, S., Ziv, M.: Simulation based medical education: an opportunity to learn from errors. Med. Teach. **27**(3), 193–199 (2005)
17. Susi, T., Johannesson, M., Backlund, P.: Serious games—an overview. Technical Report HS-IKI-TR-07-001, School of Humanities and Informatics, University of Skovde, Sweden, Feb 2007
18. Stapleton, A.J.: Serious games: serious opportunities. In: Proceedings of the 2004 Australian Game Developers' Conference, pp. 1–6. Melbourne (2004)
19. Becker, K., Parker, J.: The Guide to Computer Simulations and Games. Wiley, Indianapolis (2012)
20. Thiagarajan, S., Stolovitch, H.D.: Instructional Simulation Games. Educational Technology Publications, Englewood Cliffs (1978)
21. Tashiro, J., Dunlap, D.: The impact of realism on learning engagement in educational games. In: Proceedings of ACM Future Play 2007, pp. 113–120. Toronto (2007)
22. Farmer, E., von Rooij, J., Riemersma, J., Joma, P., Morall, J.: Handbook of Simulator Based Training. Ashgate Publishing Limited, Surrey, UK (1999)
23. Hays, R.T., Singer, M.: Simulation Fidelity in Training System Design. Springer, New York (1989)
24. Ker, J., Bradley, P.: Simulation in medical education. In: Swanwick, T. (ed.) Understanding Medical Education: Evidence, Theory and Practice, pp. 164–180. Wiley, New York (2010)
25. Rehmannand, A.J., Mitman, R.D., Reynolds, M.C.: A handbook of flight simulation fidelity requirements of human factors research. Technical report, Crew System Ergonomics Information Analysis Center, Wright-Patterson Air Force Base, Dayton, 1995
26. Muchinsky, P.M.: Psychology Applied, to Work. Wadsworth (1999)
27. Brown, J.S., Collins, A., Duguid, S.: Situated cognition and the culture of learning. Educ. Researcher **18**(1), 32–42 (1989)
28. Dalgarno, B., Lee, M.J.W.: What are the learning affordances of 3-D virtual environments? Br. J. Educ. Technol. **41**(1), 10–32 (2010)
29. Hulusic, V., Aranh, M., Chalmers, A.: The influence of cross-modal interaction on perceived rendering quality thresholds. In: Proceedings of the 16th International Conference in Central Europe on Computer Graphics, Visualization and Computer Vision, pp. 41–48. Plzen–Bory, Czech Republic, 4–7 Feb 2008
30. Blascovich, J., Bailenson, J.: Infinite Reality. Harper Collins, New York (2011)
31. Cook, D.A., Hamstra, S.J., Brydges, R., Zendejas, B., Szostek, J.H., Wang, A.T., Erwin, P.J., Hatala, R.: Comparative effectiveness of instructional design features in simulation-based education: systematic review and meta-analysis. Med. Teach. **35**(1), e867–e898 (2013)
32. Kneebone, R.L.: Practice, rehearsal, and performance: an approach for simulation-based surgical and procedure training. J. Am. Med. Assoc. **302**(12), 1336–1338 (2009)

33. Ziv, A., Wolpe, P.R., Small, S.D., Glick, S.: Simulation-based medical education: an ethical imperative. Acad. Med. **78**(8), 19–22 (2003)
34. Raths, D.: Virtual reality in the OR. Training Dev. Mag. **6**(8), 36–40 (2006)
35. Gallagher, A., Ritter, M., Champion, H., Higgins, G., Fried, M., Moses, G., Smith, D., Satava, R.: Virtual reality simulation for the operating room, proficiency-based training as a paradigm shift in surgical skills training. Ann. Surg. **241**(2), 364–372 (2005)
36. Kohls-Gatzoulis, J.A., Regehr, G., Hutchison, C.: Teaching cognitive skills improves learning in surgical skills courses: a blinded, prospective, randomized study. Can. J. Surg. **47**(4), 277–283 (2004)
37. Hull, L., Arora, S., Aggarwal, R., Darzi, A., Vincent, C., Sevdalis, N.: The impact of nontechnical skills on technical performance in surgery: a systematic review. J. Am. Coll. Surg. **214**(4), 214–230 (2012)
38. Fitts, P.M., Posner, M.I.: Human Performance. Brooks and Cole, Oxford, England (1967)
39. Rosser Jr, J.C., Lynch, P.J., Haskamp, L.A., Gentile, D.A., Yalif, A.: The impact of video games in surgical training. Arch. Surg. **142**(2), 181–186 (2007)
40. CBS News. Video gamers make good surgeons. http://www.cbsnews.com/2100-204_162-610601.html/. 9 Feb 2009
41. Seitz, A.R., van Wassenhove, V., Shams, L.: Simultaneous and independent acquisition of multisensory and unisensory associations. Perception **36**(10), 1445–1453 (2007)
42. Carlile, S.: Virtual Auditory Space: Generation and Application. R.G. Landes, Austin TX, USA (1996)
43. Mastoropoulou, G., Debattista, K., Chalmers, A., Troscianco, T.: The influence of sound effects on the perceived smoothness of rendered animations. In: Proceedings of the Symposium on Applied Perception in Graphics and Visualization 2005, pp. 9–15. La Coruna, Spain, 26–28 Aug 2005
44. Bonneel, N., Suied, C., Viaud-Delmon, I., Drettakis, G.: Bimodal perception of audio-visual material properties for virtual environments. ACM Trans. Appl. Percept. **7**(1), 1–16 (2010)
45. Angel, E.: Interactive Computer Graphics: A Top-Down Approach Using OpenGL. TPearson Education Inc., Boston (2008)
46. Hulusic, V., Debattista, K., Aggarwal, V., Chalmers, A.: Maintaining frame rate perception in interactive environments by exploiting audio-visual cross-modal interaction. Vis. Comput. **27**(1), 57–66 (2011)
47. Hulusic, V., Harvey, C., Debattista, K., Tsingos, N., Walker, S., Howard, D., Chalmers, A.: Acoustic rendering and auditory-visual cross-modal perception and interaction. Comput. Graph. Forum **31**(1), 102–131 (2012)
48. Shams, L., Kim, R.: Crossmodal influences on visual perception. Phys. Life Rev. **7**(3), 295–298 (2010)
49. Sabri, H., Cowan, B., Kapralos, B., Moussa, F., Cristancho, S., Dubrowski, A.: Off- pump coronary artery bypass surgery procedure training meets serious games. In: Proceedings of the International Symposium on Haptic Audio-Visual Environments and Games, pp. 123–127. Phoenix, 16–17 Oct 2010
50. Fann, J.I., Caffarelli, A.D., Georgette, G., Howard, S.K., Gaba, D.M., Young-blood, P., Mitchell, R.S., Burdon, T.A.: Improvement in coronary anastomosis with cardiac surgery simulation. J. Thorac. Cardiovasc. Surg. **136**(6), 1486–1491 (2008)
51. Hogue, A., Kapralosand, B., Zerebeckiand, C., Tawadrous, M., Stanfield, B., Hogue, U.: Stereoscopic 3d video games and their effects on engagement. In: Proceedings of the Stereoscopic Displays and Applications XXIII, pp. 828816–828816-7. Burlingame, CA, 23–25 Jan 2012
52. UCSF.: Medical education: Potential contributions of stereoscopic 3D. School of Medicine, Office of Educational Technology, University of California San Francisco (2013)
53. Hollander, A., Roseand, H., Kollin, J., Moss, W.: Attack of the s. Mutans!: a stereoscopic-3d multiplayer direct-manipulation behavior-modification serious game for improving oral health in pre-teens. In: Proceedings of the Stereoscopic Displays and Applications XXIII, pp. 828816–828816-7. Burlingame, CA, 23–25 Jan 2012

54. Kapralos, B., Jenkin, M., Milio, E.: Virtual audio systems. Presence: Teleoperators Virtual Env. **17**(6), 527–549 (2008)
55. Bonneel, N., Suied, C., Viaud-Delmon, I., Drettakis, G.: Inter-active sound propagation using compact acoustic transfer operators. ACM Trans. Graph. **31**, 7 (2012)
56. Nordahl, R., Turchet, L., Serafin, S.: Sound synthesis and evaluation of interactive footsteps and environmental sounds rendering for virtual reality applications. IEEE Trans. Visualizations Comput. Graph. **17**(1), 1234–1244 (2011)
57. Zhou, Z.Y., Cheok, A.D., Qiu, Y., Yang, X.: The role of 3-d sound in human reaction and performance in augmented reality environments. IEEE Trans. Syst. Man Cybern.—Part A: Syst. Hum. **37**, 262–272 (2007)
58. Makino, H., Ishii, I., Nakashizuka, M.: Development of navigation system for the blind using GPS and mobile phone communication. In: 18th Annual Meeting of the IEEE Engineering in Medicine and Biology Society, pp. 506–507. Amsterdam, the Netherlands, 31 Oct–3 Nov 2005
59. Kramer, G., Walker, B., Bonebright, T., Cook, P., Flowers, J.H., Miner, N., Neuhoff, J.: Sonification Report: Status of the Field and Research Agenda. Technical report, Department of Psychology, University of Nebraska—Lincoln, Lincoln (2010)
60. Cowan, B., Kapralos, B.: A gpu-based method to approximate acoustical reflectivity. J. Graph. GPU Game Tools **15**(4), 210–215 (2011)
61. Cowan, B., Kapralos, B.: Gpu-based real-time acoustical occlusion modeling. Virtual Reality **14**(3), 183–196 (2010)
62. Cowan, B., Kapralos, B.: Gpu-based one-dimensional convolution for real-time spatial sound generation. Loading Feature Issue: FuturePlay 2008 Ed. **3**(9) (2009)
63. Iuppa, N., Borst, T.: End-to-End Game Development: Creating Independent Serious Games and Simulations from Start to Finish. Focal Proess, Oxford (2010)

Chapter 15
The Ongoing Development of a Multimedia Educational Gaming Module

Elizabeth Stokes

Abstract This chapter gives an explanation of the development of an ongoing multimedia educational gaming module. Carrying out this module over many years resulted in establishing that pupils' on the austistic spectrum have a diverse and variance in their educational abilities, medical conditions, computer skills, likes and interests. However, generic computer games were not being developed in collaboration with practitioners with the use of a holistic approach specifically for these learners. The objective was for undergraduate students to develop personalised games, from the practitioners' completion of individualised profiles, as their assignment. The students used the profiles as a baseline, to carry out in depth research and used their imaging, sound, animation and authoring skills for developing games, whilst adhering to the practitioners' specifications stated on the profiles. The chapter concludes by demonstrating the positivity of bringing society into academia with coursework based on real users, resulting in being beneficial to all the participants.

Keywords Multimedia · Therapeutic · Educational · Games · Software · Module · Students coursework · Learners · Practitioners · Schools · Autism · Disabilities

15.1 Introduction

In view of establishing that learners with disabilities, for example, those on the autistic spectrum have a diversity of needs, generic games were not being developed in collaboration with practitioners specifically for these individuals. This chapter sets out to explain how a module [1] was developed in order to solve

E. Stokes (✉)
Department of Computer Science, School of Science and Technology,
Middlesex University, The Burroughs, Hendon, London NW4 4BT, UK
e-mail: estokes12000@yahoo.co.uk; e.stokes@mdx.ac.uk

A. L. Brooks et al. (eds.), *Technologies of Inclusive Well-Being*,
Studies in Computational Intelligence 536, DOI: 10.1007/978-3-642-45432-5_15,
© Springer-Verlag Berlin Heidelberg 2014

this problematical issue. The aim is to demonstrate how an educational gaming module was successfully developed over many years. The objectives were for a lecturer to work in collaboration with teachers (practitioners) on the completion of profiles. University students' coursework would use the profiles as a baseline to develop individualised educational games for real autistic pupils in schools. The collaboration of teachers and the lecturer resulted in merging society and academia. The realism of this coursework motivated the students with the knowledge that their efforts not only resulted in them gaining theoretical and practical technological skills (resulting in them passing the module) [1], but their efforts could be potentially of educational use and help and support the practitioners as an educational classroom tool.

15.2 The Coursework

The coursework was produced in order to assess the students' knowledge, understanding and academic attainment in relation to the learning outcomes of the module [1]. Students gained theoretical knowledge of planning, research and design methodologies, data gathering, design principles and the user interface resulting in quality control, for the development of games for real users in society. Students gained an understanding of the basic principles of multimedia and associated technologies, experiences of different software tools and an understanding of the appropriateness of multimedia elements (text, sound, animation, graphics) in relation to their End User's preference to a tactile, visual, auditory modality. This together with their research, computing, scripting, Human Computer Interaction (HCI) and multimedia practical skills resulted in the development of individualised educational games for real learners on the autistic spectrum in the society.

In the past the lecturer gave students completely hypothetical scenarios not relating to real-world situation or allowed students to choose a topic resulting in developing and a multimedia artifact as their coursework. Many students strive to achieve the best results possible through their input in time and energy on short lived, stressful, and self-profiting work. Unfortunately, after the coursework is graded, the effort is sadly soon forgotten. The coursework ends up being of pointless use to them or others, resulting in abandonment on a shelf to gather dust, stored away out of sight probably never to be looked at again and even discarded.

Therefore, in order to produce a coursework, which could continue and be of use to others, a real-life problem in society was used, with students playing a participatory role and focusing on producing educational games for real learners with a disability in schools. Researchers [2, 3] claimed that some students regarded their participation in research as being educationally beneficial to them; this was echoed by many researchers [4–6] stating that...

> ...a good number of surveys broadly demonstrate that students believe they have benefited educationally from participation and that they do find it a positive and useful experience... [6]

15.3 The Lifecycle of the Gaming Coursework

15.3.1 Case Studies (Profiles)

The importance of case studies, as some researchers [7] pointed out, is that it enabled some developers to not just produce effective products by just considering the user friendliness or work from their own personal guidelines. Instead developers using case studies are able to acquire an in depth knowledge of diverse variation of each users' needs together with a thorough appreciation of aims, objectives and problems which needs to be met in order to produce an effective product. This investigation [8, 9] was in agreement with [7] claims, as to the importance of considering diversity, when using case studies in student's assignments. Some researchers [8] claimed that the use of real case studies, in students' coursework, resulted in being an effective learning and teaching aid, with students gaining transferable skills, from applying their learning to ...real world projects [1].

This coursework was first developed and implemented in 2000 with a lecturer working in collaboration with practitioners of one school and six of their learners on the autistic spectrum [1]. The lecturer developed a blank template of a profile (case study), which was given to the practitioners to complete [9]. Several hundred students were given a full explanation of their participatory role in this research and each student were randomly given a completed and anonymous profile, and a clear set of requirements, for producing an individualised educational multimedia game, over 12 weeks, for each particular autistic learner [1]. Therefore, students got to understand the specific learners' needs, by carrying out in-depth research resulting in the production of a report based on a pupil's profile (case study) [9].

15.4 Pupils' Profile

The investigation demonstrates the importance of each learner's variance and spectrum of needs [9]. The investigation began with a pilot study using six representative pupils from a randomly chosen representative special school in the United Kingdom. The practitioners had to complete a profile of each pupil.

15.4.1 Gender

Five of the participants were male and one was female (Table 15.1). This correlates with Johnson's [10] statement that '...boys are four times more likely to develop autism than girls...'

Table 15.1 Variance in each pupil's spectrum of needs [9]

Pupils	Gender	Age	Medical conditions	Communicates
1	M	12	ASD (3)	Non-verbal no words only sounds. Uses PECS, TEACCH and gestures
2	M	14	ASD (3), ADHD (3) Allergies (3)	Verbal (5/6 year old) TEACCH
13	M	14	Autistic tendencies (5)	A few odd words in frequent use. Points, quickly becomes upset if no response
4	M	15	Demonstrates triad of difficulties necessary for diagnosis of autism	Verbal. Makes needs known through using two, three words together
5	M	13	ASD	Uses up to 100 words to communicate basic needs. Can use communication book and symbols and can gesture and lead
6	M	17	Moderate autism	No speech. Uses a form of "Makaton". Also uses pictures and photographs

15.4.2 Age

Two of the pupils were of the same age and other pupils varied in ages from 12 to 17 years old (Table 15.1).

15.4.3 Medical Conditions

Although the practitioners were asked to indicate any medical conditions, most of the practitioners did not state each pupil's medical condition. They were also asked the grade where on the autistic spectrum they regarded each pupil, with one being Serve to five being Mild. Unfortunately three of the pupils were not graded by the practitioners, however, two of the pupils were graded as being three and one was graded as one being at the mild end of the spectrum. This demonstrates a variance in the pupils' spectrum of needs [9]. Unfortunately, they did not state where on the spectrum each pupil was in relation to their triad of impairments (speech, language, communication, social interaction, imagination and rigidity of thought).

15.4.4 Communication

Some practitioners gave limited information as to the extent of each pupil's verbalisation. There was no indication as to the extent of sound some pupils produced or whether they classed the non-verbal pupils as being totally mute. They did not state on the profile the extent of the pupils use of signs, symbols, gestures and

Table 15.2 The variance in each pupil's spectrum of needs in relation to their computer and reading ability and their comprehension [9]

Pupils	Computer ability	Reading level	Comprehension
1	Limited with support can type out words but needs adults support. To develop mouse control	Pre reading level. Software to help letter matching and recognition letter sequencing (visual cues needed) learning letters	Variable—understands a limited amount of language but requires demonstration/gestures from adult
2	Good	Level B 5–14 elaborated curriculum approximate primary 2	Attention and concentration poor due to ADHD, general comprehension good
3	Minimal touch screen, mouse skills	Non reader (5)	At 18 months developmental stage in some areas of understanding (5). Good level or word structure when e.g. work first at table then out on bus
4	Can co-operate independently	Reading at 5/6 years	N/S
5	Can operate independently including use of mouse	Pre-reading 3/4 years	N/S
6	Can use mouse well. Adequate ability. Can perform 60 piece jigsaws fairly easily. Uses PC for games e.g. Tonka	Not measurable	Understands spoken word 90 % of time. Also can use tools, machines for specific jobs

facial expression and objects. The practitioners could have stated whether it would be educationally and therapeutically beneficial to include each pupil's chosen communication method in the software being developed and whether this should be animated together with text and sound (Table 15.1).

15.4.5 Computer Ability

Practitioners gave some information on the profiles in relation to each pupil's computer ability (Table 15.2); however, more information would have been very useful. The practitioners could have stated whether each pupil had a performance for a particular or combination of modalities (Visual, Auditory and Tactile) and multimedia elements (Text, Graphics, Animation and Sound).

15.4.6 Reading Level

It appears from the completed profiles that each pupil was at a different reading stage. It would have been beneficial if the practitioners could have stated if each pupil benefited from having the text in the software in upper case or lower case, the size of the words to be used and whether the text should be coloured, animated with the text sounded out. It would have been useful if the practitioner had indicated if the software developed included a story with the text highlighted and spoken.

15.4.7 Comprehension

The practitioners gave an adequate amount of information regarding each pupil's comprehension level demonstrating a variance in understanding [9].

15.4.8 Educational Area to Consider

Although the practitioners indicated numerous educational areas in relation to literacy and numeracy, once again there is a variance [9] in the different educational areas for each pupil (Table 15.3).

15.4.9 Likes

There is a huge difference in each pupil's likes (Table 15.3) [9].

15.5 The Development and Production

Therefore, the development and production of the customised games were based on the profiles given to each student. From the student's research findings based on the profiles, students produced storyboards, which resulted in the development of the games. This was as Quinn [11] stated...

> ...Software engineers must understand the needs of the users, access the strengths and weakness of current system and design modifications to the software... [11]

and as stated in Donegan et al's research... [12]

> ...Because of their individual needs and difficulties, whichever technology they use to communicate requires a high level of personalization and customization if they are to be able to use it effectively [12].

Table 15.3 The variance in each pupil's educational areas and likes [9] are considered in the software development

Pupils	Educational areas to consider	Likes
1	Software to focus on numbers 1–10, counting or sequencing 1–20 (+visual clues) counting games anything to develop mouse control—must see a purpose to tasks	Sensory materials music bubble tubes, books, puzzles singing kettle, Tweenies and Thomas the tank
2	Spelling software, single figures addition + subtraction Software to be developed to help with ¾ letter words	Likes drawing, artwork, and working on diary—How about a multimedia diary for him to keep updated? Likes music therapy likes art work likes stories flash cards, all work on computer, music therapy and art work (a special favourite) stories
3	Use concrete clues e.g. beater for music. Early stages of cause and effect needed. Keep it very simple. Instant quick feedback. A clear finished said in software. Software to teach learn to wait needed with reward from screen turn taking game. Lots of animation, sound needed	Likes music, Disney characters buzz lightyear, dvd/video
4	Independence. Communication and language. Self help in community e.g. shopping, awareness of money	Healthy eating, motivated by food (loves baking). Enjoys swimming, horse riding and music
5	Communication and language. Numbers 1–10. Self help skills (P.S.E)	Enjoys cartoons i.e. Tom and Jerry, Thomas the Tank etc. Loves sudden noise or toys that play music. Enjoys cutting and particularly "Thomas the Task" posting favourites
6	Communication development. Outside working opportunities	Outside working opportunities. Tonka originally but prefers completing task of jigsaws picture

For quality control the software was evaluated by the students' peers and modified. However, as the games were developed as an assignment, the games were graded by the academics in the university. Further quality control was carried out by the academics and further modified before submission to the school.

15.6 Feedback and the Next Cohort of Students

Quality control was also carried out by the practitioners involved through ongoing assessment, monitoring and evaluation of the games for their educational use. The quality control validated the ongoing exhaustive testing process resulting in making it easier to isolate and fix errors. The data gathered resulted in the academic gaining feedback from receiving ongoing evaluation and the updating of profiles or new profiles.

Further cohorts of students took into consideration the updated profiles of the learners' changing needs and the evaluated feedback from the practitioners, highlighting the strengths and weakness of the games developed by the previous cohort of students. This enabled them to make the appropriate modifications [11].

Niès and Pelayo's study [13] showed how collaboration helped to ...*resolve the limits of direct users involvement and usual problems pertaining to users' needs description and understanding.*

Therefore, the module [1] underwent an iterative approach, with each new cohort of students receiving updated and modified profiles, this gave the students ...*a gross expectation of the users knowledge at the beginning of the interaction* [14] and feedback from the practitioners enabled ongoing amendments, modification and improvements to be made.

The students would be able to learn from the mistakes made by their previous peers in order to perfect the games for the practitioners and ultimately for the learners. The module [1] used an iterative process through an ongoing continuous use of Human Computer Interaction (HCI) and Hierarchical Task Analysis (HTA) approach [15] and an Autistic User Centred Design-For-One approach with an ongoing validation or testing obtained from adapted and adaptive (tailored) systems.

15.7 The Responsibility of the Academic

The development and effective ongoing module [1], with the academic wholly responsible for working in collaboration with practitioners and acquiring the complete modified profiles and new profiles, which resulted in the random distribution of the profiles to the students. The lecturer was also responsible for teaching the students research, theoretical and practical multimedia skills and to understand and interpret the profiles accurately. This resulted in the development of the most appropriate individualised educational games; based on the teacher's specifications and these would be met in the outcome of each student's contribution.

15.8 Students' Useful Contribution and the Assets They Bring

Students had a great deal to contribute with a deep and caring optimistic approach to developing a game which would be of educational use to their own learner with a disability. There have been several hundred students who have carried out this coursework over 13 years (from 2000). The pupils came from different backgrounds; there is a variance in age, experience, intellectual ability, ethnicity and social class [9]. Some of the mature students brought their personal lifelong experience into the research, e.g. a retired headmaster of a school who studied the module [1]. The multicultural students had great empathy, knowledge and understanding with the equally multi-cultural learners' profiles issued to the students. This enabled the students to transfer their personal knowledge and understanding, already acquired and brought into academia, with this coursework e.g. student saying "…great I have got a child from my part of the world". Therefore, students gained a great deal whilst carrying out their studies.

15.9 Students' Gain from the Coursework

The incentive for carrying out this coursework for the students was more than just being assessed and graded for demonstrating their knowledge and skills attained from undertaking the coursework. They acquired an understanding of what they were learning at university in their 'academic-world' with the ability to relate this to real-life situations in today's society in the 'real-world'.

15.10 Partnership and Collaboration

Schuler's [16] cited in Foot and Sanford [6] comment reinforcing positive ethical advantages for students through their encouragement and adoption of a vital 'partnership' role with the academic's enthusiasm, wealth of knowledge, experience and help throughout and their indirect partnership with members of society such as the practitioners. The researcher is also in agreement with Schuler [16] who considered student's participation in a very real assignment as giving them an advantage of being less stressful as they do not have to give consideration to an autonomous assignment. Students were also in-directly collaborating with professionals in society whilst carrying out their studies. Therefore, this coursework resulted in students acquiring self-satisfaction and sense of achievement that their work could be educationally beneficial to less able learners with educational needs, thereby, making a real contribution to society.

However, the researcher is in agreement with Foot and Sanford [6] who states that… students give positive ratings to research in which they have participated (in order to bolster their original decision to participate). This is true in this case of this particular module, especially as elective students who have made the original decision to do the module, result in giving positive written and verbal feedback with praise for the assignment and the module and even resulting in a change of degree programme [1].

15.11 Theoretical Knowledge, Practical Skills and Design Techniques

Students learn the importance of research in order to implement an autistic-user-centred design-for-one approach. The theoretical knowledge and the practical skills learnt from doing the module resulted in the incentive of putting more effort into this coursework. The students, therefore, were more aware that their hard work would result in not only producing good pieces of coursework to pass the module and be awarded a grade but also, had an incentive that they were making a real contribution for a real member of society who would gain from all their efforts [1].

15.12 Relating Academia with Society and Humanity

Students were able to relate what they were learning in the academic environment to what was actually happening in society. This gave them the opportunity to play a big part in humanity by helping someone less able by providing them with a game, which could potentially help their educational needs, as well as providing the practitioners with and educational aid for the classroom. Making students active participants in their own coursework, for the good of mankind, as well as their own personal achievement, has shown to be far more rewarding for them.

15.13 Gains from the Experience Itself

Conversely, the researcher disagrees with Foot and Sanford's [6] questioning whether students gained from the experience, but agrees with them that students' research active experience helped their understanding and 'appreciation of the scientific process'.

As already been stated by other researchers [6], students particularly prefer "research which is integrated into their own educational programme or their participation is used in a concrete way to illustrate a psychological principle or an aspect of the scientific process" [6].

15.14 Students' Gains

Therefore, the students' participation in this coursework, gave them a real learner to research, in order to produce customised educational games, resulting in coercing them into a very worthwhile venture, whilst enabling them to earn their credits towards their degree course from the completion of the games.

Students are often assessed through reports, presentations, posters, essays and class tests individually, in pairs or group work. Students put a lot of effort into passing these assignments with some students finding these assignments a chore, leaving their work to the last minute and not enjoying it. However, for several years it has become apparent that the majority of students carrying out this coursework have enjoyed it as it assessed their knowledge, understanding, abilities and skills achieved. This was demonstrated through the students producing far more work than requested of them, with some students donating cards, vouchers and presents for the pupils, as well as, the games they had developed for them.

Some students take the assignment so personally as to refer to their anonymous case study (profile) as "my child". The majority of student's own written feedback, indicated that they found the coursework "very rewarding". Some students ended up continuing working with disabilities with voluntary and paid employment. A graduate demonstrated the game she developed for the module in her job interview with Hewlett Packard UK and Microsoft, this resulted in being told by that they were so impressed with the student's coursework and both companies offered the student employment [1].

A great number of students studying this module [1] resulted in carrying out further research into autism or disabilities for their final year projects and dissertations, with some going on to postgraduate (Master and PhD) levels of research.

15.15 Schools, Practitioners and Pupils Gains

As well as the free games made specifically for each learner in the school, the academic negotiated a financial deal with the student union shop and the university bookshop. As the students needed to buy a package (e.g. consisting of headphones with mic, CD, logbook etc.) from the student union shop in order to carry out the assignment and the recommended reading books from the bookshop, both shops agreed to donate a £1 on each package and book bought to the school. The money went to the schools participating at the time.

The practitioners benefited by receiving ongoing individualised educational games, developed to their specifications and for their learners with disabilities.

It enabled them to specify exactly the educational content of the games they required to be developed which would complement the areas being covered by the curriculum in the classroom and which would be of educational use for each individual learner, through the ongoing updated profiles, evaluation and feedback process. The pupils gained by obtaining their very own personalised educational games.

Therefore, this ongoing module has now been carried out by thousands of students over 13 years (from 2000), in collaboration with 44 practitioners of 148 learners with autism or other disabilities, from 9 schools across the United Kingdom [1]. All the participants in this investigation have acknowledged the benefits of the whole collaborative iterative ongoing process [17]. This real coursework is now achieving real effective, efficient, and enjoyable results for the academic, the students, the practitioners and their learners. The continuation of this work has resulted in the development of the Games for Learners on the Autistic Spectrum and/or with other Disabilities (GLAD) software development company with the development of personalised, individualised educational software. For more details email estokes12000@yahoo.co.uk [18].

15.16 Conclusion

To conclude, the ongoing development of this multimedia educational gaming module was ethically justified through the validity of this very real coursework [1]. It demonstrates that by bringing a real-life situation into academia increases the students' stimulation, motivation and interest in a positive and beneficial way for them and others in society. The importance is focused on the on going updating of the individual's case studies, (profiles), the evaluation and feedback from the partnership with all parties concerned. This chapter has shown how the assignment brought together academia and society, resulting in all parties reaping many of the benefits from this coursework and module [1].

Therefore, the pupils (learners) gained from this real assignment by having their own individualised educational games developed for them. The educationalists (teachers) and therapists (speech and language) also gain from having the appropriate educational software developed for them to use in the classroom for each pupil. The students not only gained a grade for passing the module [1], resulting in them eventually gaining a degree, but also were motivated and stimulated at the thought of producing a product that would help to solve a problem in society. This was achieved with the students developing a product, as their assignment, in order to be of educational help to the practitioners. They could also see their efforts as being of real educational use to learners with disabilities, in the real world, long after its completion and their graduation.

Therefore, in order for the students to develop a personalised educational product there was an importance to demonstrate that learners on the autistic spectrum have diverse and variance needs, [9] thereby, needing a collaboration holistic design-for-one approach, which generic computer games were not adopting. The objectives were met with students developing personalised games, from the practitioners' completion of individualised profiles, as their assignment. The students, used this, as a baseline, to carry out in depth research and use their own imaging, sound, animation and authoring skills for developing games, whilst adhering to the practitioners' specifications stated on the profiles.

To conclude, this chapter demonstrated the positivity of bringing society into academia with coursework based on real users. This resulted in being beneficial to all the participants effectively, fairly, ethically and educationally from the results of the ongoing collaboration development of this multimedia educational gaming module [1]. Therefore, this empirical investigation makes a valuable, original and academic, contribution to knowledge, in the Computing, Education and Special Needs fields.

References

1. Stokes, E.: Ongoing Development of a Multimedia Gaming Module for Speech, Language and Communication. Pervasive Health—State-of-the-Art & Beyond Springer series book to be published 2014
2. Roberts, L.D., Allen, P.J.: Student perspectives on the value of research participation. In: McCarthy, S., Dickson, K.L., Cranney, J., Trapp, A., Karandashev, V. (eds.) Teaching Psychology Around the World, vol. 3, pp. 198–212. Cambridge Scholars Publishing, Newcastle upon Tyne (2012)
3. Roberts, L.D., Allen, P.J.: A brief measure of student perceptions of the educational value of research participation. Aust. J. Psychol. **65**(1), 22–29 (2013)
4. Christensen, L.: Deception in psychological research: When is its use justified? Pers. Soc. Psychol. Bull. **14**, 664–675 (1988). Cited in Foot, H., Sanford, A.: The use abuse of student participant. The Psychologist **17**(5) May issue (2004)
5. Kimmel, A.J.: In defence of deception. Am. Psychol. **53**, 803–804 (1998). Cited in Foot, H., Sanford, A.: The use abuse of student participant. The Psychologist. **17**(5) May issue (2004)
6. Foot, H., Sanford, A.: The use abuse of student participant. The Psychologist **17**(5), 664–675 (2004)
7. Shneiderman, B., Plaisant, C.: Designing the User Interface, 4th edn. Pearson Addison Wesley, Boston (2004)
8. Elrod, C.C., Murray, S.L., Flachsbart, B.B., Burgher, K.E., Foth, D.M.: Utilizing multimedia case studies to teach the professional side of project management. J. STEM Educ. Innovations Res. (2010)
9. Stokes, E.: Profiles used in teaching research for developing multimedia games for pupils on the Autism Spectrum (AS). Int. J. Disabil. Hum. Dev. **7**, 37–49 (2008)
10. Johnsons, S., Hollis, C., Hennessy, E., Kochhar, P., Woke, D.: Screening for autism in preterm learners: Diagnostic utility of the social communication questionnaire. Arch. Dis. Child. **96**, 73–77 (2011)
11. Quinn, M.J.: Ethics for the Information Age, pp. 22–29. Pearson Addison Wesley, Boston (2004)
12. Donegan, M., Morris, J.D., Corno, F., Signorile, I., Chió, A., Pasian, V., Vignola, A., Buchholz, M., Holmqvist, E.: Understanding users and their needs. Int. J. Univers. Access Inf. Soc. **8**(4), 259–275 (2009)
13. Niès, J., Pelayo, S.: From users involvement to users' needs understanding: A case study. Int. J. Med. Inform. Hum. Factors Eng. Healthc. Appl. **79**(4), 76–82 (2010)
14. Elsom-Cook, M.: Principles of Interactive Multimedia. McGraw-Hill, London (2001)
15. Cox, D.: Task Analysis, Usability and Engagement. Human-Computer Interaction. Interaction Design and Usability Lecture Notes in Computer Science, vol. **4550**, pp. 1072–1081 (2007)
16. Schuler, H.: Ethical Problems in Psychological Research. Academic Press, New York (1982). Cited in Foot, H., Sanford, A.: The use abuse of student participant. The Psychologist, **17**(5) May issue (2004)
17. Dix, A., Finlay, J., Abowd, G., Beale, R.: Human–Computer Interaction, 3rd edn. Pearson Education, Harlow (2004)
18. Stokes, E.: Games for Learners with Autism and other Disabilities (GLAD) Personalised individualised educational software. Technology for Inclusion Conference Central Enfield City Learning Centre, July 2013

Part V
Disruptive Innovation

Chapter 16
Disruptive Innovation in Healthcare and Rehabilitation

A. L. Brooks

Abstract Disruption is a powerful body of theory that describes how people interact and react, how behavior is shaped, how organizational cultures form and influence decisions. Innovation is the process of translating an idea or invention into a product or service that creates value or for which customers will pay. Disruptive Innovation in context of the author's body of work in healthcare and rehabilitation relates to how development of a cloud-based converged infrastructure resource, similar to that conceived in a national (Danish) study titled Humanics, can act as an accessible data and knowledge repository, virtual consultancy, networking, and training resource to inform and support fields of researchers, practitioners and professionals. High-speed fiber networking, smart phone/tablet apps, and system presets can be shared whilst AI and recommendation engines support directing global networks of subscribers to relevant information including methods and products. Challenges and problems to fully realize potentials are speculated. A mature body of research acts as the vehicle to illustrate such a concept.

Keywords Disruptive innovation · Transdisciplinary convergence · Cloud-based convergent infrastructure · Societal demographics · Research · Education

16.1 Introduction

16.1.1 Disruptive Innovation

Rather than detailing Disruptive Innovation, this 'position' chapter introduces the concept by predicting a similar disruption in healthcare and rehabilitation as in

A. L. Brooks (✉)
Director of Sensorama Lab, Medialogy/AD:MT, School of ICT, Aalborg University Esbjerg,
Niels Bohrs vej 8, 6700 Esbjerg, Denmark
e-mail: tb@create.aau.dk

A. L. Brooks et al. (eds.), *Technologies of Inclusive Well-Being*, 323
Studies in Computational Intelligence 536, DOI: 10.1007/978-3-642-45432-5_16,
© Springer-Verlag Berlin Heidelberg 2014

corporate business e.g. [1–3] and education e.g. [4]. In this case the predicted disruption is via a cloud-based convergent infrastructure that acts as a resource to bring together disparate and related healthcare professionals and researchers to communicate and share knowledge and data to benefit the healthcare and related fields.

Disruption is acknowledged as a powerful body of theory that describes how people interact and react, how behavior is shaped, how organizational cultures form and influence decisions [1]. Innovation is the process of translating an idea or invention into a product or service that creates value or for which customers will pay [5].

Disruptive Innovation in context of the author's body of work relates to how development of a cloud-based converged infrastructure resource can benefit professionals in the fields of healthcare and rehabilitation. In this respect, the concept was conceived under the SoundScapes research resulting from extensive periods of institute-based experimental research and a 6-year research study across clinics, hospitals and institutes with patients and staff. First published in Japan 1999 [6], and subsequently under a national funded (Danish) healthcare project titled Humanics (1996–2002)—as reported in [7, 8], the original concept was how the Internet can be a communication and reporting network between healthcare professionals and patients who train at home via Virtual Interactive Space (VIS) [6]. The research explored serious games and creative expression (music-making, digital painting and robotics control)—all via unencumbered gesture using bespoke non-worn sensor apparatus.

In line with related advances and more efficient and speedier network infrastructures the concept has since been refined and extensions planned as a secure accessible data and knowledge repository, virtual consultancy, networking, and training resource to inform and support fields of researchers, practitioners and professionals. It is now envisaged as a growing and evolving entity where software (e.g. apps, presets, etc.) for in-session intervention and new assessment tools are shared. Cloud-based AI and recommendation engines will support directing global networks of subscribers to relevant information including methods, use-strategies, and products.

Mindful of this vision, this position chapter next introduces the concept of cloud-based infrastructures as disruptive innovation in healthcare and rehabilitation. Cloud-based in context relates to associated means of connectivity, sharing and networking via e.g. virtualization, bring your own device (BYOD), self-forming and mesh networks, and the Internet of things (IOT). Exemplifying the author's position in the field, the chapter follows with an overview of the mature research and examples of development from the author's SoundScapes/ArtAbilitation/GameAbilitation and Ludic Engagement Designs for All (LEDA) body of work. This research illustrates how serious games (structured interactions), computer-generated alternative realities and play aligned with complementary creative expression designed sessions (abstract or free interactions) can supplement and potentially even supplant traditional methods of intervention in rehabilitation and therapy. Challenges and problems to fully realize potentials of basing aspects of the work 'in the cloud' are discussed. A focus is to optimally engage and immerse the participant (or patient) in the computer-generated environment to maximize the experience toward a state of presence to realize targeted progress and outcomes. A section on these human traits follows later in the chapter.

16.2 Cloud-Based Disruptive Innovation

The history of cloud computing dates from the 1950s[1] when large main-frame computers were networked with access from various locations optimized [9, 10]. Associated is the tacit knowledge gained from the author having worked throughout the 1980s on networks of industrial control systems (ICSs) and distributed control system (DCS) in the form of large Honeywell TDC2000 mainframe computers that were accessed 24/7 from a network of remote terminals capable of monitoring and operating a physical environment.

Advantages of adopting cloud-based infrastructures in healthcare and rehabilitation are many. The client site (e.g. hospital, clinic or institute), independent healthcare professional, or even the patient, doesn't have the problem of software integration as any problems, updates, performance and maintenance are addressed by the vendor's expert team who ensure a secure service accessed via HTTP. The costs are predictable (usually a monthly subscription) and costs are distributed. Disadvantages include that whilst the service can be customized for the largest clients, the smaller customers need to accept the service as is. Another potential problem could be that as a critical service is supplied the customer has limited control and needs to rely on the vendor.

Cloud-based computing allows users to benefit from integrated convergence of technologies without the need for deep knowledge about or expertise with each one of them. The concept in the specific context of this publication aims to unite and link healthcare professionals and related communities around communication (data, information and knowledge) and training (methods, strategies, and apparatus—hard/software) repositories where secure exchanges, sharing and networking is available. Goals include increased efficiency of practice (intervention, evaluations and refinements) and investment (e.g. optimal use of purchased systems) as well as access to latest tools (e.g. digital assessment etc.). Addition benefits help the users to enable a focusing on their core business instead of being impeded by IT obstacles.

Mell and Grance [11] defined Cloud Computing via identifying "five essential characteristics", these are:

---

[1] http://en.wikipedia.org/wiki/Cloud_computing

- *On-demand self-service*. *A consumer can unilaterally provision computing capabilities, such as server time and network storage, as needed automatically without requiring human interaction with each service provider.*
- *Broad network access*. *Capabilities are available over the network and accessed through standard mechanisms that promote use by heterogeneous thin or thick client platforms (e.g., mobile phones, tablets, laptops, and workstations).*
- *Resource pooling*. *The provider's computing resources are pooled to serve multiple consumers using a multi-tenant model, with different physical and virtual resources dynamically assigned and reassigned according to consumer demand...*
- *Rapid elasticity*. *Capabilities can be elastically provisioned and released, in some cases automatically, to scale rapidly outward and inward commensurate with demand. To the consumer, the capabilities available for provisioning often appear unlimited and can be appropriated in any quantity at any time.*
- *Measured service*. *Cloud systems automatically control and optimize resource use by leveraging a metering capability at some level of abstraction appropriate to the type of service (e.g., storage, processing, bandwidth, and active user accounts). Resource usage can be monitored, controlled, and reported, providing transparency for both the provider and consumer of the utilized service.*

The virtualizing of IP and Fibre Channel storage networking allows for single console management and pooling and sharing of IT resources. Thus, rather than dedicating a set of resources to a particular computing technology, application or line of business, converged infrastructures, a pool of virtualized server, storage and networking capacity is created that is shared by multiple applications and lines of business...in this context the global healthcare community—including e.g. hospitals, institutes, clinics, General Practitioners (GPs), etc. These end users access the scalable cloud-based applications through a web browser, thin client, or mobile app while the business software and user's data are stored on servers at a remote location i.e. allocated space for a user to deploy and manage software "in the cloud". Thus, a bring your own device (BYOD) becomes feasible.

Further Mell and Grance [11] state:

A cloud infrastructure is the collection of hardware and software that enables the five essential characteristics of cloud computing.

The cloud infrastructure can be viewed as containing both a physical layer and an abstraction layer. The physical layer consists of the hardware resources that are necessary to support the cloud services being provided, and typically includes server, storage and network components. The abstraction layer consists of the software deployed across the physical layer, which manifests the essential cloud characteristics. Conceptually the abstraction layer sits above the physical layer.

Service Models:

Software as a Service (SaaS). The capability provided to the consumer is to use the provider's applications running on a cloud infrastructure. The applications are accessible from various client devices through either a thin client interface, such as a web browser (e.g., web-based email), or a program interface. The consumer does not manage or control the underlying cloud infrastructure including network, servers, operating systems, storage, or even individual application capabilities, with the possible exception of limited user-specific application configuration settings.

Platform as a Service (PaaS). The capability provided to the consumer is to deploy onto the cloud infrastructure consumer-created or acquired applications created using programming languages, libraries, services, and tools supported by the provider. The consumer does not manage or control the underlying cloud infrastructure including network, servers, operating systems, or storage, but has control over the deployed applications and possibly configuration settings for the application-hosting environment.

Infrastructure as a Service (IaaS). The capability provided to the consumer is to provision processing, storage, networks, and other fundamental computing resources where the consumer is able to deploy and run arbitrary software, which can include operating systems and applications. The consumer does not manage or control the underlying cloud infrastructure but has control over operating systems, storage, and deployed applications; and possibly limited control of select networking components (e.g., host firewalls).

Deployment Models:

Private cloud. The cloud infrastructure is provisioned for exclusive use by a single organization comprising multiple consumers (e.g., business units). It may be owned, managed, and operated by the organization, a third party, or some combination of them, and it may exist on or off premises.

Community cloud. The cloud infrastructure is provisioned for exclusive use by a specific community of consumers from organizations that have shared concerns (e.g., mission, security requirements, policy, and compliance considerations). It may be owned, managed, and operated by one or more of the organizations in the community, a third party, or some combination of them, and it may exist on or off premises.

Public cloud. The cloud infrastructure is provisioned for open use by the general public. It may be owned, managed, and operated by a business, academic, or government organization, or some combination of them. It exists on the premises of the cloud provider.

Hybrid cloud. The cloud infrastructure is a composition of two or more distinct cloud infrastructures (private, community, or public) that remain unique entities, but are bound together by standardized or proprietary technology that enables data and application portability (e.g., cloud bursting for load balancing between clouds).

Deployment models such as the private and community models above relate to the Humanics concept resulting from the author's SoundScapes research as discussed in the Sect. 16.3. This is where clinics with their teams of experts metaphorically act as the "provider" with patients' linking in as "clients" (but where the Danish welfare state pay for the service).

16.3 SoundScapes: Serious Games, Alternative Realities and Play

SoundScapes is a mature body of research and development that has been well published in that it explores bespoke behavior training interventions via ICT that are specifically tailored for individual participants undergoing rehabilitation to creatively express, play and perform. The ICT strategy integrates various elements from serious games, alternative realities and play.

Unlike certain definitions of serious games that disregard entertainment when used in healthcare and rehabilitation e.g. [12, 13], the concept of SoundScapes holds entertainment of the participant as a key in-session goal. Entertainment experience in this context is designed for and regarded as a catalyst that engages the participant through the fun, enjoyment and pleasure of participating in the activity. The activity is designed as a rehabilitation intervention in line with a healthcare team's goal for the participant's progression where 'actions' are the main unit of analysis.

The SoundScapes hypothesis is that the effect of the intervention is stronger when the participant is not focused on the rehabilitation goal or therapy situation, but rather to have fun in the moment. However, this is problematic and an ongoing challenge to achieve through for example personnel or situation influences. In line with this was how SoundScapes targeted online intervention with interactive gesture controlled games, digital painting, and music making when researching with acquired brain injured patients in the Humanics project (1996–2002)—see [7, 8]. A concept behind the author's design being that the tele-abilitation aspect would place the patient in a home-family situation where the playing or creative expression would optimize distancing from the therapy situation. However, from that distance, game interactions and results could be conveyed as shared data to a clinic for analysis for the assigned healthcare professionals (physical and occupational therapists, psychologists, neuropsychologists and speech therapists) to give feedback without the necessity of travelling and expense. Parallel to the Humanics study was a European Future Emerging Technology (FET) project that also resulted from the author's earlier research where additionally non-humanoid robotic devices were controlled by gesture (via both worn or non-worn sensors) as a healthcare intervention to supplement traditional methods—see [14, 15]. In the author's SoundScapes research interactive computer-based gesture-controlled serious games have been explored since 1998. Prior to that interactive music making and sound manipulation was primarily used.

Serious Games is briefly introduced in the Sect. 16.3.1.

16.3.1 Serious Games

The term *serious games* is used in reference to any kind of online game for learning where the games are designed with the intention of improving some specific aspect of learning that is relevant, instantly useful, and fun [16]. However, the players come to serious games with expectations of learning. It is such expectation of outcome that is considered non-optimal in SoundScapes as it has been evident that participants push their physical limitations when they are motivated through engagement to achieve the state of flow [17], present and immersed in the activity, rather than thinking of the consequences. In other words when an activity necessitates a physical input to trigger and manipulate multi-media the physical actions give proprioceptive feedback to the participant and that is where a focus begins in an interaction. The goal is to transfer the participant's attention away from the primary and to the multimedia (i.e. secondary) feedback so it is that stimuli that drive subsequent (inter)actions. This causal activity results in the human afferent efferent neural feedback loop closure.

Derryberry [16] argues that serious games "offer a powerful, effective approach to learning and skills development". Furthering her position, she states that "What sets serious games apart from the rest is the focus on specific and intentional learning outcomes to achieve serious, measurable, sustained changes in performance and behavior. Learning design represents a new, complex area of design for the game world. Learning designers have unique opportunities to make a significant contribution to game design teams by organizing game play to focus on changing, in a predefined way, the beliefs, skills, and/or behaviors of those who play the game, while preserving the entertainment aspects of the game experience" (p. 4).

Salen and Zimmerman [18] argue a game that is well-designed yields meaningful play defining it as "what occurs when the relationships between actions and outcomes in a game are both discernable *(sic)* and integrated into the larger context of the game." In context to SoundScapes and the use of games in rehabilitation and healthcare, a refinement of meaningful play is posited as *what occurs when the relationships between actions and outcomes in a game are both discernible and integrated into the larger context of the situation.* This is where 'situation' includes the targeted progression according to the participant's profile where actions and outcomes are key units of analysis in SoundScapes of both the digital media and human performance. Thus, by elaborating beyond 'the larger context of the game' to 'the larger context of the situation'.

It is acknowledged that "there is widespread consensus that games motivate players to spend time on task mastering the skills a game imparts" [19]. However, this author argues that when a game is conceived for use in healthcare, the gameplay skills in the computer generated alternative reality need to be designed balanced and aligned to the skill involved in interaction in order to increment a patient's progress target of physical improvement. In other words a focus must be on physical input or functional improvement that is mapped accordingly to stimulate repeated training to achieve in gameplay. Advances in affordable game input devices and accessible mapping software to digital content increase potentials of what is possible through using serious games in this context.

16.4 Alternative Reality

Computer-generated Alternative Reality (e.g. of a game or creating art-related virtual interactive space) can be designed according to the personal profile of the participant and the goals of the sessions. The goal is to engage the participant in activities that comprise the alternative reality. Such engagement relates to immersion, presence and flow, human traits that relate to play and useful in treatment in healthcare and rehabilitation. The next section analogizes such traits from a games perspective.

16.4.1 Analogizing from a Games Perspective

In a 2012 National Science Foundation (NSF) publication titled Interactive Media, Attention, and Well-Being,[2] the executive summary states:

> Behavioral training interventions have received much interest as potentially efficient and cost-effective ways to maintain brain fitness or enhance skilled performance with impact ranging from health and fitness to education and job training. In particular, neuroscience research has documented the importance of explicitly training (i) attentional control, in order to enhance perceptual and cognitive fitness as well as (ii) kindness and compassion, to produce changes in adaptive emotional regulation and well-being. At the same time, video game play has become pervasive throughout all layers of current society, thus providing a potentially unique vehicle to deliver such controlled training at home in a highly engaging and cost-efficient manner. Yet, several gaps remain in terms of realizing the true potential of the medium for positive impact, as developing engaging and effective research-based games anchored in neuroscientific principles that can have scalable, sustainable publishing models presents several new inter-disciplinary challenges.

A panel of international experts associated to the National Science Foundation (NSF) then identified in the same document (ibid) five main areas of focus toward improving the scientific validation of games designed to boost well-being or attention:

(i) **Better understanding of core game mechanics driving impact outcomes.** *Clearly not all games are created equal when it comes to fostering brain plasticity—some game mechanics appear more efficient than others, calling or a concerted effort in characterizing those game dynamics that are most potent in inducing brain plasticity and learning.*

(ii) **Incorporating inter-individual differences in game design.** *Recognizing that there are as many ways to play a game as there are players and experience levels, the need to acknowledge and exploit inter-individual variability was highlighted, calling for the design of individualized game experiences taking into account not only game play, but also physiological and brain markers in real time.*

(iii) **Greater focus on social and emotional skills.** *The fact that emotion and social conduct may be considered skills rather than traits, and thus like all skills can exhibit sizeable plasticity, calls for more games designed to impact affective states.*

(iv) **Clearer validation methodologies and benchmarks.** *Not a week goes by without some new claim about a new piece of software curing ADD/ADHD, or a new mini-game that slows cognitive aging. Yet, few of these statements withstand scrutiny. A hot debate about best methodological and reporting practices is thus underway in the*

---

[2] http://www.bcs.rochester.edu/games4good/GameWellBeingAttention_NSFReportpdf

field. In addition, objective demonstration of efficacy calls for larger multi-site studies, and possibly an infrastructure allowing independent evaluation of game/ intervention efficacy.

(v) **Developing sustainable, scalable publishing models.** *Translating in-lab research documenting a beneficial effect of video games on attention or well-being into a commercially-viable product that can reach many people and truly produce social change is a tall order. Lessons could be learned from the pharmaceutical industry, but alternate paths may be worth considering for behavioral interventions, such as through video games.*

Concluding the executive summary (ibid) the report states the need for "New approaches to the design, assessment, publishing and on-going optimization of video games for enhancing well-being and attention ..." where users and researchers can access and control the mechanics identified to foster brain plasticity to "thus adapt the games to their needs not only for play, but also for research." An example of such tailoring of bespoke games is shown in Fig. 16.1 from the Humanics study at the Centre for Rehabilitation of Brain Injury (CRBI) an acknowledged leading center for acquired brain injury treatment in Denmark, which is located adjacent to Copenhagen University. With Multiplayer online games already networking global players the serious games and interactive content could be easily transferred to the cloud.

Outcomes from SoundScapes, whilst positive, suggest that video games have a restrictive tree-based structure constraining suitability optimal for only certain scenarios and conditions. Abstract digital creativity via unrestricted gesture-control, in the form of music-making, painting, and robotic lighting has a differing structure and is suitable for certain scenarios and conditions. The argument is hereby made that these structures are complementary, exhibiting common and differing aspects where play and social/machine interactions are central.

The NSF report suggests how expert brain scientist evaluations would identify control points of successful games so that rigorous experimental systematic research would highlight impact on brain plasticity, learning and modification of cognitive processes. In order to achieve advancements in this necessitates acknowledgement of the involved disciplinary differences where the scientists establish key parameters and dimensions hypothesized to affect and nurture plasticity in terms comprehensible for the video game industry experts. In line with this, the argument made herein, supported by this authors' prior publications (and others), highlight how afferent efferent neural feedback closure in the way described affects and nurtures brain plasticity through advanced understandings of media plasticity manipulation [20]. The further argument of the common structures between using 'art' and games in this context widens the hypothesis so that fertile research is conducted having common societal goals.

However, many contemporary formal treatment programs using games in intervention (GameAbilitation) do not balance the structured with the abstract (ArtAbilitation). This strategy in SoundScapes is discussed further in the following sections.

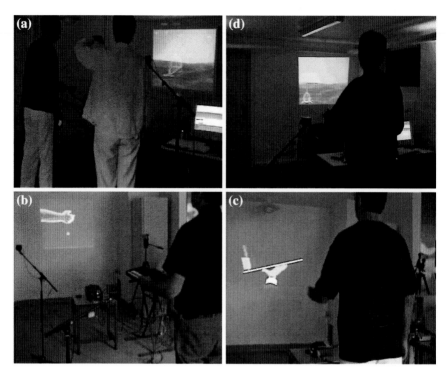

Fig. 16.1 Created gesture-controlled game examples. **a** *upper left* (dolphin catching fish) patient controlling horizontal navigation with facilitator/therapist controlling vertical, progressing to patient controlling both axis (**d**). **b** *lower left* dynamic hand/arm motion mapped to release throwing ball from hand. **c** *lower right* balancing a glass on a tray with adjustable friction to train Proprioception and Kinesthetic awareness, control and responses [7]

A challenge is thus posited for such strategies to include art-based human endeavors especially as increased aged numbers may suggest that a 'game-for-all' success is less likely than an 'art-for-all' strategy. In line with this belief the author has evolved the concept of 'Ludic Engagement Designs for All' (LEDA). This is where a compendium of both structured games and unstructured (as experienced by the participant/patient) art-related/creative play/untethered expression is advanced offering tools designed for facilitators' intervention. Just as game designers focus on creating the players' experiences and the underlying game-mechanics, settings and components are all in service of these experiences. Under the contextual abstract 'performance art' of SoundScapes, designers create, adapt and map for a participant experience. The parameters of change that influence the experience are designed as quickly retrieved presets under the model 'Zone of Optimized Motivation' (ZOOM) [21–23] where the innate mechanics, settings and components are all accessible to the designer for iterative tailoring.

While intervention using video games may involve changing difficulty levels and the ease of doing this is embedded, there are major constraints on the facilitator being able to change the abstract form in an optimal way without specific training.

In line with this a weakness of the SoundScapes research is that the author as concept originator, system designer and programmer, as well as having the role as facilitator. This is where intervention involves real-time improvisation according to the 'moment' aligned with real-time (in-action) conscious and unconscious assessment of interactions so as to change system parameters accordingly (or not) to optimize the experience within the constraints of the tools afforded by the designer. Such parameter change is established as presets that can be incremented with minimal distraction in order to keep the contact/communication with the patient/participant. Post-session evaluations equally have involved the author/designer/facilitator so that observations lead to reflections and refinement iterations for the next session—so for example refined presets. Thus, each session in a program should incrementally improve the tailoring toward achieving the desired goals from the treatment. This designed for 'patient-/facilitator-centred balance', where the units of analysis are actions and responses of the patient, optimally realizes a custom interactive installation for each individual. This further emphasizes the association to the artist as inventor and designer in SoundScapes.

SoundScapes as a European small medium enterprise (SME) has evolved in three main ways beyond apparatus and method [24]. The 'products' are now (1) a contemporary complex located in Denmark that acts as a training of trainers weekend retreat, (2) commissioned consultations and expert evaluations for professional organizations and for public bodies and research projects (including keynotes, presentations and workshops), and (3) commissions of single case study consultation and intervention that involves the creation of custom solutions for one specific end-user and where the specific trainer of the patient is educated to use the hybrid ICT solution(s). In addition the resources from the years of research in the field are being archived (with permissions) as a cloud-based resource for students of the SoundScapes approach as well as peers in healthcare. This is seen as a future sharing of data resource that is secure to professional subscribers. Online training, consulting and communication/networking via such a cloud-based infrastructure can only benefit the field as outlined in this chapter.

16.4.2 Training Trainers Retreat

It is evident that for continued development a focus is required on the 'training of trainers' in line with lifelong learning and the Danish model of social welfare and technology (Denmark is acknowledged as a leader in ICT with Cluster centers on an advanced high-tech society, featuring a world-class ICT infrastructure and the world's most e-ready population[3,4]). Consequently, a private facility has been realized in Esbjerg on the south-west coast of Denmark near the German border.

---

[3]   http://www.weforum.org/issues/global-information-technology

[4]   http://www.eiu.com/site_info.asp?info_name=ereadiness&page=noads

Contributing to the foundation of the concept of training trainers was how feedback from staff that used commissioned SoundScapes systems at institutes whilst reporting positive outcomes they consistently felt more time and training was needed to realize optimal intervention. Such time was problematic to be scheduled in their daily activities also in such a formal situation there was peer and employer pressure.

In an effort to satisfy this need SoundScapes as well as establishing the new training complex is exploring means to train and acknowledge proficiency through accreditation and licensing such that employers would pay for the service to increase staff competences with ICT tools to thus optimize investments. A side note to this is how over the many years of visiting institutes, hospitals and clinics etc., it was obvious how the technology tools were found in a 'bottom drawer' or 'remote cupboard' due to either a lack of training of use or, when training had been given, the person trained had become pregnant or changed jobs and thus was no longer at the work place that had invested in the technology. A part of the SoundScapes consultancy is to visit, compile and review such tools to see is they can be used and staff trained in how to use them before a need for further technology investment.

The training involves presenting the concept, which considers when a human's faculties are hindered through impairment and the potentials of ICT-based alternative channeling of stimuli on neural plasticity and physical efferent responses and how empowering fun 'doing' as a core of participating and actively engaging in goal-driven activities that can lead to improved quality of life. The engaged participation has embedded aspects that offer opportunities for professional healthcare assessment, critique and reflection to input to the designer of the experience toward more formal/traditional outcomes. This adoption is evident as advances in sensor-based interfaces make affordable interactive access to responsive stimuli where issues such as impairment, dexterity and strength become redundant. These issues are addressed by the designers who work first with user personas to achieve communal solutions and subsequently, using their distinct profiles, a bespoke elucidation. Each communal solution can lead to numerous bespoke elucidations, thus a growing compendium of environment tools are created to be resourced as seen fitting.

A problem has been that the fitting of a tool to a situation necessitates imagination beyond the obvious that includes a carefully planned intervention strategy focused on the emergent model titled Zone of Optimized Motivation (ZOOM) [21–23]. The facilitator is thus key personnel in driving the intervention toward the designed-for progress.

The participant experience of a SoundScapes session is of creative fun and playful achievement over challenges. Innate session social interactions between the participant and the facilitator are related to inter-subjective machine-mediated interactions. These affect participant intra-subjectively in relation to their aesthetic resonance (see later section on this), which is represented by non-verbal signifiers indicating well-being as the participant in turn responds to the responsive interactive machine-content that results from an immediately-prior feedforward action.

Human afferent efferent neural feedback loop closure results from this bespoke approach to alternative healthcare thus supplementing traditional by involving Universal Access via Ludic Engagement Designs for All (LEDA) [25].

The intervention described is subject of numerous publications since 1999. The initial article [6] reported on approximately 15 years of the concept in applied research with numerous subjects across impairment/abilities, age, and situations. A direct result was a commercial product and patent titled 'Communication method and apparatus' [24]. This patent detailed a compendium of sensor-based devices to source data from human participants and the subsequent mapping of the data to digital content to offer information for use in and across art and healthcare. This information is thus derived from an 'inhabited information space' in the form of an interactive environment consisting of responsive content that can be tailored according to a profile. This plasticity of digital media, in its flexible form in both sensing efferent feed-forward input (biofeedback data both internal and external) and stimulating feedback (across modalities), offers enormous potentials that has been evident in the SoundScapes work.

Impact is evident by wide-ranging uptake and adoption of the concept that motivates participation and thus quality of life. In art context, the 'designed-for' representation of human endeavor (performance) involved in conveyance of a profound human message is the vehicle embodying aesthetic resonance rather than the artwork (art as object) or art-product (music and/or painting). Thus, the 'designer' acts as virtual facilitator/artist who expresses through the medium of a remote human 'performer/participant' who, in turn, is empowered and motivated to express themselves through the 'goal-bespoke' design and embodied interaction. This all-encompassing transdisciplinary design/intervention role surpasses traditional by its innate applied philosophical perspective balanced with a practical knowledge and understanding, thus, realizing disruptive innovation and impact in/ across art and healthcare. In this role the designer morphs between concrete structured planner, learner/teacher/facilitator, and abstract improvising artist. From this mature body of research a physical retreat (as outlined above) as well as a cloud-based resource is being created.

16.4.3 A Cloud-Based Archive Architecture

The author's SoundScapes body of work has been active across exploring human performance in healthcare and rehabilitation aligned with human performance in the arts. For many years it has been presented at national and international scale at major events and venues. It has also been conducted at small local institutions and festivals.

To date the research archive plans to include SoundScapes data from (1) nationally in Denmark: (a) materials from a six-years study at a day-care institute for PMLD adults titled Aktivitetscentret Frederiksbjerg in Aarhus; (b) Sessions conducted at the Centre for Advanced Visualization and Interaction, CAVI,[5] Aarhus University where the author was the first artist in residence (1999–2001); (c) Hospitals and clinics including Aarhus Universitetshospital, and The Center for Rehabilitation of Brain Injury (CRBI) at the University of Copenhagen... and (2) outside of Denmark—internationally project partners of Twi-aysi[6] (1999–2000), which was part of the European Community i3 programme of technology research. A future probe funded by The European Network for Intelligent Information Interfaces; CAREHERE[7] (2000–2001), which was a European Project funded under Framework V IST Key Action 1 supporting the programme for Applications Relating to Persons with Special Needs Including the Disabled and Elderly.

Common in the above list was the use of the author's SoundScapes system (apparatus and method).

At the hospitals, institutes and clinics the concept was implemented in empowering children and adults with special needs, the elderly in long term care and people undergoing rehabilitation in hospital or at home, following for example stroke or brain injury. By giving access to affordable, appealing and readily usable state of the art technology for the improvement of their physical and cognitive skills using feedback from acoustic and visual stimuli. The project was concerned with the (re-) development of physical and cognitive skills by interaction with a responsive sound and visual environment: the improvement of motor control through direct and immediate feedback through the aural and visual senses.

At the performance art events, exhibition (e.g. museums of Modern Art) and venues (e.g. Olympics, Paralympics, Cultural Capital of Europe etc.), the concept was implemented in activities of interaction to be able to research and learn about both physical and cognitive actions, responses, and motivations (etc.) to the various computer-generated stimuli that the participant self-directed through gesture. This way of considering performance highlights the author's embedded abnormal artistic reflection and critique across forms where an influence is from being born into a artistic family including members having profound impairment, an education in engineering, and years of working in music, dance and theatre creating interactive spaces within which dancers and actors were empowered to move and control theatrical elements. The next section justifies with quotes on performance art.

16.4.4 Justifying the Transcending of Performance Art-Related Quotes

Performance became accepted as a medium of artistic expression in its own right in the 1970s. At that time, conceptual art—which insisted on an art of ideas over product, and on

---

[5] http://cavi.au.dk/

[6] http://www.bristol.ac.uk/Twi-aysi/

[7] http://www.bristol.ac.uk/carehere/

an art that could not be bought and sold—was in its heyday and performance was often a demonstration, or an execution, or those ideas.

Performance has been considered as a way of bringing to life the many formal and conceptual ideas on which the making of art is based. Live gestures have constantly been used as a weapon against conventions of established art.

...whenever a certain school, be it Cubism, Minimalism or conceptual art, seemed to have reached an impasse, artists have turned to performance as a way of breaking down categories and indicating new directions.

Performance has been a way of appealing directly to large public, as well as shocking audiences into reassessing their own notions of art and its relation to culture.

...performance has provided a presence for the artist in society. This presence, depending on the nature of the performance, can be esoteric, shamanistic, instructive, provocative, or entertaining.

...a permissive, open-ended medium with endless variables, executed by artists impatient with the limitations of more established forms, and determined to take their art directly to the public. For this reason its base has always been anarchic. By its very nature, performance defies precise or easy definition beyond the simple declaration that it is live art by artists.

...no other artistic form of expression has such a boundless manifesto, since each performer makes his or her own definition in the very process and manner of execution [26], pp. 7-9.

Extracting from these quotes from Goldberg's book one can see the transdisciplinary perspectives such that the creation of an interactive installation at an art location or conducting a session at a hospital, institute or clinic embodies similar attributes of a virtual interactive space (VIS). That is whereby *expression has such a boundless manifesto* whereby *each performer makes his or her own definition in the very process and manner of execution.* The *limitations of more established forms* can relate to traditional therapy intervention or even their frustration at their impaired body, wishing to find ways to explore it further. The *anarchic* relates to the provocation and that felt by those stuck in the established not wanting to see the benefits and opportunities. Thus, *By its very nature, performance defies precise or easy definition beyond the simple declaration that it is live art by artists* or in this case where those who are impaired are empowered to establish their own definition of what is presented as it defies current definitions.

SoundScapes is a statement reflecting how *Live gestures have constantly been used as a weapon against conventions* where in this case it demonstrates *a way of breaking down categories and indicating new directions.* Opportunities and potentials in healthcare and rehabilitation are thus illustrated of ICT used in such a direction. It is *esoteric, shamanistic, instructive, provocative, or entertaining,* therefore, and through unrestricted events, it is *a way of appealing directly to large public, as well as shocking audiences into reassessing their own notions.*

SoundScapes is an art of an *idea over product* yet where the process of creation and expression is the product and where the art of creating that Virtual Interactive Space differs little from other artists creating such interactive installations for Museums of Modern Art (as the author did—e.g. Circle of Interactive Light [COIL] 1998–1999 exhibition tour Danish MoMAs). In healthcare and Rehabilitation it is considered *an art that could not be bought and sold* as the art is internalized by the participant/patient with their abilities to achieve and perceive the associated aesthetic resonance at an individual level and with varied outcome (Fig. 16.2 where a Profoundly Multiple Learning Disabled (PMLD) person and a Learning Disabled person creatively expresses themselves (LD)).

In this body of work, ability (functional, physical, and cognitive) is in focus rather than impairment/disability. The content that is mapped to respond to the participant's input similarly is flexible to match the profile and goal. An outcome goal is an assessment of the aesthetic resonance of the patient or participant so that in-action and on-action evaluations are as valid and reliable as possible. Indications of aesthetic resonance can be individual and influenced by condition and impairment. Aesthetic Resonance was subject of European funded projects at the turn of the century—namely CARESS[8] and CAREHERE.[9] Aesthetic Resonance is introduced in the Sect. 16.4.5.

16.4.5 Aesthetic Resonance

This section discusses the concept of Aesthetic Resonance (AR), which is a term used across various art-related disciplines (Table 16.1).

In the work responsible for this chapter (i.e. SoundScapes), AR is defined as "a situation where the response to an intent is so immediate and aesthetically pleasing as to make one forget the physical movement (and often effort) involved in the conveying of the intention" [20]. A history of the term is posited. 'Aesthetic Resonation' is a related term coined by Professor Phil Ellis [27, p. 175] that evolved from his body of Sound Therapy work with children empowered to control sound through motion that resulted in their "aesthetic motivation" and "internal resonance" [28, p. 77]. The 1997 definition was "special moments experienced by individuals described as having profound and multiple learning difficulties, in which they achieve total control and expression in sound after a period of intense exploration, discovery and creation."

'Aesthetically Resonant' was coined within a European Community funded project titled Creating Aesthetically Resonant Environments in Sound (CARESS) within the i3 program of technology research. i3 (1995–2002) was the European long-term research initiative to develop intelligent information interfaces. The

---

[8] http://www.bristol.ac.uk/caress/

[9] http://www.bristol.ac.uk/carehere/

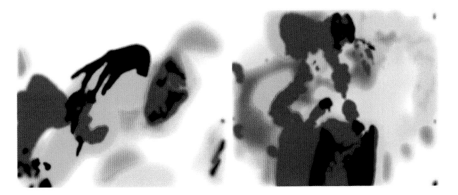

Fig. 16.2 Exemplifying free gesture in virtual interactive space (VIS) mapped to painting. *Left* PMLD participant (*hand* and *head* gesture) *Right* LD (*head, hands, torso*) [14]

Table 16.1 Examples of use of the term "aesthetic resonance" besides SoundScapes	• *Sound therapy* for the profoundly disabled [67] • Rehabilitation with audiovisual stimulus[10] • Internet performance art [68] • Reflecting artist experiences [69] • Questioning music, creative process and self-experience [70] • Sculpture art [71]

CARESS funding was under a program entitled Experimental School Environments (ESE) exploring new child centered paradigms for learning, through novel IT-based devices; artifacts and environments. Ellis [29, p. 114] subsequently adopted "Aesthetic Resonance" as a term when solely reporting the CARESS research—defining AR as: "special moments when a child achieves real control and expression after a period of intense exploration, discovery and creation—moments which can be seen to be both 'endearing' and 'touching'. In CARESS a commercial linear ultrasonic sensor device and worn encumbering sensor apparatus was used across a wide age/ability range.

Following the cessation of CARESS (c. 1999/2000) the partners from Bristol University and Sweden declared a desire to explore multimodal feedback beyond the auditory that was the focus in Ellis' Sound Therapy research. As the author (Brooks) had previously been applying his research with diverse stimuli (visual, auditory, robotic, games, VR, AV effects...etc.) and had progressed beyond encumbering "wired" solutions as used in CARESS an approach was made to research SoundScapes as a CARESS follow on funded under i3 as Twi-aysi[11] (The world is as you see it). The feasibility (future) probe used SoundScapes concepts of

---

[10] ftp://musart.dist.unige.it/Pub/Publications/2004/AISB04-Gesture.pdf

[11] www.bris.ac.uk/Twi-aysi

method and apparatus to explore AR beyond solely auditory feedback and wired sensing devices and to include gesture control of Virtual Reality and games.[12]

The success of the i3 future probe resulted in a European project titled CAREHERE[13] (Creating Aesthetic Resonant Environments for Handicapped, Elderly, and Rehabilitation). Project partners defined AR as: "an environment giving patients a visual and acoustic feedback depending on a qualitative analysis of their (full-body) movement, in order to evoke ludic aspects (and consequently introduce emotional-motivational elements)" (p. 269[14]). The ludic aspect that is referred to here was researched further within a body of work titled "Ludic Engagement Designs for All (LEDA)" resulting in an international conference e.g. [30–32]. Legacy to the original work is evident here, as is Ellis' (2004) subtle text reformulation suggesting acknowledgement of Aesthetic Resonance in Sound-Scapes occurring beyond solely auditory feedback stimulus and beyond solely participants with profound and multiple learning difficulties as in his earlier definition related to his Sound Therapy research. SoundScapes has evolved to form the movements "ArtAbilitation" and "GameAbilitation" where a wide selection of digital media content is matched to sensing means and a participant profile so that adaptive Aesthetic Resonant Virtual Interactive Spaces (VIS) evolve. Within this framework Aesthetic Resonance—the participant's representation of an 'inner quality' comprising self-agency, awareness of causality, flow, ludic engagement, from successful matching of affordances and related environment/interaction attributes. This referred to 'inner quality' is introduced in the next section, which then presents the association to beliefs in neural correlates potentiated via this concept.

16.4.5.1 Contextual Framework

'Aesthetic' derives from the Greek 'aisthesthai', to 'perceive'. In this research it refers to a participant's external representation of what is internally perceived as pleasurable. This is speculated to be where the inter-subjective interactions between human and mediating technology/digital media result in an intra-subjective stimulation such that the participant's afferent efferent neural feedback loop is closed as the participant immerses into the action. Links to Presence are reported [33].

The externalization of the perceived experience is an intuitive response to the interactions. It signifies a 'communication', a linkage, to an innate 'inner quality' that has relationship to self-achievement, self-empowerment and self-agency. This complex communication is mediated by the created environment, adaptively designed to aesthetically please as an interactive system, with its profile-matched-

---

[12] An example video is at www.youtube.com/watch?v=m5-I9NHPt2I.

[13] www.bristol.ac.uk/carehere

[14] ftp://musart.dist.unige.it/Pub/Publications/2004/AISB04-Gesture.pdf

affordances interfaced to a proficient facilitator who conducts the intervention session. The session experience as much as possible is 'participant-driven'—and a video/sensor-based data-rich archiving system for observing human performance behavior responses according to digital media content responses. In this way, the *human performance plasticity* (including physiological system attributes of brain, CNS, motoric, sensing …) and the *digital media plasticity* are inter-related. Through an understanding of this inter-related plasticity AR can be considered directly linked to neural correlates and thus to NeuroAesthetic Resonance [34].

In SoundScapes, 'Resonance' refers to the above-mentioned 'inner quality' resulting from the interactive system-mediated experience. This known quality of the human system refers to how it responds during the observation of an action resulting in 'inner-mirroring'. This is where higher order motor action plans are coded so that an internal copying of the interactions is not repeated but is, rather, used as the basis for the next motivated action [35]. Related work is that of [36] who presented a case for how mirror neurons and imitation learning are the driving force behind "the great leap forward" in human evolution; and the subsequent reports by [36] where action representation in mirror neurons is presented; [37] considering audiovisual mirror neurons and action recognition; and the [35] chapter discussing mirror neurons and imitation in relation to development and resonance attributes and the hypothesis that there is a very general, evolutionary ancient mechanism, referred to as a "resonance" mechanism, through which pictorial descriptions of motor behaviors are matched directly on the observer's motor "representations" of the same behaviors mirrored internally. According to Rizzolatti et al. [35], the resonance mechanism is a fundamental mechanism at the basis of inter-individual relations including some behaviors commonly described under the heading of "imitation." In line with this argument, the evolution of one-to-one scale digital media mirroring in SoundScapes is built upon the author's applied research as facilitator where imitation was successful as an intervention approach in rehabilitation with ICT. The mirroring with adaptive digital media—in the form of multimedia, Virtual Reality, Art, Games…, resulted in the author's hypothesis on how digital media plasticity that can be matched to a user profile to affect human performance plasticity, offers great potentials in and beyond the rehabilitation that is the focus of the research to date.

To bring into context, the 'observed action' described above in SoundScapes moved from a facilitator interaction as imitation to a mirroring via digital media not anymore of a specific motor action as in manual imitation but a mirroring whereby a continuum from abstract to figurative can be adjusted. It is rather the participant observation of the reaction of the feedback content to the controlling motor action, i.e. his or her primary action, which then acts as stimulus for the subsequent 'system-generated' motor action. Aesthetic Resonance is thus directly associated to afferent-efferent neural feedback loop closure, which is tentatively suggested as a reason why interactive multimedia is effective with disabled people as reported in e.g. [38, 39]. Evaluation of the use of the system (apparatus and method) is thus primarily (and typically) via the participant's body movement/motivated actions, including those involved in the interactive control of the multimedia, alongside verbal utterances, facial expressions, and other non-verbal behavior/communica-tion—all which can be sourced as input to the system and analyzed for behavior patterns and traits linked to the stimuli—see patent detail [24].

Further evolution of this rich VIS data environment is how amassed information can result from the interactions, which in turn can be filtered to inform a system-embedded/integrated intelligent agent (e.g. Dynamic Difficulty Adjustment DDA) such that the system automatically responds to situations that are too difficult for the participant so that a motivated engagement is promoted [21]. However, this is not the core of this chapter so therefore not elaborated further.

Concluding this section, Aesthetic Resonance is therefore posited as a com-plement to Neuroaesthetics in order to develop the concept of NeuroAesthetic Resonance where the representation of the 'inner quality' and resonance mecha-nism of the human experience is motivated by the tailored interaction that achieves for example, Flow [17], self-agency, and ludic engagement. Contributing to a grounding of this argument, Sect. 16.4.6 presents Neuroaesthetics.

16.4.6 Neuroaesthetics/Neuroesthetics

Neuroesthetics (or neuroaesthetics) is a relatively recent sub-discipline of empirical aes-thetics. Empirical aesthetics takes a scientific approach to the study of aesthetic perceptions of art and music. Neuroesthetics received its formal definition in 2002 as the scientific study of the neural bases for the contemplation and creation of a work of art. [40].

It is not the purpose of this chapter to introduce Neuroaesthetics in detail as there is a readily growing body of publications. The advances in brain imaging technologies e.g. magnetic resonance imaging (fMRI), magneto encephalography (MEG), and positron emission tomography (PET), increasingly enable analyse of human responses to experiencing stimuli. Such analysis can greatly assist the design of adaptive virtual environments where selection of input and display apparatus is flexibly mixed and matched to content according to the participant profile and responses. In other words, through understanding inner human reaction, design improvements to the 'system of empowerment' can be made. SoundScapes' conceptual framework is focused around offering a variety of means to stimulate alternative channels of Neuroplasticity such that typical association between afferent sensing and the person's movement is diverted around damaged areas of the brain. So for example auditory feedback triggered via side-to-side movement gives a feedback sense of balance and body proprioception that can assist when such sensitizing in the brain is damaged.

The potentiating of alternative synaptic pathways in this way achieves asso-ciations that are conceptualised to evolve to thus contribute to the person's improved functionality via brain plasticity (alternative sensitizing/channelling). Over time, the challenge (externally i.e. sensing parameters and mapped content) is incremented according to the motion learning (kinetic chain action-responses) to maintain and optimize the motivation. In this way, the state of Flow is targeted such that any conscious limitations maybe subdued through the interaction evoking an experience that is highly stimulating in many ways. Neuroplasticity is used in this case to refer to the ability of the brain and nervous system to change structurally and functionally as a result of input from the environment [41]. This capacity can be at various levels, ranging from cellular changes involved in learning, to large-scale changes involved in cortical remapping in response to injury.

The most widely recognized forms of plasticity are learning, memory, and recovery from brain damage.[15] Participants with impairments in line with these attributes are the main foci of designing the interactive environments, otherwise referred to as Virtual Interactive Space (VIS) as outlined earlier in this document.

Contemporary artists challenge traditional ontology through dematerialization of the object, by not being fixated on a medium, and through not seeking to produce objects of aesthetic interest. This, as artistic value, can sometimes reside in something other than its physical appearance, with which one must be in direct perceptual contact to appreciate its specific aesthetic value [42]. For further on this direct human experience of the phenomenon, see Acquaintance Principle and related [43–49].

The next sections relate art and aesthetics to the SoundScapes Virtual Interactive Space (VIS) via introducing Neuroaesthetic Resonance as a development from Aesthetic Resonance and Neuroaesthetics.

16.4.6.1 Evolution of Neuroaesthetic Resonance

Aesthetics is much broader than the field of the arts,[16] however, confusingly; traditional art theories (and non-questioning theorists) try to give 'necessary and sufficient conditions' before labelling a piece of work as art or assigning aesthetic value [50]. Evolving from the author's prior empirical work investigating Aesthetic Resonance and Aesthetic Resonant Environments introduced prior in this document within the context of SoundScapes, ArtAbilitation, GameAbilitation and Ludic Engagement Designs for All (LEDA), is an exploration of the underlying conditions that contribute to aesthetic experience, as well as aesthetic behaviors, using scientific methods applied within a socio-cultural real-world context. Thus, a participant's capacity to express/articulate aesthetic judgments, create, and to receive/respond to aesthetic stimuli are assessed as related performance conditions. This directly relates to the neural correlates of how direct and immediate interactions with adaptive digital media (i.e. its' plasticity) can influence and result in positive human performance as plasticity outcomes. This is especially needed when creating alternative rehabilitation to supplement traditional where mirroring,

---

[15] http://www.science-of-aesthetics.org

[16] http://www.science-of-aesthetics.org

creative expression, gameplay, and imitation is used in the contemporary intervention.

Grounding this argument is how Expression Theory evolved during the Romanticism movement of the nineteenth century to challenge the prior 'imitation theory' by moving from the objective oriented outer world to a subject oriented inner perception of the mind. This is fitting as the concept focused on thoughts and ideas, feelings, and cognitive faculties to evoke the human audience's emotions. Thus, through this focus on the subjective mind, the 'expression theory' is said by its advocates to be inclusive of both the artist and the audience, with innate power enabling an articulation of the communicative and educative power of the mind [50].

Other aesthetic theories, such as attributed to Collingwood and previously Croce [51], similarly relate the mind and its function, expression of emotions, and intuition. Additionally, Collingwood's theory posited that artworks act as a vehicle for people's learning, thus, the artist becomes an educator and the artwork becomes some kind of educational vehicle. In context to this body of work, the terms 'artist' and 'audience', 'learning' and 'educator' can be interpreted as applying to different entities as outlined in the SoundScapes body of research. Related to this are how the rules of user interface design similarly consider such aesthetic related to ability and human performance [52].

Reflecting these points is that innate to 'Contemporary Art Theory' is a departure from the traditional elements of artist, artwork, audience, and 'art world' institutions and instead are tensions between normalcy, i.e. being within certain limits that define the range of normal functioning, and creativity. Thus, as well as presenting how a cloud-based initiative can be seen as disruptive innovation, this chapter also wishes to provoke critical and reflective peer discussions on how the dynamic 'inner quality' of a human is beyond a passive appreciation of a static painting that is commonly associated to Neuroesthetics. The intended provocation needs to be understood in the positive manner prescribed as to question what is happening so that future contributions may utilise the outcomes to contribute to making a new generation of art that targets direct and immediate cognitive appreciation, conflict or questioning in line with this author's attempts. In this respect, by SoundScapes targeting those who are most marginalised and where it is often evident that there is an increased intensity of sensitivity, an inner learning is targeted for the participant achieved from the fun experience indicated via aesthetic resonance indices aligned with engagement, immersion and presence aspects.

In this way, aesthetic values can be questioned alongside neural correlates of human performance and responses to self as creator where ICT empowers participation and progress in healthcare and rehabilitation.

The work is underlined by a broader interpretation of what art is (as presented earlier in the chapter) contradicting opinion of those not having acquaintance to qualify judging the assigned aesthetic value. Interpretation, understanding and meaning of this specific contemporary art form is associated to hermeneutics. Hermeneutics is a (whole/part) method of analytical questioning, in this case via (and of) an emergent model [22] the perspective of Virtual Interactive Space (VIS: [6]), which was initially described in 'Science Links Japan: Gateway to Japan's Scientific and Technical Information' as:

> The VIS system was developed to research whether multimedia feedback through movement in virtual interactive spaces is capable of enhancing current methods of rehabilitation therapy. Multidimensional Infrared light and linear ultrasonic sources create the invisible interactive spaces that are capable of translating natural 3D movement into digital information that is mappable within a computer system to give the desired multimedia feedback via a digital interface. Interactive programs for each participant are designed depending on the facility and the therapists' desired goal for the treatment. Physically limited participants are trained to explore the interactive space and to focus on the multimedia element(s), for example, sounds, robotics, and/or images, which are directly manipulated as a result of movement. The zones are programmable to give results up to twelve meters from the source therefore allowing close (fine) and distant (gross) exercises for movement analysis. Neglect, co-ordination and balance training were programmed for brain-damaged participants, while stress-relieving exercises were programmed for Profound and Multiple Learning Disabled (PMLD), Cerebral Palsy, Down syndrome, and similar participants. Both groups and therapists showed an enthusiastic response to using the system with an accelerated learning curve over traditional methods achieved. Participants were able to measure their own progress through the feedback of sounds or images within a specific program. It was also observed that certain groups experienced reduced spasm attacks and stress related disorders as a result of the use of the virtual interactive space over traditional input devices. The virtual interactive space (VIS) is a non-intimidating interface that gives immediate results to both therapist and user and provides foundation for further research and development. [6].

The 1999 article [6] reported the author's prior research at the Centre for Recovery of Brain Injury (CRBI), Copenhagen University, Denmark and was published in the Proceedings of the World Congress for Physical Therapy when hosted in Yokohama, Japan. The article informs of how VIS enables direct and immediate human experiences open for analysis. In this case in creating an alternative rehabilitation strategy model realized as a communication method and apparatus patent from 2000. Subsequent resulting commercial product was realized via the Personics Company, which was established as a direct result from the research. The company was funded through the Humanics and CAREHERE projects that resulted from the author's research. Thus, the *foundation for further research and development* directly applied.

16.4.7 Games, Interactive Technology, and 'Alternative' Environments in Healthcare

Increased adoption of ICT in the form of games, interactive technology, and virtual reality 'alternative' environments is evident in rehabilitation, healthcare and therapy [53]. Hence activities of play and creativity are prevalent. Each instance of treatment requires an individual patient profile to enable optimal addressing of individual differences. From the design side theoretically related constructs need to be assessed especially immersion, presence, and flow related to the patient's session experience. The selected system, intervention, and physical set-up need to "fit" the patient and the targeted goal(s) from the session/program. Ideally adaptive potentials are designed into the designed intervention construct. Games that include an alternative (virtual) reality that embodies gameplay that is structured and defined are used to supplement traditional methods. Alternatively, and used complimentary to the games, creative (art) expression where a patient is empowered to make music, digitally paint or control robots in an open, improvised and abstract manner is empowered.

One attraction of contemporary games and creative expression using the latest hardware is the magic of interactive technology using touch or gesture-control— with devices often referred to as Natural User Interfaces or Perceptual Controllers e.g. [54]. A benefit in healthcare and rehabilitation is the cost-to-benefit factor and the mobility potentials for the patient to take a system home to self-train with family support. This strategy is detailed previously in this chapter regarding the author's Humanics project—with acquired brain injured patients. Under this scenario, observations and evaluations of patient session-activity assists diagnosis and therapeutic processes and can hint at system refinements as well as clinical intervention decisions and directed home training. Data that enables the system to operate/respond, both human and technical, triangulates to offer quantitative assessment to correlate to the qualitative observations. Measured progress is thus attainable.

What is clear is that to design an optimal interaction (game, creative expression, etc.,) for use in healthcare one must consider all aspects of user experience—thus as well as for gameplay, one should not ignore other aspects of the experience presence, engagement, immersion, affordances, agency, play, etc. Equally important is the means of presenting the multimedia content e.g. visual (screen size), auditory (sound system), etc.—as well as the situation and the intervention framework. It is complex and how to do this is the challenge. Addition challenges are how to evaluate systematically and consistently with validity and reliability, sensitivity, robustness, non-intrusiveness, and convenience due to the multi-dimensionality construct involving psychological processes. Confounding this are the different definitions and many researchers questioning what presence is and how to measure it. Examples are elaborated in the next section.

16.4.8 Presence

Minsky [55] is credited with coining the term *Presence* having derived it from his Telepresence concept, which described a participant's sense of being physically present at a remote location where they interact with a system's human interface. Sheridan's [56] conception of Presence argues it is "closely related to the phenomenon of distal attribution or externalization, which refers to the referencing of our perceptions to an external space beyond the limits of the sensory organs themselves." He posited that when "perception is mediated by a communication technology, one is forced to perceive two separate environments simultaneously", thus one experiences a physical presence in the natural environment and a virtual Telepresence in the computer-generated environment. Since then there have been numerous definitions elaborated including the concept's innate aspects such as social presence, spatial presence etc., e.g. [57, 58].

Advances in contemporary technologies have more than ever enabled a subjective experience of presence at a computer-generated distant locale while being situated locally in a physical space—in other words a sense of "being present" in the computer-generated virtual (or synthetic) situation [59–61]. Reviews of measurement questioning various dimensions of and phenomena related to presence and immersion have been compiled e.g. [62–64]. Reflecting presence and associated complexities innate to human experience of virtual situations are explorative studies such as Presenccia - an Integrated Project funded under the European Sixth Framework Program, Future and Emerging Technologies (FET), Contract Number 27731 (http://www.presenccia.org).

Objective measures include physiological variables, i.e. biofeedback—e.g. heart rate (ECG), GSR (galvanic skin response aka skin conductance response), brainwave activity (EEG), respiration rate and peripheral skin temperature. However, variance in lag between the differing responses to stimuli is problematic as is the fact that interaction within a computer-generated situation can affect in itself regards arousal and emotion and positive/negative feelings. Inherent subjective and objective attributes of "measurement" contribute to this complex challenge as outlined [65].

Serious games and the next-generation means of experiencing the games such that immersive engagement (presence, flow, aesthetic resonance etc.) is achieved in alternative realities where advanced, alternative and expended 'play therapy' (or rather gameplay therapy) beyond current methods is the intervention. In context, the 'next-generation means' refers to new affordable hardware that, more than ever, offers an increased experience of '*Place Illusion and Plausibility*' [66]. This next-generation of affordable hardware is building a creative culture through communities of developers that are sharing rather than protecting. Reflecting this is how the move from the industrial age to the information age has fostered a catalyst in the form of the Digital Revolution. Associated to how non-contextual systems are now used in the field of healthcare and rehabilitation, there is evidence of a growing community culture of entrepreneurs that are creating products that the author considers contributing to future Disruptive Innovation in this field. Linked with this is the use of Online resources to gain investments to realize their vision without sacrificing shareholding or control (e.g. Kickstarter[17]). Such strategies will likely in the future develop ICT that will enable the patient to remain

---

[17] http://www.kickstarter.com/

home instead of needing to attend hospital or clinic for treatment as in the Humanics scenario. More recently this policy of patient empowerment has been aligned with demographics of aged, related care service providers and available resources (e.g. hospital beds)—see for example the large multi-funded Danish coordinated project "Patient@Home"[18] whose leadership had knowledge of SoundScapes and the Humanics project.

16.5 Conclusions

The chapter posits how cloud-based resources can act as a disruptive innovation in healthcare and rehabilitation. The position is argued by exemplifying via the author's mature body of research coined as SoundScapes and bringing in some of the complex entities that need to be considered when dealing with humans and ICT. Reflecting on SoundScapes initiatives within the field at the turn of the century it can be concluded that the timing is now right for the incremental move to realize robust tele-health and tele-rehabilitation solutions that include alternative strategies to supplement traditional intervention via a cloud-based product. The healthcare sector is now ready unlike in 2000. However, insight to weaknesses of the concept still needs to be addressed regarding use of the technology as staff expectations of competence need aligning with each company's investment in their training. Thus, SoundScapes has evolved to be a training and consultation hub rather than a hardware product inventor, manufacturer and producer/retailer as in the past. The vision is to be a contributor to the cloud-based resource through accredited and licensed training franchises and consulting use of ICT across the healthcare and rehabilitation fields.

The Cloud-based convergent infrastructure proposed is envisioned as a repository, a centralized hub of resources and an optimized communication network. For example, included would be e-learning and training courses; a secure and accessible taxonomy of global and local research archives including recommendation engine support—to technology/products/peers and their results/apps and presets; optimization of shared patient data (local/distant), and more. Gatekeepers would ensure secure access to professional subscribers. In line with this, alignment to the National Institute of Standards and Technology's defined "five essential characteristics" of On-demand self-service; Broad network access; Resource pooling; Rapid elasticity; and Measured service would prove a first step toward the envisioned disruptive innovation.

---

[18] http://patientathome.dk/

References

1. Christensen, C.M.: The Innovator's Dilemma: When New Technologies Cause Great Firms to Fail. Harvard Business School Press, Boston (1997)
2. Dyer, J., Gregersen, H., Christensen, C.M.: The Innovator's DNA: Mastering the Five Skills of Disruptive Innovators. Harvard Business School Press, Boston (2011)
3. Christensen, C.M., Raynor, M.: Innovator's Solution, Revised and Expanded: Creating and Sustaining Successful Growth. Harvard Business Review Press, Boston (2013)
4. Christensen, C.M., Johnson, C.W., Horn, M.B.: Disrupting Class: How Disruptive Innovation Will Change the Way the World Learns. McGraw-Hill, New York (2010)
5. Drucker, P.F.: Innovation and Entrepreneurship. HarperCollins, New York (1993/1999) (originally published by Harper & Row Publishers, NY, 1985)
6. Brooks, A.L.: Virtual interactive space (V.I.S.) as a movement capture interface tool giving multimedia feedback for treatment and analysis. Science Links. Japan. http://sciencelinks.jp/jeast/article/200110/000020011001A0418015.php (1999)
7. Brooks, A.L.: Humanics 1: a feasibility study to create a home internet based telehealth product to supplement acquired brain injury therapy. In: Proceedings of 5th International Confrence on Disability, Virtual Reality, and Associated Technologies, pp. 191–198 (2004)
8. Brooks, A.L., Petersson, E.: Humanics 2: Human Computer Interaction in acquired brain injury rehabilitation. In: Stephanidis, C. (ed.) Proceedings of the 11th International Conference on Human Computer Interaction. [CD-ROM] Lawrence Erlbaum Associates, Incorporated (2005)
9. J. Softw. Eng. Appl.5(11A), Nov 2012 (Special Issue on Cloud Computing)
10. J. Cloud Comput.: Adv., Syst. Appl. Springer. http://www.journalofcloudcomputing.com/
11. Mell, P., Grance, T.: The NIST Definition of Cloud Computing: Recommendations of the National Institute of Standards and Technology, Computer Security Division, Information Technology Laboratory, U.S. Department of Commerce. http://csrc.nist.gov/publications/nistpubs/800-145/SP800-145.pdf (2011)
12. Rego, P. Moreira, P.M., Reis, L.P.: Serious games for rehabilitation: a survey and a classification towards a taxonomy. In: 2010 5th Iberian Conference on Information Systems and Technologies (CISTI), (2010)
13. Rego, P. Moreira, P.M., Reis, L.P.: Natural user interfaces in serious games for rehabilitation. Information Systems and Technologies (CISTI) (2011)
14. Brooks, A.L., Hasselblad, S.: CAREHERE—Creating aesthetically resonant environments for the handicapped, elderly and rehabilitation: Sweden. In: Proceedings of International Conference Disability, Virtual Reality and Associated Technologies, pp. 191–198, University of Reading (2004)
15. Brooks, A.L.: Robotic synchronized to human gesture as a virtual coach in (re)habilitation therapy. In: Proceedings of 3rd International Workshop on Virtual Rehabilitation (IWVR2004), VRlab, EPFL, Lausanne, Switzerland (2004)
16. Derryberry, A.: Serious games: online games for learning (White Paper). © Adobe Systems Incorporated. Available from http://www.adobe.com/resources/elearning/pdfs/serious_games_wp.pdf (2007)
17. Csíkszentmihályi, M.: The flow experience and its significance for human psychology. In: Csikszentmihalyi, M. (ed.) Optimal Experience: Psychological Studies of Flow in Consciousness, pp. 15–35. Cambridge University Press, Cambridge (1988)
18. Salen, K., Zimmerman, E.: Rules of Play: Game Design Fundamentals. The MIT Press, Cambridge (2004)
19. Dondlinger, M.J.: Educational video game design: a review of the literature. J. Appl. Educ. Technol. 4(1), 21–31 (2007)
20. Brooks, A.L., Camurri, A., Canagarajah, N., Hasselblad, S.: Interaction with shapes and sounds as a therapy for special needs and rehabilitation. In: Proceedings of 4th International

Conference on Disability, Virtual Reality and Associated Technologies, University of Reading Press. Veszprém, Hungary, pp. 205–212 (2002)

21. Brooks, A.L.: Intelligent Decision-Support in Virtual Reality Healthcare and Rehabilitation. In: Brahnam, S., Jain, L.C. (eds.) Advanced Computational Intelligence Paradigms in Healthcare 5, Studies in Computational Intelligence, vol. 326, pp. 143–169. Springer, Berlin (2011)

22. Brooks, A.L.: SoundScapes: The evolution of a concept, apparatus, and method where ludic engagement in virtual interactive space is a supplemental tool for therapeutic motivation. University of Sunderland, UK [PhD dissertation] (2006/2011)

23. Brooks, A.L., Petersson, E.: Recursive reflection and learning in raw data video analysis of interactive 'play' environments for special needs health care. In: Proceedings of Healthcom2005, 7th International Workshop on Enterprise Networking and Computing in the Healthcare Industry, pp. 83–87 (2005)

24. Brooks, A.L., Sorensen, S.: Communication method & apparatus. US Patent 6,893,407 (2005)

25. Petersson, E.: Non-formal learning through ludic engagement within interactive environments. University of Malmo, Sweden [PhD dissertation] (2006)

26. Goldberg, R.: Performance Art: From Futurism to the Present. Thames & Hudson, London (2001)

27. Ellis, P.: The music of sound: a new approach for children with severe and profound multiple learning difficulties. Br. J. Music Educ. **14**(2), 173–186 (1997)

28. Ellis, P.: Developing abilities in children with special needs: a new approach. Child. Soc. **9**(4), 64–79 (1995)

29. Ellis, P.: Caress—an endearing touch. In: Developing New Technologies for Young Children. Trentham Books, pp. 113–137 (2004)

30. Brooks, A.L.: GameAbilitation+ artabilitation: ludic engagement designs for all (LEDA). In: Proceedings of Keynote for AIRtech 2011: Accessibility, Inclusion and Rehabilitation using Information Technologies including Simposio sobre Aplicaciones de la Informática y la Matemática en Rehabilitación y Salud. AIMRS, Cuba. Available at http://vbn.aau.dk/en/publications/gameabilitation–artabilitation (2011)

31. Petersson, E.: Editorial: Ludic Engagement Designs for All. In: Digital Creativity [special issue], vol. 19(3), pp. 141–144. Routledge, London (2008)

32. Petersson, E., Brooks, A.L.: LEDA: The 1st International Symposium on Ludic Engagement Designs for All (LEDA), Aalborg University Esbjerg. Proceedings. Aalborg, Denmark: Aalborg University Press, 28–30 Nov 2007

33. Brooks, A.L., Petersson, E.: Play therapy utilizing the Sony. EyeToy. In: Proceedings of 8th Annual International Workshop on Presence, pp. 303–314 (2005)

34. Brooks, A.L.: Neuroaesthetic resonance. Arts and Technology. Lecture Notes of the Institute for Computer Sciences, Social Informatics and Telecommunications Engineering, vol. 116, pp 57–64 (2013)

35. Rizzolatti, G. et al.: From mirror neurons to imitation: facts and speculations. In: The Imitative Mind: Development, Evolution, and Brain Bases. Cambridge Press, Cambridge (2002)

36. Kohler, E., et al.: Hearing sounds, understanding actions: action representation in mirror neurons. Science **297**, 846–848 (2002)

37. Keysers, C., et al.: Audiovisual mirror neurons and action recognition. Exp. Brain Res. **153**(4), 628–636 (2003)

38. Van Leeuwen, L. Ellis, P.: Living sound: human interaction and children with autism. In: Resonances with music in education, therapy and medicine. Verlag Dohr, pp. 33–55 (2002)

39. Gehlhaar, R.: SOUND = SPACE, the interactive musical environment. Cont. Music Rev. **6**, 1 (1991)

40. Nalbantian, S.: Neuroaesthetics: neuroscientific theory and illustration from the arts. Interdiscip. Sci. Rev. **33**(4), 357–368 (2008)

41. Shaw, C., McEachern, J. (eds.): Toward a Theory of Neuroplasticity. Psychology Press, London (2001)

42. Davies, S., Higgins, K.M., Hopkins, R., Stecker, R., Cooper, D.E.: A Companion to Aesthetics. Wiley-Blackwell, Chichester (2009)
43. Wollheim, R.: Art and its Objects. Cambridge University Press, Cambridge (1980)
44. Budd, M.: The acquaintance principle. Br. J. Aesthetics 43(4), 386–392 (2003)
45. Hayner, P.: Knowledge by acquaintance. Philos. Phenomenological Res. 29(3), 423–431 (1969)
46. Kieran, M., Lopes, D.: Knowing Art: Essays in Aesthetics and Epistemology. Springer, Dordrecht (2004)
47. Lamarque, P.: Philosophy and Conceptual Art. Oxford University Press, Oxford (2007)
48. Lamarque, P., Olsen, S.H.: Aesthetics and the Philosophy of Art: The Analytic Tradition: An Anthology. Blackwell, Oxford (2004)
49. Sibley, F.: Aesthetic and nonaesthetic. J. Philos. Rev. 74, 135–159 (1965)
50. Sheppard, A.: Aesthetics: An Introduction to the Philosophy of Art (OPUS). Oxford University Press, Oxford (1987)
51. Kemp, G.: Croce's Aesthetics. Available from http://plato.stanford.edu/archives/fall2009/entries/croce-aesthetics (2009)
52. Johnson, J.: Designing with the Mind in Mind: Simple Guide to Understanding User Interface rules. Morgan Kaufmann, Burlington (2010)
53. Gamberini, L., Barresi, G., Maier, A., Scarpetta, F.: A game a day keeps the doctor away: a short review of computer games in mental healthcare. J. Cybertherapy Rehabil. 1(2), 127–145 (2008)
54. Wigdor, D., Wixon, D.: Brave NUI World: Designing Natural User Interfaces for Touch and Gesture. Morgan Kaufmann, Burlington (2011)
55. Minsky, M.: Telepresence. MIT Press Journals, pp. 45–51 (1980)
56. Sheridan, T.B.: Presence: Teleoperators Virtual Environ. 1, 120–126 (1992)
57. Lombard, M., Ditton, T.: At the heart of it all: the concept of presence. JCMC 3(2), 4 (1997)
58. Schubert, T., Friedmann, F., Regenbrecht, H.: The experience of presence: factor analytic insights. Presence 10, 266–281 (2001)
59. Barfield, W., Hendrix, C.: The effect of update rate on the sense of presence within virtual environments. Virtual Reality: J. Virtual Reality Soc. 1(1), 3–16 (1995)
60. Slater, M., Wilbur, S.: A framework for immersive virtual environments: speculations on the role of presence in virtual environments. Presence 6, 603–616 (1997)
61. Slater, M., Usoh, M., Steed, A.: Depth of presence in virtual environments. Presence 6, 130–144 (1994)
62. International Society for Presence Research (ISPR): The ISPR-measures-compendium. Available at: http://www.temple.edu/ispr/docs/frame_measure_t.htm (2000)
63. Youngblut, C.: Experience of presence in virtual environments. Institute for Defence Analyses. Available from: http://www.dtic.mil (2003)
64. Van Baren, J., IJsselsteijn, W.: Measuring presence: a guide to current measurement approaches. Available from: http://www.immersive-medien.de/sites/default/files/biblio/PresenceMeasurement.pdf (2004)
65. IJsselsteijn, W.A., de Ridder, H., Freeman, J., Avons, S.E.: Presence: concept, determinants and measurement. In: Proceedings of the SPIE, vol. 3959, pp. 520–529 (2000)
66. Slater, M.: Philos. Trans. R. Soc. Lond. B Biol. Sci. 364(1535), 3549–3557 (2009)
67. Ellis, P.: Caress—an endearing touch. In: Developing New Technologies for Young Children. Trentham Books, 113–137 (2004)
68. Broeckmann, A.: Reseau/resonance: connective processes and artistic practice. Leonardo 37(4), 280–286 (2004)
69. Hagman, G.: Aesthetic Experience: Beauty, Creativity, and the Search for the Ideal. Rodopi, Amsterdam (2005)
70. Hagman, G.: The musician and the creative process. J. Am. Acad. Psychoanal. 33, 97–118 (2005)
71. Collins, B.: Geometries of curvature and their aesthetics. In: Emmer, M. (ed.) The Visual Mind II, pp. 141–158. MIT Press, Cambridge (2005)

Glossary

Adaptive architecture Buildings that can change their properties to adapt to different environments or users

ADHD Attention Deficit Hyperactive Disorder—A behavioural condition that makes focusing on everyday requests and routine challenging

ADL Activities of Daily Living

Animated architecture Buildings that can change their properties in real time according to input from users or the surrounding environment

Artificial intelligence Artificial Intelligence—the study and design of intelligent systems (agents) able to achieve goals through intelligent behaviour

AS Algorithmic Strategy—a combination of elementary functions needed to express behaviour

Assistive music-technology Where *assistive technology* refers to technology that is designed to enable a user to engage with activities that might ordinarily be challenging due to individual needs, *assistive music-technology* refers to those assistive technologies that are focused on music making

Bubble-tube A common sensory device—essentially a tall cylinder of water with a stream of air bubbles rising from the bottom. Bubble-tubes are often equipped with coloured lighting

CP Cerebral Palsy

DIY/hacker musician Someone who adapts and reconfigures audio-technologies to create new and unusual sound-generators/instruments

DJ An abbreviation for Disc Jockey. Originally referring to the person who would select and play the music at a disco, the term has more recently broadened to included elements of music performance where the DJ will mix, adapt and create music within a live environment

DMI Digital musical instrument

A. L. Brooks et al. (eds.), *Technologies of Inclusive Well-Being*,
Studies in Computational Intelligence 536, DOI: 10.1007/978-3-642-45432-5,
© Springer-Verlag Berlin Heidelberg 2014

Ecological validity To establish the ecological validity of a neuropsychological measure, the neuropsychologist focuses upon demonstrations of either (or both) verisimilitude and veridicality. By verisimilitude, ecological validity researchers are emphasizing the need for the data collection method to be similar to real life tasks in an open environment. For the neuropsychological measure to demonstrate veridicality, the test results should reflect and predict real world phenomena

EEG Electroencephalography—recording of electrical activity in neurons in cortex through electrodes placed on the scalp

Experience-based plasticity The ability of the nervous system to respond to intrinsic or extrinsic stimuli through a reorganization of its internal structure

Fibromyalgia A long term condition characterized by chronic widespread pain throughout the body. The pain is allodynic (a heightened and painful response to pressure). The condition also often includes a range of other symptoms and as a result is often referred to as Fibromyalgia Syndrome

fMRI Functional Magnetic Resonance Imaging is a technique that detects increased blood flow in regions of the brain, the increase in which is associated with increased neural activity

fMRI Functional Magnetic Resonance Imaging. A brain imaging technique used to depict brain activity by measuring metabolism in the brain

Frontostriatal system The frontalstriatal system is responsible for executive functions and supervisory attentional processing. In neurodevelopmental disorders that disrupt executive functioning, a heterogeneous pattern of deficits emerges, including: impulsivity, inhibition, distractibility, perseveration, decreased initiative, and social deficits. These cognitive symptoms are characteristic of pervasive developmental disorders such as attention-deficit hyperactivity disorder and autism

GSR Galvanic Skin Response involves the analysis of the skin conductivity which provides an indication of psychological or physiological arousal. Changes in the skins moisture level resulting from sweat glands are controlled by the sympathetic nervous system, as a result of this GSR can provide a rapid indication of a subjects stress levels

Haptic Relating to active tactile-interaction of the kind that might exist within a human computer interface

HMD A Head Mounted Display is a display system worn on the head that presents views to one or both eyes. In Immersive Virtual Reality this is both eyes, often using stereoscopic displays to create the illusion of depth through the use of parallax. Other HMD systems exist that present 'augmented reality where views of the real and virtual worlds are combined

Infinity tunnel A sensory device that uses a combination of LEDs and mirrors to create an illusion of a never ending tunnel of lights

Intelligent architecture Like animated architecture. But it also has a set or short and/or long term goals that it bases its actions on

Intensive interaction As defined by the Intensive Interaction Institute, "Intensive interaction is an approach to teaching the pre-speech fundamentals of communication to children and adults who have severe learning difficulties and/or autism and who are still at an early stage of communication development"

Inverse kinematics A mathematical system used to calculate the position of various joints and limbs relative to the position of a particular part of a 'body'; such techniques allow animators to move the hand of a 3D human model to a desired position and orientation and following this an algorithm selects the appropriate angles and positions for the wrist. elbow, and shoulder joints

IVE Immersive virtual environments. Environments that immerse their users in virtual simulations

IVR Immersive Virtual Reality—this can take a number of forms including single user and multi user systems; however in each instance rather than an image being presented on a screen as a window upon another world users occupy the virtual world either through the projection of that world onto surfaces surrounding the viewer (such as CAVES) or by a user wearing technology such as a Head Mounted Display

Life like architecture Buildings that can change their properties in real time according to input from users or the surrounding environment. similar to a living organism

Light wheel Originally made for early discotheques, a light wheel uses a rotating disc of coloured lighting gels to project constantly changing patterns onto a suitable surface e.g. a white wall

MSE Multisensory environment

Neglect An attention deficit characterized by an inability to respond to or orient towards objects in the contralesional space which cannot be attributed to visual impairments

Neuropsychological assessment A neuropsychological assessment typically evaluates multiple areas of cognitive and affective functioning. In addition to measures of intelligence and achievement, it examines a number of areas of functioning that also have an impact on performance in activities of daily living

PAT Prism Adaptation Therapy—a therapy for patients suffering from the impairment neglect. The patient is exposed to prism-induced distortion of visual input during pointing activity

Perimetry Tests designed to measure the function of the visual field of the eye excluding the central field of vision (Fovea)

PMLD Profound and multiple learning difficulties

PTSD Post Traumatic Stress Syndrome—a psychological reaction that occurs after experiencing a highly stressing event out-side the range of normal human experience and that is usually characterized by depression, anxiety, flashbacks, recurrent nightmares, and avoidance of reminders of the event

PVC Polyvinyl chloride—a commonly produced type of plastic that is available in both exible and rigid forms

Rebound room An area designed to accommodate rebound therapy which typically includes a sunken trampoline surrounded by soft-furnishings

REF Reorganization of Elementary Functions—a model of the possible mechanisms behind recovery of function in reha-bilitation

Repurposed technology A term used to describe technology that is being used in a way that it was not originally designed e.g. a gaming controller being used within an electronic musical instrument

Resonance board A at wooden board that amplifies sounds as someone explores the surface with their hands e.g. scratching, tapping

Responsive space Space that has similar qualities to animated architecture

Sensory space A generic term of reference for an area that is designated for sensory activities, also described as a multisensory envirnoment

SNE Special needs education

Snoezelen Commercial realisation of the sensory room as originally conceived by Hulsegge and Verheul

Soundbeam A non-contact approach to triggering and manipulating sound using one or more ultrasound beams. Originally created to enable dancers to produce sound based on their own movements, Soundbeam is an item of assistive music technology that is commonly found in special needs schools in the UK

VBI Vision-Based Interfaces

Velcro Registered trade name of main manufacturer of hook-and-loop fastener as used for rapid fastening

Visuomotor Eye-to-hand activity and coordination

Virtual human Virtual humans consist of artificially intelligent graphically rendered characters that have realistic appearances, can think and act like humans, and can express themselves both verbally and non-verbally. Additionally, these virtual humans can listen and understand natural language and see or track limited user interactions with speech or vision systems

Virtual reality An advanced form of human–computer interaction, in which users are immersed in an interactive and ecologically valid virtual environment

VJ An abbreviation for Video Jockey, someone who creatively mixes, adapts and controls video projections in a live performance environment

VR Virtual Reality

VR Virtual reality

Wand An interface device used in virtual environments that allows the tracking of the wand device in space. The wand has a series of but-tons that are used to trigger interactions, akin to a mouse that can be tracked in the third dimension

About the Editors

Anthony Lewis Brooks acknowledged as a third culture thinker[1] and "a world pioneer in digital media and its use with the disabled community",[2] *Brooksy* (as he is affectionately referred) resides on the south-western coast of Denmark in the city of Esbjerg where, under Aalborg University, he is director of the SensoramaLab, a VR, HCI, and Creative Design Innovation (Ludic Engagement Designs for All—LEDA) complex. He is also a leader–lecturer of the hugely successful Medialogy education—being a member of the original founding team from over a decade ago. He is a EU expert examiner under Future Emerging Technologies and has a long list of global keynote credits. His research has been responsible for national, international (EU) projects, patents, creative industry initiatives, and commercial products. National and International awards have been awarded for his work. He runs his own not-for-profit SME (Small Medium Enterprise)—a consultancy and residential spa "training trainers" beachside resort with a focus on societal impact for life long learning of ICT under welfare technology/education. As a digital media artist, his work has been presented at major events including two Olympic/Paralympic Games (1996/2000), European Culture City (1996/2000), Danish NeWave New York (1999) -... and numerous Museum of Modern Art exhibitions. Over 150 publications contribute to his concepts on *Plasticity—Digital Media and Human Performance* that align with his view on how performance art intersects with healthcare as a *Transdisciplinary convergence resulting in Disruptive Innovation in Research and Education.*

---

[1] John Brockman (1991) "… third-culture thinkers tend to avoid the middlemen and endeavor to express their deepest thoughts in a manner accessible to the intelligent reading public." An artist "rendering visible the deeper meaning of lives" http://www.edge.org/3rd_culture.

[2] http://community.tes.co.uk/tes_special_educational_needs/b/weblog/archive/2013/08/07/the-latest-on-human–computer-interaction-and-special-needs.aspx.

A. L. Brooks et al. (eds.), *Technologies of Inclusive Well-Being*,
Studies in Computational Intelligence 536, DOI: 10.1007/978-3-642-45432-5,
© Springer-Verlag Berlin Heidelberg 2014

Sheryl Brahnam is the Director/Founder of Missouri State University's infant COPE (Classification Of Pain Expressions) project. Her interests focus on face recognition, face synthesis, medical decision support systems, embodied conversational agents, computer abuse, human computer interaction, mediated psychotherapy, and artificial intelligence. Dr. Brahnam has published articles related to medicine and culture in such journals as *Artificial Intelligence in Medicine*, *Expert Systems with Applications*, *Journal of Theoretical Biology*, *Amino Acids, AI and Society, Bioinformatics, Pattern Recognition, Human Computer Interaction, Neural Computing and Applications*, and *Decision Support Systems*.

Lakhmi C. Jain is with the Faculty of Education, Science, Technology and Mathematics at the University of Canberra, Australia and University of South Australia, Australia. He is a Fellow of the Institution of Engineers Australia.

Dr. Jain founded the KES International for providing a professional community the opportunities for publications, knowledge exchange, cooperation, and teaming. Involving around 5,000 researchers drawn from universities and companies world-wide, KES facilitates international cooperation and generate synergy in teaching and research. KES regularly provides networking opportunities for professional community through one of the largest conferences of its kind in the area of KES. http://www.kesinternational.org.

His interests focus on the artificial intelligence paradigms and their applications in complex systems, security, e-education, e-healthcare, unmanned air vehicles, and intelligent agents.